# Lecture Notes in Compu

71

Edited by G. Goos, J. Hartmanis an

Advisory Board: W. Brauer   D. Gι      ....ι

**Springer**

*Berlin*
*Heidelberg*
*New York*
*Barcelona*
*Budapest*
*Hong Kong*
*London*
*Milan*
*Paris*
*Tokyo*

E. Thomas Schubert  Phillip J. Windley
James Alves-Foss (Eds.)

# Higher Order Logic
# Theorem Proving
# and Its Applications

8th International Workshop
Aspen Grove, UT, USA
September 11-14, 1995
Proceedings

Springer

Series Editors

Gerhard Goos, Karlsruhe University, Germany

Juris Hartmanis, Cornell University, NY, USA

Jan van Leeuwen, Utrecht University, The Netherlands


Volume Editors

E. Thomas Schubert
Department of Computer Science, CMPS, Portland State University
P.O. Box 751, Portland, OR 97207-0751, USA

Phillip J. Windley
Department of Computer Science, TMCB 3370, Brigham Young University
Provo, UT 84602-6576, USA

James Alves-Foss
Department of Computer Science, University of Idaho
Moscow, ID 83844-1010, USA

Cataloging-in-Publication data applied for

Die Deutsche Bibliothek - CIP-Einheitsaufnahme

**Higher order logic theorem proving and its applications** : ...
international workshop ; proceedings. - Berlin ; Heidelberg ;
New York ; London ; Paris ; Tokyo ; Hong Kong ; Barcelona ;
Budapest : Springer.
8. Aspen Grove, UT, USA, September 11 - 14, 1995. - 1995
  (Lecture notes in computer science ; Vol. 971)
  ISBN 3-540-60275-5 (Berlin ...)
NE: GT


CR Subject Classification (1991): F.4.1, I.2.2-3, B.6.3, B.7.2, D.2.2, D.4.6,
F.3.1


ISBN 3-540-60275-5 Springer-Verlag Berlin Heidelberg New York

© Springer-Verlag Berlin Heidelberg 1995
Printed in Germany

Typesetting: Camera-ready by author
SPIN 10485472     06/3142 – 5 4 3 2 1 0     Printed on acid-free paper

# Preface

This volume is the proceedings of the international workshop on *Higher Order Logic Theorem Proving and its Applications*, held at Aspen Grove, Utah, USA during September 11-14, 1995. The workshop is the eighth in a series of annual meetings that brings together researchers and practitioners to explore issues related to higher order logic theorem proving technology and the use of higher order logic as a basis for formal methods reasoning. Though the original focus of the workshop was the HOL theorem proving system, the scope has since broadened to include the development and use of other higher order logic mechanized theorem provers. Each of the thirty-five papers submitted this year was fully refereed by at least three reviewers selected by the program committee. The program committee accepted twenty-six papers for presentation and publication in the proceedings.

The papers selected fall into three general categories: representation of formalisms in higher order logic; applications of mechanized higher order logic; and enhancements to the HOL and other theorem proving systems. Papers in the first category discuss embedding a variety of formalisms in higher order logic, including ZF set theory, graph theory, TLA, pi-calculus, and VHDL.

Papers in the second category describe applications of higher order logic. Several papers describe hardware verification efforts, including pipeline verification, multiprocessor memory protocol verification, and formal circuit synthesis. Other papers discuss floating point verification, cryptographic protocol analysis, and reasoning about distributed programming languages.

The final category concerns higher order logic theorem proving infrastructure. Topics covered by papers in this area include proof methods, decision procedures, proof engineering support, user interface tools, definition mechanisms, transformation techniques, and proof checking.

The conference continued its tradition of providing an open venue for the discussion and sharing of preliminary results. Eight researchers were invited to present their work during an informal poster session. The workshop was sponsored by the Departments of Computer Science at Brigham Young University, Portland State University, and the University of Idaho. The conference organizers would also like to thank Paul Black, Kelly Hall, Michael Jones, Rosina Bignall, Trent Larson, and Robert Beers, all of whom helped ensure a successful conference.

More information about future conferences and the HOL system can be found on the World Wide Web at http://lal.cs.byu/lal/hol-documentation.html.

September, 1995

*Tom Schubert*
*Phil Windley*
*Jim Alves-Foss*

# Conference Organization

**Workshop Chair**
Dr. Phillip Windley
Dept. of Computer Science TMCB 3370
Brigham Young University
Provo, Utah 84602-6576
e-mail: windley@cs.byu.edu

**Program Chair**
Dr. Thomas Schubert
Dept. of Computer Science
P.O. Box 751
Portland State University
Portland, Oregon 97207-0751
e-mail: schubert@cs.pdx.edu

**Special Sessions Chair**
Dr. Jim Alves-Foss
Dept. of Computer Science
University of Idaho
Moscow, Idaho 83844-1010
e-mail: jimaf@cs.uidaho.edu

## Program Committee

Jim Alves-Foss (Idaho)
Flemming Andersen (TDR)
Richard Boulton (Cambridge)
Albert Camilleri (HP)
Shui-Kai Chin (Syracuse)
Elsa Gunter (AT&T)
John Herbert (SRI)
Ramayya Kumar (FZI)
Miriam Leeser (Cornell)

Tim Leonard (DEC)
Karl Levitt (UC Davis)
Paul Loewenstein (SUN)
Tom Melham (Glasgow)
Tom Schubert (Portland State)
David Shepherd (SGS-THOMSON)
Joakim von Wright (Åbo Akademi)
Phil Windley (BYU)

## Additional Reviewers

Brian R. Becker, A. Rosina Bignall, Paul E. Black, Michael Butler, Surekha Ghantasala, Jim Grundy, Kelly M. Hall, John Harrison, Mark Heckman, Shahid Ikram, Michael Jones, Jang Dae Kim, Thomas Långbacka, Munna, John O'Leary, Stacey Son, Donald Syme, Wai Wong, Cui Zhang

# Contents

# Mechanizing a $\pi$-calculus equivalence in HOL

Otmane AÏT MOHAMED*

CRIN-CNRS & INRIA-Lorraine, BP 239, Vandœuvre-lès-Nancy Cedex, France

**Abstract.** The $\pi$-calculus is a relatively simple framework in which the semantics of dynamic creation and transmission of channels can be described nicely. In this paper we consider the issue of verifying *mechanically* the equivalence of $\pi$-terms in the context of bisimulation based semantics while relying on the general purpose theorem prover HOL. Our main contribution is the presentation of a proof method to check early equivalence between $\pi$-terms. The method is based on $\pi$-terms rewriting and an operational definition of bisimulation. The soundness of the rewriting steps relies on standard algebraic laws which are formally proved in HOL. The resulting method is implemented in HOL as an automatic tactic.

## 1 Introduction

The $\pi$-calculus [12] is an extension of CCS [10] based on the idea that processes can communicate channel names. This possibility dramatically increases the expressive power of the calculus, for instance it allows one to model networks with a dynamically changing topology [14], and reasonable encodings of the $\lambda$-calculus and of higher-order process calculi have been proposed [11, 2, 15].

This paper reports on work concerning the *mechanical verification of equivalence* between $\pi$-terms in the context of bisimulation based semantics within the general purpose theorem prover HOL. This work is based on the one described in [3], and is constructed on top of our mechanization of the $\pi$-calculus theory in the HOL system [1].

The embedding of the $\pi$-calculus in HOL is inspired by previous works on the mechanization of process algebra in HOL, namely, the mechanization of CSP by Camilleri [6], the mechanization of CCS by Nesi [13], and, recently, the mechanization of the $\pi$-calculus by Melham [9].

Our general goal is two fold: firstly, we want to develop formally the theory of the $\pi$-calculus and secondly, we want to apply this framework to the verification of applications specified in the $\pi$-calculus.

In the proof construction process it is of the utmost importance to have tactics that carry out simple parts of the proof automatically. In particular our goal here is to define a tactic that can solve automatically the equivalence problem for *finite* terms (a term is said to be *finite* if it contains only finite summations, and no recursion). More precisely, the main contribution of this

---

* email:amohamed@loria.fr

paper is to describe a method of checking early equivalence between *finite* $\pi$-terms, and its implementation in HOL (the method can be applied to general $\pi$-terms as well but in this case termination is not guaranteed). This method is composed of two basic parts:

- In the first part a $\pi$-term is rewritten into a *prefixed form* where it is possible to read directly one of its next actions (if any). More precisely, a process $P$ is said to be in a prefixed form if it is $Nil$, or it is in one of the two forms: $\alpha.P$ or $\alpha.P + Q$, where $Nil$ is the terminated process, $\alpha$ is a suitable prefix, and $+$ is the non-deterministic sum. The rewriting rules are obtained by forcing a suitable orientation of basic algebraic laws of the $\pi$-calculus, typically we apply a suitable form of the expansion theorem and a certain number of rules concerning the commutation of restriction with the other operators.

  Since the algebraic laws have been formally derived in our HOL's formalisation of the $\pi$-calculus [1] we are able to derive easily the soundness of the rewriting process. Let us anticipate that, in order to represent certain intermediate states of the computation we employ a few new operators. These auxiliary operators give an equational characterization of parallel composition. They were introduced for the first time by Bergstra and Klop in their finite axiomatization of strong bisimulation equivalence over ACP [4, 5]. This in turn leads to transformational proof techniques for showing that a process implementation meets its specification. These operators are introduced in HOL's $\pi$-calculus theory by following the *definitional* principle, in order to ensure the consistency of this extension. Related algebraic laws are derived formally and applied to the rewriting process.
- In the second part two prefixed forms are compared according to a suitable set of rules. These rules are based on the definition of the bisimulation relation. Their soundness is shown within the HOL system.

In general, our tactic alternates the computation of prefixed forms (part 1) and their comparison according to the rules of bisimulation (part 2).

The structure of the paper is as follows: In section 2 we give a brief presentation of the $\pi$-calculus: its syntax, its semantics and the definition of the strong early equivalence. In section 3 we recall some aspects of the formalisation of the $\pi$-calculus in HOL. In section 4 we give the definition of our proof method, followed by its representation in the HOL system. We conclude with some remarks and some directions for future work.

## 2   $\pi$-calculus

In this section, we present the syntax and the semantic of the $\pi$-calculus, as well as the definition of the strong early equivalence. For further details on these topics we refer to [12].

The syntax of the $\pi$-calculus is given by the following BNF grammar:

$$P ::= Nil \mid X \mid \alpha.P \mid [x = y]P \mid (\nu c)P \mid (P + P) \mid (P \mid P) \mid Rec\ X\ P$$
$$\alpha ::= c(x) \mid \bar{c}d \mid \tau$$

Where $Nil$ is the inactive process. $c(x).P$ is a process that receives an arbitrary channel $d$ at $c$ and then it behaves like $P[d/x]$. $\bar{c}d.P$ sends the channel $d$ along $c$ and then behaves like $P$. $\tau.P$ performs a silent action $\tau$ and then behaves like $P$. The process $[x = y]P$ behaves like $P$ if $x$ and $y$ are identical, and otherwise like $Nil$. $P_1 + P_2$ represents nondeterministic choice. $P_1 \mid P_2$ represents two processes acting in parallel with the possibility of communication. The term $(\nu c)P$ behaves like $P$ except that the channel $c$ is local to $P$. The actions at channels $c$ are prohibited (but communication between components of $P$ along the channel $c$ are not). The process $Rec\ X\ P$ specifies a process that has a recursive behaviour. We abbreviate the process $(\nu d)\bar{c}d.P$ by $\bar{c}(d).P$, i.e, the process that sends a new local channel to its environment.

The definitions of *free* and *bound* names are standard. (In $c(x).P$ and in $(\nu x)P$ the variable $x$ is bound). We denote by $Fn(P)$ and $Fn(\alpha)$ the set of free names of $P$ and $\alpha$ and by $Bn(P)$ and $Bn(\alpha)$ the set of bound names of $P$ and $\alpha$. $N(P)$ and $N(\alpha)$ denote the set of names occurring in $P$ and $\alpha$. We shall identify two processes that differ only by their bound names. A substitution $\sigma$ maps channel names to channel names. We denote by $P\sigma$ the process obtained by replacing by $x\sigma$ each free occurrence of a variable $x$ in $P$. The operational semantics of the calculus is given via a labelled transition system, which is displayed in Fig. 1. We have omitted the symmetric versions of the rules *sum1*, *par1*, *com1* and *close1*.

| | | |
|---|---|---|
| *in* | $w \notin Fn(\nu x)P$ | $\Rightarrow c(x).P \xrightarrow{c(w)} P[w/x]$ |
| *out* | | $\Rightarrow \bar{c}d.P \xrightarrow{\bar{c}d} P$ |
| *tau* | | $\Rightarrow \tau.P \xrightarrow{\tau} P$ |
| *match* | $P \xrightarrow{\alpha} P'$ | $\Rightarrow [c = c]P \xrightarrow{\alpha} P'$ |
| *par1* | $P \xrightarrow{\alpha} P' \wedge Bn(\alpha) \cap Fn(Q) = \emptyset$ | $\Rightarrow P \mid Q \xrightarrow{\alpha} P' \mid Q$ |
| *sum1* | $P \xrightarrow{\alpha} P'$ | $\Rightarrow P + Q \xrightarrow{\alpha} P'$ |
| *rec* | $P[Rec\ X\ P/X] \xrightarrow{\alpha} P'$ | $\Rightarrow Rec\ X\ P \xrightarrow{\alpha} P'$ |
| *com1* | $P \xrightarrow{c(w)} P' \wedge Q \xrightarrow{\bar{c}d} Q'$ | $\Rightarrow P \mid Q \xrightarrow{\tau} P'[d/w] \mid Q'$ |
| *close1* | $P \xrightarrow{c(w)} P' \wedge Q \xrightarrow{\bar{c}(w)} Q'$ | $\Rightarrow P \mid Q \xrightarrow{\tau} (\nu w)(P' \mid Q')$ |
| *nu* | $P \xrightarrow{\alpha} P' \wedge c \notin N(\alpha)$ | $\Rightarrow (\nu c)P \xrightarrow{\alpha} (\nu c)P'$ |
| *open* | $P \xrightarrow{\bar{d}c} P' \wedge d \neq c \wedge w \notin Fn((\nu c)P)$ | $\Rightarrow (\nu c)P \xrightarrow{\bar{d}(w)} P'[w/c]$ |

**Fig. 1.** $\pi$-calculus transition system

**Strong early bisimulation**

**Definition 1.** A binary relation $R$ between process terms is a strong early simulation if whenever $(P, Q) \in R$ then,

- if $P \xrightarrow{\alpha} P'$ and $\alpha \equiv \tau$ or $\alpha \equiv \bar{c}d$ then $Q \xrightarrow{\alpha} Q'$ for some $Q'$ with $(P', Q') \in R$.
- if $P \xrightarrow{c(x)} P'$ and $x \notin N(P \mid Q)$ then, for all $y$, $Q \xrightarrow{c(x)} Q'$ for some $Q'$ with $(P'\{y/x\}, Q'\{y/x\}) \in R$.
- if $P \xrightarrow{\bar{c}(d)} P'$ and $d \notin N(P \mid Q)$ then $Q \xrightarrow{\bar{c}(d)} Q'$ for some $Q'$ with $(P', Q') \in R$.

$R$ is an early bisimulation if $R$ and $R^{-1}$ are early simulations. Two processes $P$ and $Q$ are early bisimilar, written $P \overset{.}{\sim} Q$, if $(P, Q) \in R$, for some early bisimulation $R$.

The relation $\overset{.}{\sim}$ is not preserved by input prefix; the full congruence is obtained using name instantiation. Thus two processes $P$ and $Q$ are early congruent, written $\sim$ if $P\sigma \overset{.}{\sim} Q\sigma$, for every substitution $\sigma$.

## 3   $\pi$-calculus in HOL

In this section we recall briefly some aspects of the formalization of the $\pi$-calculus in HOL; the reader should refer to [1] for more details. Our mechanization is very similar to the one proposed by Melham [9]. The main differences can be summarized as follows: (i) our $\pi$-calculus theory is not polymorphic, (ii) channels are represented by the HOL built in type string, and (iii) the recursive behaviour is specified by using the Rec operator rather than bang(!). More precisely, we have defined a concrete data type Pr to represent the set of processes in HOL by using a built-in facility for automatically defining concrete recursive data types from a specification of their syntax [7]. The type definition for the $\pi$-calculus can be given as follows:

```
Pr =  Nil
      |Var string
      |In Ch Ch Pr
      |Out Ch Ch Pr
      |Tau Pr
      |Sum Pr Pr
      |Par Pr Pr
      |Nu Ch Pr
      |Rec string Pr
```

where Nil, Var, In, Out, Tau, Sum, Par, Nu, Rec are distinct constructors, and Ch is a type abbreviation of the logical built-in type string. The type definition package returns a theorem which characterises the type Pr and allows reasoning about this type.

⊢ ∀e f0 f1 f2 f3 f4 f5 f6 f7 f8 f9.
$\exists_{unique}$ fn.
  (fn Nil = e) ∧
  (∀s. fn(Var s) = f0 s) ∧
  (∀s0 s1 P. fn(In s0 s1 P) = f1(fn P)s0 s1 P) ∧
  (∀s0 s1 P. fn(Out s0 s1 P) = f2(fn P)s0 s1 P) ∧
  (∀P. fn(Tau P) = f4(fn P)P) ∧
  (∀s0 s1 P. fn(Match s0 s1 P) = f5(fn P)s0 s1 P) ∧
  (∀P1 P2. fn(Sum P1 P2) = f6(fn P1)(fn P2)P1 P2) ∧
  (∀P1 P2. fn(Par P1 P2) = f7(fn P1)(fn P2)P1 P2) ∧
  (∀s P. fn(Nu s P) = f8(fn P)s P) ∧
  (∀s P. fn(Rec s P) = f9(fn P)s P)

The elementary syntactic theory of the $\pi$-calculus contains the definitions of free names, bound names and the definitions of substitution and $\alpha$-equivalence. All these definitions, are easily formalized in HOL by using the ML function new_recursive_definition and the built-in facility for defining inductive relations [8].

The next step in the formalization is the mechanization of the operational semantics of the $\pi$-calculus, the definition of strong early equivalence, and the derivation of its algebraic laws. The transition system is based on four actions. These are represented in the logic by values of type Act which is specified as follows:

```
Act = tau
    |out Ch Ch
    |in Ch Ch
    |oute Ch Ch
```

Here tau,out,in,oute are distinct constructors. The definition is automatic using the same mechanism employed in the Pr type definition.

The transition system is defined in HOL by using the derived principle of *inductive predicate definition* [8], which proves automatically the existence of a relation inductively defined by a user specification. The HOL system offers a variety of general purpose proof tools associated with the derived rule of inductive predicate definition. These tools allow one to reason about inductively defined relations. Several tactics permit one to perform goal-directed inductive proofs which are related to the labelled transition system. It is possible to prove that a certain labelled transition between $\pi$-terms holds by calling a tactic which is constructed automatically from the rules of the labelled transition system. Moreover there is also an automatic proof procedure for performing case analysis on the transition system.

The definitions of early strong bisimulation and early strong congruence can be directly represented in higher order logic:

```
⊢ ∀R.
    Strong_Sim R =
    (∀P Q.
     R P Q ⇒
     (∀P'. Trans P tau P' ⇒  (∃Q'. Trans Q tau Q' ∧ R P' Q')) ∧
     (∀c x P'.
       Trans P(in c x)P' ∧ ¬x IN ((N_Pr P) UNION (N_Pr Q)) ⇒
        (∀y.
         ∃Q'.
          Trans Q(in c x)Q' ∧ R(Subt1_Pr P'(x,y))(Subt1_Pr Q'(x,y)))) ∧
     (∀x y P'.
       Trans P(out x y)P' ⇒  (∃Q'. Trans Q(out x y)Q' ∧ R P' Q')) ∧
     (∀x y P'.
       Trans P(oute x y)P' ∧ ¬y IN ((N_Pr P) UNION (N_Pr Q)) ⇒
       (∃Q'. Trans Q(oute x y)Q' ∧ R P' Q')))
```

```
⊢ ∀P Q. Strong_Equiv P Q =
        ∃R. R P Q ∧ Strong_Sim R ∧ Strong_Sim(λx y. R y x)
```

```
⊢ ∀P Q. Strong_Cong P Q = (∀s. Strong_Equiv(Subt_Pr P s)(Subt_Pr Q s))
```

Up to this point we have mimicked Melham's work modulo the fact that we adopt early equivalences (the largest ones) while Melham concentrates on the later ones. It is in the development of the algebraic theory of the $\pi$-calculus in HOL that our work goes well beyond that of Melham (he only proved the algebraic laws of the sum operator). This turns out to be a non trivial task. As a typical example one may consider the proofs related to parallel composition which are long and require the utilization of a technique known as "bisimulation up to" (introduced in [12]). For instance to prove the commutativity of parallel composition, $P \mid Q \sim Q \mid P$, we must prove first that $P \mid Q \sim_{up\_nu} Q \mid P$, for a suitable definition of $\sim_{up\_nu}$ and the theorem $\forall PQ. P \sim_{up\_nu} Q \Rightarrow P \sim Q$. The relation $\sim_{up\_nu}$ is introduced in HOL as:

```
⊢_def ∀P Q. EQUIV_UP_NU P Q =
        ∃R. R P Q ∧ Strong_Sim_UP_NU R ∧ Strong_Sim_UP_NU(λx y. R y x)
```

The definition of Strong_Sim_UP_NU is similar to the definition of Strong_sim except for the tau clause and it is defined as: (we represent by three dots the common part of the two definitions)

$\vdash_{def} \forall R.$
    Strong_Sim_UP_NU R =
    ($\forall$P Q. R P Q $\Rightarrow$
      ($\forall$P'.Trans P tau P' $\Rightarrow$
        ($\exists$Q'.Trans Q tau Q' $\wedge$
          (R P' Q' $\vee$
           ($\exists$P'' Q'' w. (P' = Nu w P'') $\wedge$ (Q' = Nu w Q'') $\wedge$ R P'' Q'')))) $\wedge$
      ...

The algebraic laws are available as a collection of equational theorems in HOL. We have used them to show the correctness of the $\pi$-calculus specification of the addition of integers in HOL [1]. In this paper, these theorems were used in the proof of the algebraic laws of our extended $\pi$-calculus, which are useful in the computation of the prefixed form. Concerning the expansion theorem, we have implemented it as a conversion since it represents a schematic axiom, following the work of Nesi in her CCS formalization of this theorem.

In the method that we propose in this paper, we use, an alternative formulation of the expansion theorem based on two auxiliary operators $|_l$ and $|_s$. In this formulation the expansion theorem can be finitely axiomatized. Another advantage of this approach is that we can define a strategy to reduce any $\pi$-term to a form that is weaker than the sum of prefixed forms. The *lazy* evaluation of $\pi$-terms is important in the definition of the verification tactic.

## 4 Equivalence proof method

In this section, we describe an effective method of proving that two finite $\pi$-terms are strong early equivalent and its representation in the HOL system. We will consider a simple language which is an extension of finite, and matching free $\pi$-calculus, and which will be called $\pi_e$-calculus. Its syntax is given by the following grammar :

$$A ::= \mathbf{H} \, P \mid \gamma \bullet A \mid (A \mid_s A) \mid (A \mid_l A) \mid (A \oplus A) \mid (A \| A) \mid (\nu_e c) A$$
$$\gamma ::= \mathbf{h} \, \alpha \mid c\langle z \rangle$$

where $P$ and $\alpha$ range respectively over the type Pr and Act. We use the constructors $\mathbf{H}$ and $\mathbf{h}$ to mark these values but we will omit them whenever this fact is clear from the context. Let us continue with the description of the remaining operators: $\alpha \bullet A$ represents a process which is able to execute the action $\alpha$ and then behaves like $A$. $c\langle z \rangle \bullet A$ represents a process that has received a channel $z$ along a channel $c$ and is ready to become $A$. The operator $|_s$ forces a synchronisation between its arguments and the operator $|_l$ forces its left operand to perform the first action. The operators $\bullet$, $\oplus$, $\|$, and $\nu_e$ have exactly the same semantics as their corresponding operator $., +, |$, and $\nu$. They are introduced here to support intermediate steps in the computation of the prefixed form. Note that $\pi_e$-calculus contains the language generated by the $\pi$-calculus grammar. For this reason, we have to make sure that the introduction of these operators preserves the semantics of the $\pi$-calculus. More precisely, if $P$ and $Q$ are $\pi$-terms and $\sim_{ext}$ represents the equivalence relation over the extended $\pi$-calculus, then the following equations must hold:

$$\mathbf{h}\ \alpha \bullet \mathbf{H}\ P\ \sim_{ext}\ \mathbf{H}\ (\alpha.P)$$
$$\mathbf{H}\ P \oplus \mathbf{H}\ Q\ \sim_{ext}\ \mathbf{H}\ (P+Q)$$
$$\mathbf{H}\ P\|\mathbf{H}\ Q\ \sim_{ext}\ \mathbf{H}\ (P\mid Q)$$
$$(\nu_e c)\mathbf{H}\ P\ \sim_{ext}\ \mathbf{H}\ ((\nu c)P)$$

This is ensured by a suitable definition of the extended equivalence ($\sim_{ext}$) and the extended transition relation.

## 4.1 Extended labelled transition system

We define the relation $\rightsquigarrow$ by extending the transition relation $\longrightarrow$ and adding the appropriate rules which give meaning to the new notations. The extended transition system is given in Fig. 2. We have omitted the symmetric versions of the rules *choice1,pars1,pars2,pars1_ cl,pars2_ cl,par1,com1,com2,close1,close2*.

## 4.2 Extended strong early bisimulation

**Definition 2.** A binary relation $R$ between process terms is an (extended) strong early simulation if whenever $(A, B) \in R$ then,

- if $A \overset{\alpha}{\rightsquigarrow} A'$ and $\alpha \equiv \tau$ or $\alpha \equiv \bar{c}d$ then $B \overset{\alpha}{\rightsquigarrow} B'$ for some $B'$ with $(A', B') \in R$
- if $A \overset{c(x)}{\rightsquigarrow} A'$ and $x \notin Ext\_N(A\|B)$ then, for all $y$, $B \overset{c(x)}{\rightsquigarrow} B'$ for some $B'$ with $(A'\{y/x\}, B'\{y/x\}) \in R$ or $B \overset{c(y)}{\rightsquigarrow} B'$ for some $B'$ with $(A'\{y/x\}, B')$.
- if $A \overset{\bar{c}(d)}{\rightsquigarrow} A'$ and $d \notin Ext\_N(A\|B)$ then $B \overset{\bar{c}(d)}{\rightsquigarrow} B'$ for some $B'$ with $(A', B') \in R$.
- if $A \overset{c(z)}{\rightsquigarrow} A'$ then $B \overset{c(z)}{\rightsquigarrow} B'$ for some $B'$ with $(A', B') \in R$ or $B \overset{c(x)}{\rightsquigarrow} B'$ and $x \notin Ext\_N(A\|B)$ for some $x, B'$ with $(A', B'\{z/x\}) \in R$.

$R$ is an (extended) early bisimulation if $R$ and $R^{-1}$ are (extended) early simulations. Two processes $A$ and $B$ are extended early bisimilar, written $A \sim_{ext} B$, if $(A, B) \in R$, for some (extended) early bisimulation $R$.

**Proposition 3 (Congruence Law).** *The relation $\sim_{ext}$ is a congruence with respect to the $\oplus, |_l, |_s$ operators. That is,*
$$A \sim_e B \Rightarrow A \oplus A' \sim_e B \oplus A', A \mid_l A' \sim_e B \mid_l A', A \mid_s A' \sim_e B \mid_s A'.$$

**Proposition 4 (Exp_Law).** $A\|B \sim_{ext} A \mid_l B + B \mid_l A + B \mid_s A$

The proof of proposition 3 and proposition 4 is performed in the HOL system. First a suitable relation is found and then one uses the definition of the extended transition system to show that this relation is an extended transition system.

| | | | |
|---|---|---|---|
| $Pr$ | $P \xrightarrow{\alpha} P'$ | $\Rightarrow$ | $\mathbf{H}\, P \overset{h\cdot\alpha}{\leadsto} \mathbf{H}\, P'$ |
| $inp$ | | $\Rightarrow$ | $c\langle d\rangle \bullet A \overset{c\langle d\rangle}{\leadsto} A$ |
| $Tau$ | | $\Rightarrow$ | $\tau \bullet A \overset{\tau}{\leadsto} A$ |
| $Out$ | | $\Rightarrow$ | $\bar{c}d \bullet A \overset{\bar{c}d}{\leadsto} A$ |
| $In$ | $w \notin (Ext\_Fn\_Pr((\nu_e x)A))$ | $\Rightarrow$ | $c(x) \bullet A \overset{c(w)}{\leadsto} A[w/x]$ |
| $choice1$ | $A \overset{\gamma}{\leadsto} A'$ | $\Rightarrow$ | $A \oplus B \overset{\gamma}{\leadsto} A'$ |
| $parl$ | $A \overset{\gamma}{\leadsto} A' \wedge Ext\_Bn\_Act(\gamma) \cap Ext\_Fn\_Pr(B) = \emptyset$ | $\Rightarrow$ | $A \mid_l B \overset{\gamma}{\leadsto} A'\|B$ |
| $pars1$ | $A \overset{c(w)}{\leadsto} A' \wedge B \overset{\bar{c}d}{\leadsto} B'$ | $\Rightarrow$ | $A \mid_s B \overset{\tau}{\leadsto} A'[d/w]\|B'$ |
| $pars2$ | $A \overset{c\langle d\rangle}{\leadsto} A' \wedge B \overset{\bar{c}d}{\leadsto} B'$ | $\Rightarrow$ | $A \mid_s B \overset{\tau}{\leadsto} A'\|B'$ |
| $pars1\_cl$ | $A \overset{c(w)}{\leadsto} A' \wedge B \overset{\bar{c}(w)}{\leadsto} B'$ | $\Rightarrow$ | $A \mid_s B \overset{\tau}{\leadsto} (\nu_e w)(A'\|B')$ |
| $pars2\_cl$ | $A \overset{c(w)}{\leadsto} A' \wedge B \overset{\bar{c}(w)}{\leadsto} B'$ | $\Rightarrow$ | $A \mid_s B \overset{\tau}{\leadsto} (\nu_e w)(A'\|B')$ |
| $parl$ | $A \overset{\gamma}{\leadsto} A' \wedge Ext\_Bn\_Act(\gamma) \cap Ext\_Fn\_Pr(B) = \emptyset$ | $\Rightarrow$ | $A\|B \overset{\gamma}{\leadsto} A'\|B$ |
| $com1$ | $A \overset{c(w)}{\leadsto} A' \wedge B \overset{\bar{c}d}{\leadsto} B'$ | $\Rightarrow$ | $A\|B \overset{\tau}{\leadsto} A'[d/w]\|B'$ |
| $com2$ | $A \overset{c\langle d\rangle}{\leadsto} A' \wedge B \overset{\bar{c}d}{\leadsto} B'$ | $\Rightarrow$ | $A\|B \overset{\tau}{\leadsto} A'\|B'$ |
| $close1$ | $A \overset{c(w)}{\leadsto} A' \wedge B \overset{\bar{c}(w)}{\leadsto} B'$ | $\Rightarrow$ | $A\|B \overset{\tau}{\leadsto} (\nu_e w)(A'\|B')$ |
| $close2$ | $A \overset{c(w)}{\leadsto} A' \wedge B \overset{\bar{c}(w)}{\leadsto} B'$ | $\Rightarrow$ | $A\|B \overset{\tau}{\leadsto} (\nu_e w)(A'\|B')$ |
| $nu$ | $A \overset{\gamma}{\leadsto} A' \wedge c \notin Ext\_N\_Act(\gamma)$ | $\Rightarrow$ | $(\nu_e c)A \overset{\gamma}{\leadsto} (\nu_e c)A'$ |
| $open$ | $A \overset{\bar{d}c}{\leadsto} A' \wedge d \neq c \wedge w \notin Ext\_Fn\_Pr((\nu_e c)A)$ | $\Rightarrow$ | $(\nu_e c)A \overset{\bar{d}(w)}{\leadsto} A'[w/c]$ |

**Fig. 2.** Extended $\pi$-calculus transition system

# 5 Implementation

We concentrate now on the issues related to the formalisation of our extended calculus in HOL . The extended languages of processes and actions are introduced in the logic as new types: Ext_Pr and Ext_Act by using the built in facility for defining recursive types:

```
Ext_Pr =  H Pr
         |Pref Ext_Act Ext_Pr
         |Pars Ext_Pr Ext_Pr
         |Parl Ext_Pr Pr
         |Choice Ext_Pr Ext_Pr
         |Comp Ext_Pr Ext_Pr
         |Res Ch Ext_Pr
```

where H,Pref,Pars,Parl,Comp,Choice,Res are distinct constructors.

```
Ext_Act = h Act
         |inp Ch Ch
```

where h,inp are distinct constructors.

The syntactic theory of the extended $\pi$-calculus is developed according to the one developed for the $\pi$-calculus. All the definitions are given in a straightforward manner.

The extended relation `Ext_Trans` is introduced in the logic by using the inductive definition package in the same way as for the formalization of the transition relation `Trans`. These two relations are related via the following theorems:

**From Trans to Ext_Trans**

$$\forall P \; a \; P'. \; \text{Trans } P \; a \; P' \Rightarrow \; \text{Ext\_Trans } (H \; P) \; (h \; a) \; (H \; P')$$

**From Ext_Trans to Trans**

$$\forall P \; a \; A'. \; \text{Ext\_Trans } (H \; P) \; (h \; a) \; A' \Rightarrow \; \exists P'. \text{Trans } P \; a \; P' \wedge (A' = H \; P')$$

The definitions of the `Ext_sim`, `Ext_bisim`, and `Ext_equiv` are defined in the HOL system as:

```
⊢def ∀R.
    Ext_sim R =
    (∀A B.
     R A B ⇒
     (∀A'. Ext_Trans A(h tau)A' ⇒
       (∃B'. Ext_Trans B(h tau)B' ∧ R A' B')) ∧
     (∀c d A'. Ext_Trans A(h (out c d))A' ⇒
       (∃B'. Ext_Trans B(h (out c d))B' ∧ R A' B')) ∧
     (∀c d A'.Ext_Trans A(h (oute c d))A' ∧
        ¬d IN ((Ext_N_Pr A) UNION (Ext_N_Pr B)) ⇒
       (∃B'. Ext_Trans B(h (oute c d))B' ∧ R A' B')) ∧
     (∀c x A'. Ext_Trans A(h (in c x))A' ∧
        ¬x IN ((Ext_N_Pr A) UNION (Ext_N_Pr B)) ⇒
       (∀y.∃B'. Ext_Trans B(h (in c x))B' ∧
         R(Ext_subt1 A'(x,y))(Ext_subt1 B'(x,y)) ∨
         Ext_Trans B(inp c y)B' ∧ R(Ext_subt1 A'(x,y))B')) ∧
     (∀c d A'.Ext_Trans A(inp c d)A' ⇒
       (∃B'.Ext_Trans B(inp c d)B' ∧ R A' B' ∨
         (∃x.Ext_Trans B(h (in c x))B' ∧
           ¬x IN ((Ext_N_Pr A) UNION (Ext_N_Pr B)) ∧
           R A'(Ext_subt1 B'(x,d)))))))
```

$\vdash_{def}$ ∀R. Ext_bisim R = Ext_sim R ∧ Ext_sim(λx y. R y x)

$\vdash_{def}$ ∀A B. Ext_equiv A B = (∃R. R A B ∧ Ext_bisim R)

**From Strong_Sim to Ext_sim**

Let `R` be a strong early simulation. From this relation we can construct an extended strong early simulation, the idea amounts to adding the label **H** to each element of the relation `R`. This is done in HOL by the combinator `G` which is defined as follows:

$\vdash_{def}$ G = ($\lambda$R A B. $\exists$P Q.((($A = H$ P) $\wedge$ ($B = H$ Q)) $\rightarrow$ R P Q | F))

Hence, we can prove the theorem:

$\vdash$ $\forall$R. Strong_Sim R $\Rightarrow$ Ext_sim(G R)

This theorem states that every early strong simulation is also an extended early simulation.

## From Ext_sim to Strong_Sim

Conversely, from an extend strong early simulation, we construct a strong early simulation, by means of the combinator G$^{,}$ which is defined as follows:

$\vdash_{def}$ G$'$ = ($\lambda$R P Q. R(H P)(H Q))

The derived theorem is:

$\vdash$ $\forall$R . Ext_Sim R $\Rightarrow$ Strong_Sim (G$'$ R)

From these theorems, we can prove the following theorem (Ext_equiv_equiv) which asserts that the two equivalences defined on Pr and Ext_Pr coincide.

$\vdash$ $\forall$P Q. Strong_Equiv P Q = Ext_equiv (H P) (H Q)

The proof of this theorem uses the previous results and some properties of the combinators G and G$^{,}$.

Thus, in order to prove that two $\pi$-terms are early equivalent we will use the extended relation. The advantage of this relation is that it can be verified by an effective method based on the algebraic laws and the operational definition. The verification task is then decomposed as follows:

1. Let $P \sim Q$ be the equivalence we want to prove.
2. Apply the theorem Ext_equiv_equiv. So we get the goal: $H$ $P$ $\sim_e$ $H$ $Q$
3. Compute the prefixed form of $H$ $P$ and $H$ $Q$.
4. Compare these prefixed forms according to the early equivalence rules. If the processes are not equivalent then we may fail otherwise we recursively call the step 3 until we reach the terminated process.

Since the $\pi$-terms are finite, termination is guaranteed. Thus we get a decision procedure which decides the bisimulation of finite $\pi$-calculus processes.

In the sequel, we present the most important algebraic laws which allow one to reduce any $\pi$-term to a prefixed form. We also specify a strategy for their application. In practice, this is implemented as a conversion which, given a $\pi$-term A computes its prefixed forms C such that Ext_equiv A C. Then, we present a set of suitable proof rules which allow us to derive that two $\pi$-terms in a prefixed forms are early equivalent.

## Prefixed form

The prefixed form plays a crucial role as the behaviour of the remaining operators is specified on them. It is defined formally as:

$$C ::= \mathbf{H}\ Nil \mid \alpha \bullet A \mid \alpha \bullet A \oplus B$$
$$\alpha ::= \tau \mid c(x) \mid \overline{c}d \mid \overline{c}(d)$$

## Parallel Left

$$\mathbf{H}\ Nil \mid_l A \sim_{ext} \mathbf{H}\ Nil$$

$$\tau \bullet A \mid_l B \sim_{ext} \tau \bullet (A\|B)$$

$$\overline{c}d \bullet A \mid_l B \sim_{ext} \overline{c}d \bullet (A\|B)$$

$$c(x) \bullet A \mid_l B \sim_{ext} c(x) \bullet (A\|B) \ \ if \ x \notin Ext\_Fn\_Pr(B)$$

$$\overline{c}(d) \bullet A \mid_l B \sim_{ext} \overline{c}(d) \bullet (A\|B) \ \ if \ d \notin Ext\_Fn\_Pr(B)$$

$$(\alpha \bullet A \oplus A') \mid_l A'' \sim_{ext} \alpha \bullet A \mid_l A'' \oplus A' \mid_l A''$$

## Synchronisation

$$\mathbf{H}\ Nil \mid_s C' \sim_{ext} \mathbf{H}\ Nil \qquad C' \mid_s \mathbf{H}\ Nil \sim_{ext} \mathbf{H}\ Nil$$

$$c(x) \bullet A \mid_s \overline{c}d \bullet B \sim_{ext} \tau \bullet (A[d/x]\|B) \quad (and\ sym)$$

$$c(x) \bullet A \mid_s \overline{c}(d) \bullet B \sim_{ext} \tau \bullet (\nu_e d)(A[d/x]\|B)) \ if \ (d \notin Ext\_Fn\_Pr(A)) \quad (and\ sym)$$

$$c(x) \bullet A \mid_s \overline{c'}d \bullet B \sim_{ext} (\mathbf{H}\ Nil) \ if \ (c \neq c') \quad (and\ sym)$$

$$c(x) \bullet A \mid_s \overline{c'}(d) \bullet B \sim_{ext} (\mathbf{H}\ Nil) \ if \ (c \neq c') \quad (and\ sym)$$

$$\alpha \bullet A \mid_s (\beta \bullet A' \oplus A'') \sim_{ext} (\alpha \bullet A \mid_s \beta \bullet A') \oplus (\alpha \bullet A \mid_s A'')$$

$$(\alpha \bullet A \oplus A') \mid_s (\beta \bullet B \oplus B') \sim_{ext} \alpha \bullet A \mid_s (\beta \bullet B \oplus B') \oplus A' \mid_s (\beta \bullet B \oplus B')$$

## Choice equation

$$A \oplus \mathbf{H}\ Nil \sim_{ext} A, \ A \oplus B \sim_{ext} B \oplus A, \ A \oplus (A' \oplus A'') \sim_{ext} (A \oplus A') \oplus A''$$

## Restriction

$$(\nu_e c)\mathbf{H}\ Nil \sim_{ext} \mathbf{H}\ Nil \qquad\qquad (\nu_e c)c(x) \bullet A \sim_{ext} \mathbf{H}\ Nil$$
$$(\nu_e c)\overline{c}d \bullet A \sim_{ext} \mathbf{H}\ Nil \qquad\qquad (\nu_e c)\overline{c}(d) \bullet A \sim_{ext} \mathbf{H}\ Nil$$
$$(\nu_e c)\tau \bullet A \sim_{ext} \tau \bullet (\nu_e c)A$$
$$(\nu_e c)d(x) \bullet A \sim_{ext} d(x) \bullet (\nu_e c)A \quad if \quad c \neq d \wedge c \neq x$$
$$(\nu_e c)\overline{d}e \bullet A \sim_{ext} \overline{d}e \bullet (\nu_e c)A \quad\ \ if \quad c \neq d \wedge c \neq e$$
$$(\nu_e c)\overline{d}(e) \bullet A \sim_{ext} \overline{d}(e) \bullet (\nu_e c)A \quad if \quad c \neq d \wedge c \neq e$$
$$(\nu_e c)(A \oplus B) \sim (\nu_e c)A \oplus (\nu_e c)B$$

We present now a method of bringing a $\pi$-term to its prefixed form. A more effective and flexible approach consists in using the algebraic laws for the extended $\pi$-calculus as rewriting rules and implementing a control which guides their application. We propose a formal system in which we can derive judgements: $\vdash A \mapsto C$. This is presented in Fig. 3.

| | |
|---|---|
| | $\Rightarrow C \mapsto C$ |
| $A \mapsto C$ | $\Rightarrow (\mathbf{H}\ Nil) \oplus A \mapsto C$ |
| $A \mapsto C', C' \oplus B \mapsto C$ | $\Rightarrow A \oplus B \mapsto C$ |
| | $\Rightarrow (\alpha \bullet A \oplus A') \oplus B \mapsto \alpha \bullet A \oplus (A' + B)$ |
| | |
| | $\Rightarrow (\nu_e c)C \mapsto C' \qquad (C ::= \mathbf{H}\ Nil \mid \alpha \bullet A)$ |
| | $\quad C'$is obtained according to the restriction law |
| $(\nu_e c)(\alpha \bullet A) \oplus (\nu_e c)B \mapsto C$ | $\Rightarrow (\nu_e c)(\alpha \bullet A \oplus B) \mapsto C$ |
| $A \mapsto C', (\nu_e c)C' \mapsto C$ | $\Rightarrow (\nu_e c)A \mapsto C$ |
| | |
| $A \mid_l B \oplus B \mid_l A \oplus A \mid_s B \mapsto C$ | $\Rightarrow A\|B \mapsto C$ |
| | $\Rightarrow C' \mid_s C'' \mapsto C$ |
| | $\quad C$ is obtained according to the synchronisation law |
| $A \mapsto C', B \mapsto C'', C' \mid_s C'' \mapsto C$ | $\Rightarrow A \mid_s B \mapsto C$ |
| | |
| | $\Rightarrow C' \mid_l B \mapsto C$ |
| | $\quad C$ is obtained according to the left composition law |
| $A \mapsto C', C' \mid_l B \mapsto C$ | $\Rightarrow A \mid_l B \mapsto C$ |

**Fig. 3.** Prefixed form rewriting strategy

Now, we assert that the rewriting relation is sound: every rewriting step applies a valid equivalence of $\sim_{ext}$, thus preserving extended equivalence between terms.

**Proposition 5 (Correction).** *If $A \mapsto C$ then $A \sim_{ext} C$*

# 6 Proof system for Early bisimulation

Finally, we present a set of rules which allow one to derive that two processes are related by extended early bisimulation. To achieve this we use an auxiliary relation $sim_l$ which roughly checks only half of the bisimulation definition.

$$A \sim_l B = (A \xrightarrow{\tau} A' \Rightarrow \exists B'.B \xrightarrow{\tau} B' \wedge A' \sim_{ext} B') \qquad \vee$$
$$(A \xrightarrow{\bar{c}d} A' \Rightarrow \exists B'.B \xrightarrow{\bar{c}d} B' \wedge A' \sim_{ext} B') \qquad \vee$$
$$(A \xrightarrow{\bar{c}(d)} A' \wedge d \notin Ext\_N\_Pr(A\|B) \Rightarrow \exists B'.B \xrightarrow{\bar{c}(d)} B' \wedge A' \sim_{ext} B') \qquad \vee$$
$$(A \xrightarrow{c(x)} A' \wedge x \notin Ext\_N\_Pr(A\|B) \Rightarrow \forall y.\exists B'.(B \xrightarrow{c(x)} B' \wedge A'[y/x] \sim_{ext} B'[y/x]\vee$$
$$B \xrightarrow{c(y)} B' \wedge A'[y/x] \sim_{ext} B')) \qquad \vee$$

$$(A \xrightarrow{c(z)} A' \Rightarrow \exists B'.(B \xrightarrow{c(z)} B' \wedge A' \sim_{ext} B')\vee$$
$$(\exists x.B \xrightarrow{c(x)} B' \wedge x \notin Ext\_N\_Pr(A\|B) \wedge A' \sim_{ext} B'[z/x]))$$

**The proof system**

**Equiv_Rule**

$$\frac{A \sim_l B \wedge B \sim_l A}{A \sim_{ext} B}$$

**Equiv_1_Rule**

$$\overline{\mathbf{H} \; Nil \sim_l A}$$

$$\frac{A \sim_{ext} B}{\tau \bullet A \sim_l \tau \bullet B} \qquad \frac{A \sim_{ext} B}{\bar{c}d \bullet A \sim_l \bar{c}d \bullet B}$$

$$\frac{A \sim_{ext} B}{\bar{c}(d) \bullet A \sim_l \bar{c}(d) \bullet B}$$

$$\frac{z \in Ext\_Fn\_Pr(c(x) \bullet A\|B) \Rightarrow c\langle z \rangle \bullet A[z/x] \sim_l B}{c(x) \bullet A \sim_l B}$$

$$\frac{A \sim_{ext} B[z/y]}{c\langle z \rangle \bullet A \sim_l c(y) \bullet B}$$

$$\frac{\gamma \bullet A \sim_l \alpha \bullet B}{\gamma \bullet A \sim_l \alpha \bullet B \oplus B'} \qquad \frac{\gamma \bullet A \sim_l B'}{\gamma \bullet A \sim_l \alpha \bullet B \oplus B'}$$

$$\frac{\alpha \bullet A \sim_l \beta \bullet B \wedge A' \sim_l \beta \bullet B}{\alpha \bullet A \oplus A' \sim_l \beta \bullet B} \qquad \frac{\alpha \bullet A \sim_l \beta \bullet B \oplus B' \wedge A' \sim_l \beta \bullet B \oplus B'}{\alpha \bullet A \oplus A' \sim_l \beta \bullet B \oplus B'}$$

$$\frac{A \mapsto C' \qquad B \mapsto C'' \qquad C' \sim_l C'''}{A \sim_l B}$$

# 7 Conclusion

We have presented an implementation of a tactic to check early equivalence for a recursion free $\pi$-calculus. This tactic is based on a $\pi$-calculus proof environment we have constructed within the theorem prover HOL. The version of the $\pi$-calculus represented in HOL is that described by Milner & al. in [12] and our encoding follows quite closely Melham's encoding while extending it by deriving within HOL many algebraic laws of the $\pi$-calculus. To deal with the expansion theorem, we have used an alternative formulation based on two auxiliary operators which have allowed us to give an equational presentation of the expansion theorem. They were introduced in the $\pi$-calculus theory by extending the syntax of $\pi$-terms. We have shown that this extension is conservative:

$$\forall P, Q \in Pr. P \sim Q \quad iff \quad P \sim_{ext} Q.$$

Having this theorem, the algebraic laws for the extended $\pi$-calculus are easily derived and the correctness of our tactic is ensured. In future work we hope to extend our approach to deal with all the operators of the $\pi$-calculus. In the presence of the Rec operator our tactic can be applied but we can not guarantee its termination. We intend to add a simple form of induction to deal with recursively defined processes, namely *Unique Fixpoint Induction*:

$$\frac{P = Q[P/x]}{P = Rec\ X\ Q}$$

(this rule is only sound for "guarded processes" [10]).

# Acknowledgments

Thanks are due to my supervisor Roberto Amadio for all his help. I also want to thank anonymous referees for their comments and valuable suggestions.

# References

1. O Aït-Mohamed. Vérification de l'équivalence du $\pi$-calcul dans HOL. Research Report 2412, Institut National de Recherche en Informatique et Automatique, Novembre 1994.
2. R Amadio. On the reduction of chocs bisimulation to $\pi$-calculus bisimulation. In SLNCS 715, editor, *CONCUR93*, pages 112–126, 1993. Also appeared as Research Report Inria-Lorraine 1786, October 1992.
3. R Amadio and O Aït-Mohamed. An analysis of $\pi$-calculus bisimulation. Technical Report 94-2, ECRC, 1994.
4. J A Bergstra and J W Klop. Process algebra for synchronous communication. *Information and Control*, 60:109–137, 1984.
5. J A Bergstra and J W Klop. Algebra of communicating processes with abstraction. *Theoretical Computer Science*, 33:77–121, 1985.

6. A J Camilleri. Mechanizing CSP trace theory in Higher Order Logic. *IEEE Transactions on Software Engineering*, 16(9):993–1004, 1990.

7. T F Melham. Automating recursive type definitions in higher order logic. In G. Birtwistle and P. Subrahmanyam, editors, *Current Trends in Hardware Verification and Automated Theorem Proving*, pages 341–386. Springer-Verlag, 1989.

8. T F Melham. A package for inductive relation definitions in HOL. In P.J. Windly, M. Archer, K.N. Levitt, and J.J Joyce, editors, *Proceedings of the 1991 International Workshop on the HOL Theorem Proving System and its Applications*, pages 350–357. IEEE Computer Society Press, 1992.

9. T F Melham. A mechanized theory of $\pi$-calculus in HOL. *Nordic Journal of Computing*, 1(1):50–76, 1994.

10. R Milner. *Communication and Concurrency*. Prentice Hall, 1989.

11. R Milner. Functions as processes. *Journal of Mathematical Structures in Computer Science*, 2(2):119–141, 1992.

12. R Milner, J Parrow, and D Walker. A calculus of mobile process, part 1-2. *Information and Computation*, 100(1):1–77, 1992.

13. M Nesi. A formalisation of the CCS process algebra in Higher Order Logic. Technical Report 278, Computer Laboratory, University of Cambridge, December 1992.

14. Frederik Orava and Jaochim Parrow. An algebraic verification of a mobile network. *Formal Aspects of Computing*, 4(6):497–543, 1992.

15. D Sangiorgi. *Expressing mobility in process algebras: first-order and higher order paradigms*. PhD thesis, University of Edinburgh, September 1992.

# Non-primitive Recursive Function Definitions

Sten Agerholm

University of Cambridge Computer Laboratory
New Museums Site, Pembroke Street
Cambridge CB2 3QG, UK

**Abstract.** This paper presents an approach to the problem of introducing non-primitive recursive function definitions in higher order logic. A recursive specification is translated into a domain theory version, where the recursive calls are treated as potentially non-terminating. Once we have proved termination, the original specification can be derived easily. A collection of algorithms are presented which hide the domain theory from a user. Hence, the derivation of a domain theory specification has been automated completely, and for well-founded recursive function specifications the process of deriving the original specification from the domain theory one has been automated as well, though a user must supply a well-founded relation and prove certain termination properties of the specification. There are constructions for building well-founded relations easily.

## 1 Introduction

In order to introduce a recursive function in the HOL system, we are required to prove its existence as a total function in higher order logic (see [9], page 263). While this has been automated for certain primitive recursive functions in the type definition package [10], the HOL system does not support the definition of recursive functions which are not also primitive recursive.

This paper presents an approach to this problem via a simple formalization of basic concepts of domain theory in higher order logic. The overall idea is to take a recursive specification, stated in higher order logic and provided by a user, and translate it automatically into a domain theory version where the recursive function is defined by the fixed point operator. This recursive specification differs from the original one since it contains constructs of domain theory to handle the potentially non-terminating recursive calls. This kind of undefinedness is represented by adding a new undefined value to types, a standard construction of domain theory (called lifting).

The original function specification can be derived by eliminating undefinedness, i.e. by proving all recursive calls terminate. One approach is to prove that the argument in each recursive call decreases with respect to some well-founded relation. A binary relation is well-founded if it does not allow any infinite decreasing sequences of values. Hence, by well-founded induction, the recursive calls will terminate eventually, and the specified function is total.

The definition of well-founded recursive functions has been automated such that the details of domain theory and the well-founded induction never appear to the user; she just supplies a recursive specification, a well-founded relation, and a theorem list of termination properties of the specification. In addition, most well-founded relations that occur in practice can be proved very easily using a number of pre-proven constructions.

Our approach is rather pragmatic. While we on one hand want to be able to treat as many recursive specifications as possible, we also want the formalization of domain theory and the automated algorithms to be as simple and efficient as possible. We have therefore identified a number of restrictions of syntactic form which both simplify the formalization and the algorithms. The formalization is designed to allow very smooth and easy transitions between higher order logic and domain theory such that fairly simple tool support can make the domain theory essentially invisible. A similar methodology of having domain theory behind the scenes may be useful for other purposes.

The rest of the paper is organized as follows. The formalization of domain theory is introduced in Section 2. Automation for recursive function definitions in domain theory is presented in Section 3. Section 4 shows how well-founded recursive functions can be obtained from their domain theory versions. Section 6 treats a well-known example and finally, Section 7 contains the conclusions and related work.

**Note** This paper is a condensed version of [5], which contains more examples and a full list of theorems and tools provided by an implementation in HOL88.

## 2   Domain Theory

The basic concepts of domain theory (see e.g. [13]) can be formalized in higher order logic as follows. A partial order is a binary relation $R : \alpha \to \alpha \to bool$ which is reflexive, transitive and antisymmetric:

$$\text{po } R \stackrel{\text{def}}{=}$$
$$(\forall x.\ R\ x\ x) \wedge$$
$$(\forall xyz.\ R\ x\ y \wedge R\ y\ z \Rightarrow R\ x\ z) \wedge (\forall xy.\ R\ x\ y \wedge R\ y\ x \Rightarrow (x = y))\ .$$

A complete partial order is a partial order that contains the least upper bounds of all non-decreasing chains of values:

$$\text{cpo } R \stackrel{\text{def}}{=} \text{po } R \wedge (\forall X.\ \text{chain } R\ X \Rightarrow (\exists x.\ \text{islub } R\ X\ x))\ ,$$

where

$$\text{isub } R\ X\ x \stackrel{\text{def}}{=} (\forall n.\ R(X\ n)x)$$
$$\text{islub } R\ X\ x \stackrel{\text{def}}{=} \text{isub } R\ X\ x \wedge (\forall y.\ \text{isub } R\ X\ y \Rightarrow R\ x\ y)$$
$$\text{chain } R\ X \stackrel{\text{def}}{=} (\forall n.\ R(X\ n)(X(\text{SUC } n)))\ .$$

Note that we do not require that cpos have a least value, but the concrete cpos we use later always have one.

Also essential to domain theory is the notion of continuous function, which is a monotonic function that preserves least upper bounds of chains:

$$\textbf{cont } f(R, R') \overset{\text{def}}{=}$$
$$(\forall xy. \ R \ x \ y \Rightarrow R'(f \ x)(f \ y)) \wedge$$
$$(\forall X. \ \textbf{chain } R \ X \Rightarrow (f(\textbf{lub } R \ X) = \textbf{lub } R'(\lambda n. \ f(X \ n)))) \ ,$$

where lub is defined using the choice operator:

$$\textbf{lub } R \ X \overset{\text{def}}{=} (\epsilon x. \ \textbf{islub } R \ X \ x) \ .$$

Compared to the more powerful formalization presented in [3], a main simplification above is the formalization of partial orders as just relations instead of pairs of sets and relations. This simplification is possible since we restrict ourselves to consider only one special case of the cpo construction on continuous functions, also called the continuous function space. This is important since the general version of this construction would force us to consider subsets of HOL types; not all HOL functions are continuous on arbitrary cpos. As it appeared in [3], this in turn induces the need for partially specified functions, which are specified on such subsets only. In turn, a new $\lambda$-abstraction must be defined to make these partially specified functions determined by their action on the subsets, by ensuring that they yield a fixed arbitrary value outside the subsets. Otherwise, it is not possible to show that continuous functions constitute a cpo with the pointwise ordering.

We are able to manage the entire development with just two different cpo relations, called lrel and frel, which both have simple definitions:

$$\textbf{lrel } x \ y \overset{\text{def}}{=} (x = \textbf{bot}) \vee (x = y)$$
$$\textbf{frel } f \ g \overset{\text{def}}{=} (\forall a. \ \textbf{lrel}(f \ a)(g \ a)) \ .$$

Here bot is a constructor of a new datatype of syntax, written $(\alpha)lift$, which may be specified as follows:

$$v : (\alpha)lift ::= \textbf{bot} \mid \textbf{lift} \ (a : \alpha) \ .$$

Note that 'lift' is the name of both the type being specified and of one of the two constructors, though we use different fonts. The relation lrel ensures that bot is a bottom element, i.e. a least value which can be used to represent undefinedness in partial functions, and behaves as the discrete ordering on lifted values. The relation frel is the pointwise ordering on functions and works on functions with a lifted range type (and an unlifted domain type). It defines a cpo whose bottom element is the everywhere undefined function, i.e. the constant function that sends all values to bot.

Recursive functions can be defined as fixed points of continuous functionals. In the present approach, we restrict ourselves to consider only functions that can be defined as fixed points of continuous functionals on frel, which is the only instance of the continuous function space that we shall use here. Therefore, the fixed point operator is defined in the following special case

$$\texttt{fix } f \overset{\text{def}}{=} \texttt{lub frel}(\lambda n.\ \texttt{power } n\ f)\ ,$$

which is not parameterized over a cpo as it generally would be. The variable $f$ has type $(\alpha \rightarrow (\beta)lift) \rightarrow (\alpha \rightarrow (\beta)lift)$ and **power** is defined by

$$\texttt{power } 0\ f \overset{\text{def}}{=} (\lambda x.\ \texttt{bot})$$
$$\texttt{power}(\texttt{SUC } n)f \overset{\text{def}}{=} f(\texttt{power } n\ f)\ .$$

The fixed point theorem is also stated in a special case:

$$\vdash \forall f.\ \texttt{cont } f(\texttt{frel}, \texttt{frel}) \Rightarrow (\texttt{fix } f = f(\texttt{fix } f))\ .$$

This theorem is essential to the automation in Section 3 where it allows a recursive specification to be derived from the fixed point definition of a function.

Note that a recursive function defined as a fixed point has a type of the form $\alpha \rightarrow (\beta)lift$, where the range type is lifted. This means that recursive calls in its specification cannot be used directly with other HOL terms, which would expect an unlifted term of type $\beta$. In order to solve this problem, we introduce a construction **ext**, called function extension, which can be used to extend HOL functions in a strict way:

$$\texttt{ext } f\ \texttt{bot} \overset{\text{def}}{=} \texttt{bot}$$
$$\texttt{ext } f(\texttt{lift } x) \overset{\text{def}}{=} f\ x\ .$$

For instance, the term $\texttt{ext}(\lambda x.\ x + 5)$ extends addition to a strict function in its first argument. In the automation presented next, we shall use function extension to isolate recursive calls from pure HOL terms.

## 3  Automation for Fixed Point Definitions

The purpose of the above formalization is to serve as a basis for defining recursive functions in higher order logic. Given a recursive specification $g\ x = rhs[g, x]$, where $g$ has a type of the form $\alpha_1 \times \ldots \times \alpha_n \rightarrow \beta$ and therefore is a paired (uncurried) function, we translate the $rhs$ (right-hand side) into a domain theory functional $G$, which has type $(\alpha_1 \times \ldots \times \alpha_n \rightarrow (\beta)lift) \rightarrow (\alpha_1 \times \ldots \times \alpha_n \rightarrow (\beta)lift)$ and is a function on a variant of $g$ with a lifted range type and on $x$. More precisely, $G$ stands for $\lambda g'x.\ rhs'[g', x]$, where $g'$ has type $\alpha_1 \times \ldots \times \alpha_n \rightarrow (\beta)lift$ and $rhs'$ stands for a (lifted) domain theory version of $rhs$. It is constructed by introducing function extension to separate recursive calls (of lifted type) from real HOL terms (of unlifted type). Once we have proved that $G$ is continuous, we can then use the fixed point theorem to obtain $\vdash \texttt{fix } G = G(\texttt{fix } G)$. From this, a domain theory version of the recursive specification is obtained immediately:

$$\vdash \texttt{fix } G\ x = rhs'[\texttt{fix } G, x]\ .$$

A domain theory version of $g$ can be defined as $\texttt{fix}\ G$, but we usually do not do that.

This section describes how we can automate these steps for a large and widely used class of recursive specifications that are written as a list of conditionals:

$$f\,x = (b_1[f,x] \rightarrow h_1[f,x] \mid \ldots \mid b_n[f,x] \rightarrow h_n[f,x] \mid h_{n+1}[f,x])\ .$$

The use of let-terms nested with the conditionals is also supported, for instance:

$$f\,x =$$
$$(b_1[f,x] \rightarrow h_1[f,x] \mid$$
$$\texttt{let}\ y = g[f,x]\ \texttt{in}\ b_2[f,x,y] \rightarrow h_2[f,x,y] \mid h_3[f,x,y])\ .$$

The choice of this kind of a "backbone" of conditionals (possibly mixed with let-terms) is a pragmatic one and not necessary from the viewpoint of domain theory or the automation of this section (though it is exploited). Most recursive programs can be conveniently written using conditional control. In the next section, the conditions are also used as "context" information to prove termination of recursive calls.

Not all HOL terms of the above form are allowed. We also require:

- Recursive occurrences of $g$ in $rhs$ must be recursive calls, i.e. $g$ must always be applied to an argument.
- No recursive call is allowed to appear in the body of a $\lambda$-abstraction (unless it is part of a let-term[1]).

Both restrictions are introduced to avoid the need for more complex cpos of continuous functions than those supported by $\texttt{frel}$, and to allow recursive occurrences of a function to be separated from real HOL terms easily. Due to the restrictions, we avoid function types with a lifted range type in unexpected places (allowed in arguments of $\texttt{ext}$ only). A functional which take a function of this type as an argument is not necessarily continuous.

In the rest of this section we first describes an algorithm for generating the domain theory functional and then an algorithm for proving continuity of the functional; this property is necessary to exploit the fixed point theorem.

## 3.1 Generating the Functional

As explained above the goal is to generate a domain theory version of the right-hand side $rhs[g,x]$. This is done by two recursive algorithms, one for the backbone conditionals and one for branches and conditions. We imagine the backbone algorithm is called first with the right-hand side of a specification.

In the description below, we use primes to indicate that a term has been transformed, and therefore has a lifted type. In particular, the function variable $g : \alpha_1 \times \ldots \times \alpha_n \rightarrow \beta$ is replaced by the primed variable $g' : \alpha_1 \times \ldots \times \alpha_n \rightarrow (\beta)\mathit{lift}$ with a lifted range type. Once $rhs$ has been transformed to $rhs'$ the desired functional called $G$ is obtained by abstracting over $g'$ and $x$.

---

[1] Recall that the let-term $\texttt{let}\ a = t\ \texttt{in}\ t'[a]$ parses to the internal syntax $\texttt{LET}(\lambda a.\ t'[a])t$.

In order to be able to follow the descriptions below more easily, the reader may wish to try them on the following specification of the Ackermann function (see also Section 6):

$ACK(m, n) =$
$((m = 0) \rightarrow \text{SUC } n \mid$
$\quad (n = 0) \rightarrow ACK(\text{PRE } m, 1) \mid ACK(\text{PRE } m, ACK(m, \text{PRE } n)))$ ,

which yields the following translated term:

$((m = 0) \rightarrow \text{lift}(\text{SUC } n) \mid$
$\quad (n = 0) \rightarrow ACK'(\text{PRE } m, 1) \mid$
$\quad \text{ext}(\lambda a. \, ACK'(\text{PRE } m, a))(ACK'(m, \text{PRE } n)))$ .

**Algorithm for Backbone** The input is either a conditional, a let-term, or the last branch of the backbone conditional:

**Conditional:** The input term has the form $(b \rightarrow t_1 \mid t_2)$. The branch $t_2$, which may be a new condition or let-term in the backbone, is transformed recursively, and $t_1$ is transformed using the branch and condition algorithm described below. If the condition does not contain $g$ then the result is $(b \rightarrow t'_1 \mid t'_2)$. Otherwise, $b$ is transformed using the branch and condition algorithm and the result is $\text{ext}(\lambda a. \, (a \rightarrow t'_1 \mid t'_2))b'$, where the condition $b$ has been separated from the conditional using function extension.

**Let-term:** The input has the form $\text{let } a = t_1 \text{ in } t_2$, which may use a list of bindings separated by **and**'s. Transform $t_2$ recursively and use the branch algorithm on $t_1$. The result has the form $\text{ext}(\lambda a. \, t'_2)t'_1$. Lists of bindings are transformed into nested uses of function extension.

**Otherwise:** The term is considered to be the last branch of the backbone and therefore transformed using the branch algorithm.

**Algorithm for Branches and Conditions** The input has no particular form. The purpose of the algorithm is to lift terms that do not contain recursive calls and to isolate recursive calls using function extension in the terms that do.

**No recursive call:** If the variable $g$ does not appear in a free position in the input term $t$, then return $\text{lift } t$.

**Recursive call:** Assume the input term is a recursive call $g(t_1, \ldots, t_n)$. Each $t_i$ that contains $g$ must be transformed recursively. Separate these from the argument pair of $g$ using function extension and replace $g$ with $g' : \alpha_1 \times \ldots \times \alpha_n \rightarrow (\beta)\text{lift}$. Assuming for illustration that $g$ takes four arguments of which the first and the third ones contain $g$, then the result has the form

$$\text{ext}(\lambda a_1. \, \text{ext}(\lambda a_3. \, g'(a_1, t_2, a_3, t_4))t'_3)t'_1 \ .$$

**Let-term:** The input has the form $\text{let } a = t_1 \text{ in } t_2$, which may use a list of bindings separated by **and**'s. Transform $t_1$ and $t_2$ recursively. The result has the form $\text{ext}(\lambda a. \, t'_2)t'_1$. Lists of bindings are transformed into nested uses of function extension.

**Combination:** The term has the form $t\ t_1\ \ldots\ t_n$, where $t$ is not a combination (or an abstraction containing $g$). Each argument of $t$ that contains $g$ is transformed recursively and these arguments are separated from the combination using nested function extensions. The combination in the body of the function extensions is lifted. Assuming for illustration that the input is $t\ t_1\ t_2\ t_3\ t_4$, and that $t_1$ and $t_3$ contain $g$, then the result has the form

$$\texttt{ext}(\lambda a_1.\ \texttt{ext}(\lambda a_3.\ \texttt{lift}(t\ a_1\ t_2\ a_3\ t_4))t_3')t_1'\ .$$

For a simple example consider $5 + g(2,3)$ which is transformed into the term $\texttt{ext}(\lambda a.\ \texttt{lift}(5 + a))(g'(2,3))$.

## 3.2 The Continuity Prover

The most complicated part of the automation is perhaps the continuity prover. Given the functional $G$ constructed above, it must prove the continuity statement: $\texttt{cont}\ G(\texttt{frel}, \texttt{frel})$.

Recall that $G$ is the abstraction $\lambda g'x.\ rhs'[g', x]$. We first prove

$$\vdash \forall x.\ \texttt{cont}(\lambda g'.\ rhs'[g', x])(\texttt{frel}, \texttt{lrel})$$

and then establish the desired result using the continuity-abstraction theorem, which is stated as follows:

$$\vdash \forall h.\ (\forall x.\ \texttt{cont}(\lambda f.\ h\ f\ x)(\texttt{frel}, \texttt{lrel})) \Rightarrow \texttt{cont}(\lambda fx.\ h\ f\ x)(\texttt{frel}, \texttt{frel})\ .$$

To prove the first theorem, we let the conditional and **ext** term structure of $rhs'$ guide our action in a recursive traversal. At each stage of the recursion, we have one of the following four cases (selected top-down):

**No function call:** The term does not contain any free occurrences of $g'$. The desired continuity theorem (up to $\alpha$-conversion) is obtained by instantiating

$$\vdash \forall t.\ \texttt{cont}(\lambda f.\ t)(\texttt{frel}, \texttt{lrel})\ .$$

**Function call:** The term is a function application $g'(t_1, \ldots, t_n)$. Instantiate the following theorem with $(t_1, \ldots, t_n)$ and do an $\alpha$-conversion:

$$\vdash \forall t.\ \texttt{cont}(\lambda f.\ f\ t)(\texttt{frel}, \texttt{lrel})\ .$$

**Conditional:** The term is a conditional $(b \to t_1 \mid t_2)$. Traverse the branches recursively, yielding

$$\vdash \texttt{cont}(\lambda g'.\ t_1)(\texttt{frel}, \texttt{lrel})$$
$$\vdash \texttt{cont}(\lambda g'.\ t_2)(\texttt{frel}, \texttt{lrel})\ .$$

Note that the boolean guard $b$ cannot depend on $g'$ since such dependency would have been removed when the functional was generated. The desired result is obtained essentially by instantiating the following theorem (and using modus ponens):

$$\vdash \forall f_1 f_2.$$
$$\texttt{cont}\ f_1(\texttt{frel}, \texttt{lrel}) \Rightarrow \texttt{cont}\ f_2(\texttt{frel}, \texttt{lrel}) \Rightarrow$$
$$(\forall b.\ \texttt{cont}(\lambda f.\ (b \to f_1\ f \mid f_2\ f))(\texttt{frel}, \texttt{lrel}))\ .$$

**Function extension:** The term is an **ext** term of the form $\textbf{ext}(\lambda a.\ t_1)t_2$. The terms $t_1$ and $t_2$ are traversed recursively, yielding

$$\vdash \textbf{cont}(\lambda g'.\ t_1)(\textbf{frel},\textbf{lrel})$$
$$\vdash \textbf{cont}(\lambda g'.\ t_2)(\textbf{frel},\textbf{lrel})\ .$$

Next, generalizing over $a$, the first of these and the continuity-abstraction theorem can be used to deduce $\vdash \textbf{cont}(\lambda g'a.\ t_1)(\textbf{frel},\textbf{frel})$. The desired result is obtained essentially by instantiating the following theorem:

$$\vdash \forall f_1 f_2.$$
$$\textbf{cont}\ f_1(\textbf{frel},\textbf{frel}) \Rightarrow \textbf{cont}\ f_2(\textbf{frel},\textbf{lrel}) \Rightarrow$$
$$\textbf{cont}(\lambda f.\ \textbf{ext}(f_1\ f)(f_2\ f))(\textbf{frel},\textbf{lrel})\ .$$

This completes the description of the continuity prover.

# 4 Automation for Well-founded Recursive Definitions

In the previous section, it was shown how we can translate a recursive specification in higher order logic to a version where domain theory constructs appear to separate the potentially non-terminating recursive calls from real HOL terms. We must prove recursive calls terminate to eliminate the domain theory. Both systematic and ad hoc approaches to such termination proofs can be employed. This section shows how to automate one of the more powerful systematic ones, which is based on well-founded induction.

A large class of total recursive functions have well-founded recursive specifications. This means that the argument in each recursive call decreases with respect to some well-founded relation. A relation is well-founded if it does not allow any infinite decreasing sequences of values. Hence, by well-founded induction, recursive calls will terminate eventually, and the specified function is total.

A user must supply a well-founded relation and prove the proof obligations for termination, which are the statements saying that arguments in recursive calls decrease. The well-founded induction and the derivation of the original specification from the domain theory one can be automated as described in this section. Well-founded relations are introduced thoroughly in Section 5 below. In particular, a number of constructions are presented to make the proofs of well-foundedness essentially trivial for most relations that appear in practice.

## 4.1 Deriving the Specification

Recall that we derived the domain theory specification

$$\vdash \textbf{fix}\ G\ x = rhs'[\textbf{fix}\ G, x]$$

in the previous section, where $G$ is a meta-variable which stands for the functional $\lambda g'x.\ rhs'[g',x]$ and where $rhs'$ is a domain theory version of the right-hand side of the original higher order logic specification $g\ x = rhs[g,x]$.

In order to derive the original specification, we first prove

$$\vdash \exists g. \forall x. \mathtt{lift}(g\ x) = \mathtt{fix}\ G\ x$$

by well-founded induction; this proof is described in Section 4.3. Constant specification then yields a constant $g$ that satisfies $\vdash \forall x. \mathtt{lift}(g\ x) = \mathtt{fix}\ G\ x$. Rewriting the right-hand side with the domain theory specification above, we obtain $\vdash \forall x. \mathtt{lift}(g\ x) = rhs'[\mathtt{fix}\ G, x]$. We then prove $\vdash rhs'[\mathtt{fix}\ G, x] = \mathtt{lift}(rhs[g, x])$, by straight-forward case analyzes on the conditional backbone of $rhs$, and by exploiting the definition of $g$. Finally, using that the constructor $\mathtt{lift}$ is one-one, we arrive at the original and desired specification:

$$\vdash \forall x.\ g\ x = rhs[g, x]\ .$$

## 4.2   Generating Proof Obligations for Termination

It is easy to generate the proof obligations for termination by traversing the conditional structure of the specification. If a recursive call $g\ y$ appears in the $i$'th branch and the conditions are labeled $p_1, \ldots, p_i$, then the proof obligation computed for that recursive call is $\neg p_1 \wedge \ldots \wedge \neg p_{i-1} \wedge p_i \Rightarrow R\ y\ x$. The first and last branches are obvious special cases. Nested recursive calls are treated by replacing each nested call by a new variable[2].

## 4.3   The Well-founded Induction

We wish to prove the statement $\exists g. \forall x. \mathtt{lift}(g\ x) = \mathtt{fix}\ G\ x$ by well-founded induction. A user supplies a theorem stating some relation, $gR$ say, is well-founded and a theorem list of termination properties of the original specification.

The principle of well-founded induction is stated as follows (see Section 5):

$$\vdash \forall R.\ \mathbf{wf}\ R = (\forall P.\ (\forall x.\ (\forall y.\ R\ y\ x \Rightarrow P\ y) \Rightarrow P\ x) \Rightarrow (\forall x.\ P\ x))\ .$$

Since our induction proofs always have the same structure, it is advantageous to derive the desired instance of this theorem once and for all:

$$\vdash \forall R.$$
$$\mathbf{wf}\ R \Rightarrow$$
$$(\forall f.$$
$$\mathbf{cont}\ f(\mathtt{frel}, \mathtt{frel}) \Rightarrow$$
$$(\forall x.$$
$$(\forall x'.\ R\ x'\ x \Rightarrow (\exists y.\ \mathtt{fix}\ f\ x' = \mathtt{lift}\ y)) \Rightarrow$$
$$(\exists y.\ f(\mathtt{fix}\ f)x = \mathtt{lift}\ y)) \Rightarrow$$
$$(\exists g.\ \forall x.\ \mathtt{lift}(g\ x) = \mathtt{fix}\ f\ x))\ .$$

This is obtained by a few trivial manipulations. The induction predicate of the previous theorem is instantiated with $\lambda x.\ \exists y.\ \mathtt{fix}\ f\ x = \mathtt{lift}\ y$. Then the

---

[2] It may be necessary to restrict the range of the variable in order to prove the proof obligation. This can be done by introducing a condition in the backbone to express some property of the nested recursive call, and hence the variable (see [5]).

consequent of the theorem is skolemized, which means that the existential $\exists y$ is moved outside the $\forall x$ where it becomes $\exists g$; note that $y$ is a value while $g$ is a function. Symmetry of equality is also used on the consequent. Then the continuity assumption is used to obtain the term $\exists y.\ f(\text{fix } f)x = \text{lift } y$ instead of $\exists y.\ \text{fix } f\ x = \text{lift } y$ in the induction proof (i.e. the third antecedent); the fixed point theorem justifies this.

Returning to the (high-level) example specification of $g$, the first two assumptions of the previous theorem are discharged by the user-supplied theorem $\vdash$ wf $gR$ and the continuity prover (see Section 3.2), respectively. The last assumption yields the induction proof:

$$\forall x.$$
$$(\forall x'.\ gR\ x'\ x \Rightarrow (\exists y.\ \text{fix } G\ x' = \text{lift } y)) \Rightarrow$$
$$(\exists y.\ G(\text{fix } G)x = \text{lift } y)\ .$$

The proof of this is guided by the syntactic structure of the term $G(\text{fix } G)x$, which by $\beta$-conversion is equal to $rhs'[\text{fix } G, x]$. A case analysis is done for each condition of the conditional backbone. For each recursive call, there must be a termination theorem in the user-supplied theorem list. This allows us to use the induction hypothesis, i.e. the antecedent above. Hence, from the hypothesis and some proof obligation we derive that each recursive call terminates, which is the same as saying that it is equal to some lifted value. In this way, we become able to reduce away all occurrences of **ext**, due to the way it behaves on lifted values, and arrive at statements of the form $\exists y.\ \text{lift } t = \text{lift } y$, which hold trivially.

## 5 Well-founded Relations

A binary relation is defined to be well-founded on some type if all non-empty subsets of the type have a minimal element with respect to the relation:

$$\text{wf } R \overset{\text{def}}{=} (\forall A.\ A \neq (\lambda x.\ \text{F}) \Rightarrow (\exists x.\ A\ x \wedge \neg(\exists y.\ A\ y \wedge R\ y\ x)))\ .$$

The HOL theory of well-founded relations presented here was obtained by developing a special case of the theory presented in [1], which was based on a chapter of the book by Dijkstra and Scholten [8].

In general, it can be non-trivial to prove a given relation is well-founded. It is therefore useful to have standard ways of combining well-founded relations to build new ones easily. The theory provides the following standard constructions on well-founded relations:

**Less-than on numbers:** ML name wf_less:

$$\vdash \text{wf } \$<\ .$$

**Product:** ML name wf_prod:

$$\vdash \forall R.\ \text{wf } R \Rightarrow (\forall R'.\ \text{wf } R' \Rightarrow \text{wf}(\text{prod}(R, R')))\ .$$

Defined by

$$\text{prod}(R, R')b\ c \overset{\text{def}}{=} R(\text{FST } b)(\text{FST } c) \wedge R'(\text{SND } b)(\text{SND } c)\ .$$

**Lexicographic combination:** ML name `wf_lex`:

$$\vdash \forall R. \text{ wf } R \Rightarrow (\forall R'. \text{ wf } R' \Rightarrow \text{wf}(\text{lex}(R, R'))) \ .$$

Defined by

$$\text{lex}(R, R')b \ c \overset{\text{def}}{=}$$
$$R(\text{FST } b)(\text{FST } c) \vee (\text{FST } b = \text{FST } c) \wedge R'(\text{SND } b)(\text{SND } c) \ .$$

**Inverse image:** ML name `wf_inv_gen`:

$$\vdash \forall R. \text{ wf } R \Rightarrow (\forall R'f. \ (\forall xy. \ R' \ x \ y \Rightarrow R(f \ x)(f \ y)) \Rightarrow \text{wf } R') \ .$$

A useful special case of the construction is (ML name `wf_inv`):

$$\vdash \forall R. \text{ wf } R \Rightarrow (\forall f. \ \text{wf}(\text{inv}(R, f))) \ .$$

Defined by

$$\text{inv}(R, f)x \ y \overset{\text{def}}{=} R(f \ x)(f \ y) \ .$$

Most well-founded relations that appear in practice can be obtained easily by instantiating these constructions.

When these built-in constructions do not suffice, a relation can be proved to be well-founded immediately from the definition of `wf`, or, which is often more convenient, either from the theorem

$$\vdash \forall R. \text{ wf } R = \neg(\exists X. \ \forall n. \ R(X(\text{SUC } n))(X \ n)) \ ,$$

which states that a relation is well-founded if and only if there are no infinite decreasing sequences of values, or from the principle of well-founded induction:

$$\vdash \forall R. \text{ wf } R = (\forall P. \ (\forall x. \ (\forall y. \ R \ y \ x \Rightarrow P \ y) \Rightarrow P \ x) \Rightarrow (\forall x. \ P \ x)) \ ,$$

which states that a relation is well-founded if and only if it admits mathematical induction. Note that this theorem can be used both to prove a relation is well-founded by proving it admits induction and to perform an induction with a relation which is known to be well-founded.

# 6 Example

The theories and algorithms presented above have been implemented in HOL88. In this section, we illustrate the use of the implemented tools on a famous example of a well-founded recursive function: the (binary) Ackermann function. The theorems and tools of the implementation are described in greater detail in [5]. In particular, this section only illustrates the automation for well-founded recursive function definitions. It does not mention the support for generating the intermediate domain theory specification from which one may then proceed towards the original specification by any method of proof at hand.

Often the Ackermann function is specified by a collection of recursion equations like

$$A(0, y) = y + 1$$
$$A(x + 1, 0) = A(x, 1)$$
$$A(x + 1, y + 1) = A(x, A(x + 1, y)) \ ,$$

which are equivalent to the following conditional style of specification in HOL:

```
"ACK(m,n) =
 ((m = 0) => SUC n |
  (n = 0) => ACK(PRE m,1) | ACK(PRE m,ACK(m,PRE n)))" .
```

In this term, called ack_tm below, ACK is a variable; but we wish to obtain a constant ACK that satisfies the recursive specification. Note that the Ackermann function is not primitive recursive since it cannot be defined using the syntax of primitive recursive specifications.

In addition to the recursive specification, we must supply a well-founded relation and a list of termination properties of the specification. An ML function calculates the proof obligations for termination:

```
#calc_prf_obl ack_tm;;
["~(m = 0) /\ (n = 0) ==> R(PRE m,1)(m,n)";
 "~(m = 0) /\ ~(n = 0) ==> R(PRE m,k0)(m,n)";
 "~(m = 0) /\ ~(n = 0) ==> R(m,PRE n)(m,n)"]
: term list
```

These are constructed by looking at the arguments of each recursive call as described in Section 4.2. The variable k0 is introduced due to the nested recursive call in the last branch. Note that the ML function does not guess a well-founded relation but uses instead a variable R. We must find a proper instantiation for R and prove each resulting term is a theorem. It is easy to see that a suitable well-founded relation in this example is a lexicographic combination of the less-than ordering on natural numbers with itself. Proving that this relation is well-founded is trivial since lexicographic combination and the less-than ordering are standard constructions on well-founded relations (see Section 5):

```
#let wf_ack = MATCH_MP (MATCH_MP wf_lex wf_less) wf_less;;
wf_ack = |- wf(lex($<,$<))
```

Hence, we substitute lex($<,$<) for the variable R in the proof obligations above (there is a separate tool for this). We shall omit the proofs here but assume the proven termination properties have been saved in the ML variable obl_thl.

The Ackermann function can now be defined automatically using a derived definition tool called new_wfrec_definition. This introduces a new constant ACK and proves that it satisfies the recursive specification presented above:

```
#let ACK_DEF = new_wfrec_definition 'ACK_DEF' wf_ack obl_thl ack_tm;;
ACK_DEF =
|- !m n.
    ACK(m,n) =
    ((m = 0) => SUC n |
     (n = 0) => ACK(PRE m,1) | ACK(PRE m,ACK(m,PRE n)))
```

# 7  Conclusions and Related Work

This work was motivated by previous work on formalizing domain theory in HOL [3]. The Ackermann example was also considered there, though it was only treated manually and in a much more complicated domain theoretic framework than here. Since the present approach exploits domain theory in a very precise and concrete way, we have been able to instantiate the theory considerably. By tying domain theory up closely with higher order logic, we become able to restrict our use to involve only two different cpos and one kind of continuous functional for recursive definitions via the fixed point operator. We also avoid the need for a dependent λ-abstraction for writing functions, which was a main reason for complication in [3]. These simplifications and the design and engineering of proper tools were the main challenges of this work.

A goal was to make domain theory as invisible as possible. Indeed, in defining well-founded recursive functions the user never sees any domain theory, and in other cases the domain theory constructs are introduced automatically and have a very simple form. Their purpose is to separate the potentially non-terminating recursive calls from real HOL terms, which do not support a notion of undefinedness directly.

There might be recursive definitions that cannot be introduced by the present approach. A main restriction might be that a recursive call is not allowed to appear in the body of a λ-abstraction (unless it is part of a let-expression). The problem occurs when the recursive call uses the variable of the abstraction, in other cases the call can be moved outside the body. It might be possible to implement support for abstractions, but this would probably complicate the domain theory and the associated algorithms considerably. Another potential restriction is that all recursive occurrences of a function in the right-hand side of a specification must be applied to arguments, i.e. we do not support unapplied occurrences of functions. Finally, functions must be specified using conditionals, possibly nested with let-expressions. This conditional style is fairly powerful but we have no evidence that it will work for all applications in practice.

Konrad Slind has developed a similar package for well-founded recursive function definitions (in HOL90), but this does not support other recursive functions. Its implementation is based on the well-founded recursion theorem, which gives a more direct and efficient implementation, since all domain theory is avoided. Further, the well-founded induction is performed once and for all in the proof of the well-founded recursion theorem, whereas in the present approach an induction is performed for each definition. However, the advantage of domain theory is that it allows a version of a recursive function to be defined directly without proving whether it is total or not. Sometimes, recursive functions are undefined for some arguments due to non-terminating recursive calls, or the proof of termination may depend on correctness properties of the function; in both cases it is advantageous that we can define a version of the function and reason about this before deriving the desired function.

Mark van der Voort describes another approach to introducing well-founded recursive function definitions in [12], inspired by the one employed in the Boyer-Moore prover [6]. Like Slind, he also avoids domain theory and does not treat more general recursive functions. Though he supports well-founded recursive functions, he supplies a natural number measure with each definition instead of a well-founded relation (following Boyer-Moore). A recursive call must reduce this measure with respect to the less-than ordering. It seems more direct to use well-founded relations rather than a measure which destroys the structure of data. Further, a consequence is that an induction principle must be derived with each recursive definition.

Tom Melham's package for inductive relation definitions [11, 7] could be used to define many recursive functions as well. This would require a recursive specification to be translated into a set of inference rules that gives an inductive definition of a relation representation of the function. The recursive function could then be extracted from the inductively defined relation by a uniqueness proof, showing that the relation specifies a (potentially partial) function, and a definedness proof, showing that the relation specifies a total function. It is difficult to say whether or not such an approach would be simpler than the present one.

## Acknowledgements

The research described here was supported by an HCMP fellowship under the EuroForm network, and partly supported by BRICS[3]. I would like to thank the anonymous referees for many useful suggestions for improvements.

## References

1. S. Agerholm, *Mechanizing Program Verification in HOL*. M.Sc. thesis, Aarhus University, Computer Science Department, Report IR-111, April 1992. See also: *Proceedings of the 1991 International Workshop on the HOL Theorem Proving System and Its Applications*, IEEE Computer Society Press, 1992.
2. S. Agerholm, 'Domain Theory in HOL'. In the *Proceedings of the 6th International Workshop on Higher Order Logic Theorem Proving and its Applications*, Jeffrey J. Joyce and Carl-Johan H. Seger (Eds.), LNCS 780, Springer-Verlag, 1994.

---

[3] Basic Research in Computer Science, Centre of the Danish National Research Foundation.

3. S. Agerholm, *A HOL Basis for Reasoning about Functional Programs*. Ph.D. Thesis, BRICS RS-94-44, University of Aarhus, Department of Computer Science, December 1994.

4. S. Agerholm, 'LCF Examples in HOL'. The Computer Journal, Vol. 38, No. 2, 1995.

5. S. Agerholm, 'A Package for Non-primitive Recursive Function Definitions in HOL'. Technical Report No. 370, University of Cambridge Computer Laboratory, 1995.

6. R.S. Boyer and J.S. Moore, *A Computational Logic*. Academic Press, 1979.

7. J. Camilleri and T.F. Melham, 'Reasoning with Inductively Defined Relations in the HOL Theorem Prover'. Technical Report No. 265, University of Cambridge Computer Laboratory, August 1992.

8. E.W. Dijkstra and C. Scholten, *Predicate Calculus and Program Semantics*. Springer-Verlag, 1990.

9. M.J.C. Gordon and T.F. Melham (Eds.), *Introduction to HOL: A Theorem Proving Environment for Higher Order Logic*. Cambridge University Press, 1993.

10. T.F. Melham, 'Automating Recursive Type Definitions in Higher Order Logic'. In G. Birtwistle and P.A. Subrahmanyam (Eds.), *Current Trends in Hardware Verification and Theorem Proving*, Springer-Verlag, 1989.

11. T. Melham, 'A Package for Inductive Relation Definitions in HOL'. In the *Proceedings of the 1991 International Workshop on the HOL Theorem Proving System and Its Applications*, IEEE Computer Society Press, 1992.

12. M. van der Voort, 'Introducing Well-founded Function Definitions in HOL'. In *Higher Order Logic Theorem Proving and its Applications* (HOL workshop proceedings 1992), L.J.M. Claesen and M.J.C. Gordon (Eds.), IFIP Transactions A-20, North-Holland, 1993.

13. G. Winskel, *The Formal Semantics of Programming Languages*. The MIT Press, 1993.

# Experiments with ZF Set Theory in HOL and Isabelle

Sten Agerholm and Mike Gordon

University of Cambridge Computer Laboratory
New Museums Site, Pembroke Street
Cambridge CB2 3QG, UK

**Abstract.** Most general purpose proof assistants support versions of typed higher order logic. Experience has shown that these logics are capable of representing most of the mathematical models needed in Computer Science. However, perhaps there exist applications where ZF-style set theory is more natural, or even necessary. Examples may include Scott's classical inverse-limit construction of a model of the untyped $\lambda$-calculus ($D_\infty$) and the semantics of parts of the Z specification notation. This paper compares the representation and use of ZF set theory within both HOL and Isabelle. The main case study is the construction of $D_\infty$. The advantages and disadvantages of higher-order set theory versus first-order set theory are explored experimentally. This study also provides a comparison of the proof infrastructure of HOL and Isabelle.

## 1 Introduction

Set theory is the standard foundation for mathematics and for formal notations like Z [30], VDM [14] and TLA+ [15]. However, most general purpose mechanised proof assistants support typed higher order logics (type theories). Examples include Alf [17], Coq [7], EHDM [19], HOL [12], IMPS [9], LAMBDA [10], LEGO [16], Nuprl [5], PVS [26] and Veritas [13]. For many applications type theory works well, but there are certain classical constructions, like the definition of the natural numbers as the set $\{\emptyset, \{\emptyset\}, \{\emptyset, \{\emptyset\}\}, \{\emptyset, \{\emptyset\}, \{\emptyset, \{\emptyset\}\}\}, \cdots \}$, that are essentially untyped. Furthermore, the development of some branches of mathematics, e.g. abstract algebra, are problematical in type theory. In the long term it might turn out that such areas can be satisfactorily developed in a type-theoretic setting, but research is needed to establish this. For immediate practical applications it seems hard to justify not just taking ordinary (i.e. set-theoretic) mathematics 'off the shelf'.

Several proof assistants for set theory are available. In the formal methods community[1] the best known are probably Isabelle [23], a generic system that supports various kinds of type theory as well as ZF set theory, and EVES [28]. Both of these have considerable automation and are capable of proving difficult theorems. The Mizar system from Poland [27] is a low level proof checker based

---

[1] In other communities (e.g. artificial intelligence) there is related work on theorem proving for set theory (e.g. Ontic [18]).

on set theory that has been used to check enormous amounts of mathematics – there is even a journal devoted to it [8]. Mizar has only recently become well-known in the theorem proving community. Another active group focusses on metatheory and decision procedures for fragments of set theory [4]. Like Mizar, this work is not well known in the applied verification community.

Work with Isabelle provides particular insight into type theory versus set theory because it supports both. Users of Isabelle can choose between either ZF or various styles of type theory (including HOL-like higher order logic and Martin Löf type theory), depending on which seems most appropriate for the application in hand. However, the theories do not interface to each other, so one cannot move theorems back and forth between them (except by manually porting proofs). Anecdotal evidence from Isabelle users suggests that, for equivalent kinds of theorems, proof in higher order logic is usually easier and shorter than in set theory, but that certain general constructions (e.g. defining models of recursively defined datatypes) are easier to do in set theory. One reason why set-theoretic proofs can be more tedious is because they may involve the verification of set membership conditions; the corresponding conditions in higher order logic being handled automatically by type checking.[2]

It is hoped that the experiments reported here will clarify the strengths and weaknesses of type theory versus set theory as frameworks for mechanising the mathematics needed to support computer system verification. A long term goal is to develop a way of getting the best of both worlds in a single framework.

The rest of this paper is organised as follows: Section 2 presents a version of ZF set theory axiomatised in the HOL logic (this will be referred to as HOL-ST). This is then explored via a case study: the construction of $D_\infty$, Scott's model of the $\lambda$-calculus (Section 3). In Section 4, Isabelle's set theory and proof infrastructure is discussed in relation to HOL-ST. The final section (Section 5) summarises our conclusions.

# 2 A ZF-like Set Theory in HOL

This section is condensed from a more leisurely and detailed exposition presented elsewhere [11].

## 2.1 Standard Axioms of Set Theory

HOL-ST is an axiomatic theory based on ZF, but formulated in higher order logic. The theory, called ST, introduces a new type $V$ and a new constant $\in$ : $V \times V \to bool$, together with higher-order versions of the ZF axioms.

Because they are formulated in higher order logic, the axioms described below are strictly stronger than ZF. Furthermore, although the axiom of choice is not

---

[2] Note, however, that explicit proof of membership conditions in set theory results in more being formally proved than with type checking, which is typically done by programs outside the logic; i.e. what is formally proved in the former case is just calculated in the latter.

stated explicitly, it is easily proved in HOL using the choice function ($\varepsilon$-operator). An axiom asserting the existence of unordered pairs is also not needed as this follows from Replacement (pairs are images of $\mathbb{P}\ \emptyset$). The axioms of ST in the HOL logic are as follows.

**Extensionality :**  $\forall s\ t.\ (s = t) = (\forall x.\ x \in s = x \in t)$

**Empty set :**  $\exists s.\ \forall x.\ \neg(x \in s)$

**Union :**  $\forall s.\ \exists t.\ \forall x.\ x \in t = (\exists u.\ x \in u \wedge u \in s)$

**Power sets :**  $\forall s.\ \exists t.\ \forall x.\ x \in t = x \subseteq s$

**Separation :**  $\forall p\ s.\ \exists t.\ \forall x.\ x \in t = x \in s \wedge p\ x$

**Replacement :**  $\forall f\ s.\ \exists t.\ \forall y.\ y \in t = \exists x.\ x \in s \wedge (y = f\ x)$

**Foundation :**  $\forall s.\ \neg(s = \emptyset) \Rightarrow \exists x.\ x \in s \wedge (x \cap s = \emptyset)$

**Infinity :**  $\exists s.\ \emptyset \in s \wedge \forall x.\ x \in s \Rightarrow (x \cup \{x\}) \in s$

These axioms use the auxiliary constants $\subseteq, \emptyset, \cup, \cap$ and singleton sets, which are easily defined prior to their use.

## 2.2  Auxiliary Set-theoretic Notions

Familiar set-theoretic notions can be defined from the axioms of ST using the definitional mechanisms of HOL. Some useful constants are listed below (see [11] for more explanation):

**Empty set :**  $\forall x.\ \neg(x \in \emptyset)$

**Subset relation :**  $s \subseteq t = \forall x.\ x \in s \Rightarrow x \in t$

**Union :**  $x \in (s \cup t) = x \in s \vee x \in t$

**Big union :**  $x \in \bigcup s\ =\ (\exists u.\ x \in u \wedge u \in s)$

**Power set :**  $x \in \mathbb{P}\ s\ =\ x \subseteq s$

**Set abstraction (separation) :**  $x \in \text{Spec}\ s\ p\ =\ x \in s \wedge p\ x$

**Simple abstraction notation :**  $\{x \in s \mid p[x]\}\ =\ \text{Spec}\ s\ (\lambda x.\ p[x])$

**Generalized abstraction notation** :
  $\{t[x_1,\ldots,x_n] \in s \mid p[x_1,\ldots,x_n]\}$ =
    Spec s $(\lambda x. \exists x_1\ldots x_n. (x = t[x_1,\ldots,x_n]) \wedge p[x_1,\ldots,x_n])$

**Image of a (logical) function** :
  $y \in$ Image f s $= \exists x. x \in s \wedge (y = f\ x)$

**Indexed union** :  $\bigcup_{x \in X} f(x)$  $=$  $\bigcup(\text{Image f X})$

**Infinite set** :
  $\emptyset \in$ InfiniteSet $\wedge\ \forall x. x \in$ InfiniteSet $\Rightarrow (x \cup \{x\}) \in$ InfiniteSet

**Finite sets** : $\{x_1, x_2, \ldots, x_n\}$  $=$  $\{x_1\} \cup (\{x_2\} \cup \ldots (\{x_n\} \cup \emptyset))$

**Ordered pairs** : $\langle x,y \rangle$  $=$  $\{\{x\}, \{x,y\}\}$

**Maplet notation** : $x \mapsto y$  $=$  $\langle x,y \rangle$

**Cartesian products** :
  $X \times Y$  $=$  $\{\langle x_1,x_2 \rangle \in \mathbb{P}(\mathbb{P}(X \cup Y)) \mid x_1 \in X \wedge x_2 \in Y\}$

**Relations** : $X \leftrightarrow Y$  $=$  $\mathbb{P}(X \times Y)$

**Partial functions** :
  $X \nrightarrow Y$ =
  $\{f \in X \leftrightarrow Y \mid \forall x\ y_1\ y_2. x \mapsto y_1 \in f \wedge x \mapsto y_2 \in f \Rightarrow (y_1 = y_2)\}$

**Total functions** :
  $X \rightarrow Y$ = $\{f \in X \nrightarrow Y \mid \forall x. x \in X \Rightarrow \exists y. y \in Y \wedge \langle x,y \rangle \in f\}$

**Set function application** : $f \diamond x$  $=$  $\varepsilon y. x \mapsto y \in f$

**Set function abstraction** :
  $\lambda x \in X. f(x)$  $=$  $\{x \mapsto y \in X \times$ Image f X $\mid y = f\ x\}$

**Natural numbers** :
  num2Num 0  $=$  $\emptyset$
  num2Num(SUC n)  $=$  num2Num n $\cup$ {num2Num n}

  Num  $=$  $\{x \in$ InfiniteSet $\mid \exists n. x =$ num2Num n$\}$

  $(\forall n.$ Num2num(num2Num n)  $=$  n$)\ \wedge$
  $(\forall x. x \in$ Num  $=$  (num2Num(Num2num x) = x$))$

Note that functions of higher order logic are different from functions 'inside' set theory, which are just certain kinds of set of ordered pairs. The former will be called *logical functions* and the latter *set functions*. Logical functions have a type in HOL of the form $\alpha \to \beta$ and are denoted by terms of the form $\lambda x.\ t[x]$, whereas set functions have type $V$ and are denoted by terms of the form $\lambda x \in X.\ s[x]$. The application of a logical function $f$ to an argument $x$ is denoted by $f\ x$, whereas the application of a set function $s$ to $x$ is denoted by $s \diamond x$.

## 2.3   Identity and Composition

The identity and composition functions can be defined in set theory. The identity function is defined by:

$$\mathsf{Id}\, X \;=\; \{\langle x, y\rangle \in X \times X \mid x = y\}.$$

Note that it would not be possible to define a set for $\mathsf{Id}$ without the set argument $X$ since, to avoid paradoxes, sets must be built from existing ones.

The composition function could be defined in the same way, but we can exploit the presence of the (set) function arguments to avoid further arguments:

$$f \, O \, g \;=\; \{\langle x, z\rangle \in \mathsf{domain}\, g \times \mathsf{range}\, f \mid \exists y.\ \langle x, y\rangle \in g \wedge \langle y, z\rangle \in f\}.$$

Here, domain and range are defined using Image by taking the first and second components of each pair in a set function. Note that the composition function is a logical function, it has type $V \to V \to V$.

## 2.4.   Dependent Sum and Product

The dependent sum is a generalisation of Cartesian products $X \times Y$, the set of all pairs of elements in $X$ and $Y$. In the dependent sum, the second component may depend on the first:

$$\sum_{x \in X} Y\, x \;=\; \bigcup_{x \in X} \bigcup_{y \in Y\, x} \{\langle x, y\rangle\}.$$

Here, $Y\, x$ can be any term of type $V$ containing $x$ or not. In the same way, the dependent product is a generalisation of $X \to Y$:

$$\prod_{x \in X} Y\, x \;=\; \{f \in \mathbb{P}(\sum_{x \in X} Y\, x) \mid \forall x \in X.\ \exists! y.\ \langle x, y\rangle \in f\}.$$

Clearly, both $\times$ and $\to$ are special cases of the dependent sum and the dependent product, respectively. In fact, the function abstraction introduced above yields an element of the dependent product:

$$\forall f X Y.\ (\forall x.\ x \in X \Rightarrow f\, x \in Y\, x) \Rightarrow (\lambda x \in X.\, f\, x) \in \prod_{x \in X} Y\, x.$$

The following application of HOL-ST will make essential use of the dependent product to overcome limitations in the type system of HOL's version of higher order logic.

# 3   Case Study: the Inverse Limit Construction

This section, which is condensed from a longer report [1], presents a case study of the application of HOL-ST to domain theory. This work was motivated by previous work on formalising domain theory in HOL [2] where it became clear that the inverse limit construction of solutions to recursive domain equations cannot be formalised in HOL easily. The type system of higher order logic is not rich enough to represent Scott's inverse limit construction directly, though other methods of solving recursive domain equations can be encoded [24]. However, the construction can be formalised in HOL-ST directly, because set theory supports general dependent products whereas higher order logic does not.

The inverse limit construction can be used to give solutions to any recursive domain (isomorphism) equation of the form

$$D \cong \mathcal{F}(D)$$

where $\mathcal{F}$ is an operator on domains, such as sum, product or (continuous) function space, or any combination of these. Thus, it might be possible to implement tools based on the present formalisation that support very general recursive datatype definitions in HOL-ST.

This will be illustrated with the construction of a domain $D_\infty$ satisfying:

$$D_\infty \cong [D_\infty \to D_\infty],$$

where $[D \to E]$ denotes the domain of all continuous function from $D$ to $E$ (we shall always use $[\ \to\ ]$ for continuous functions and $\to$ for ordinary functions). Hence, $D_\infty$ provides a model of the untyped $\lambda$-calculus.

The version of the inverse limit construction employed here is based on categorical methods using embedding-projection pairs, see e.g. [29, 25]. This was suggested by Plotkin as a generalisation of Scott's original inverse limit construction of a model of the $\lambda$-calculus in the late 60's. The formalisation is based on Paulson's accessible presentation in the book [20] but Plotkin's [25] was also used in part (in fact, Paulson based his presentation on this).

## 3.1   Basic Concepts of Domain Theory

Domain theory is the study of complete partial orders (cpos) and continuous functions between cpos. This section very briefly introduces the semantic definitions of central concepts of domain theory in HOL-ST.

A partial order is a pair consisting of a set and a binary relation such that the relation is reflexive, transitive and antisymmetric on all elements of the set:

po $D$ =
$\forall x \in$ set $D$. rel $D\ x\ x\ \wedge$
$\forall xyz \in$ set $D$. rel $D\ x\ y \wedge$ rel $D\ y\ z \Rightarrow$ rel $D\ x\ z\ \wedge$
$\forall xy \in$ set $D$. rel $D\ x\ y \wedge$ rel $D\ y\ x \Rightarrow x = y$.

The constants set and rel equal FST and SND respectively. The constant po has type $V \times (V \to V \to bool) \to bool$.

A complete[3] partial order is defined as a partial order in which all chains have least upper bounds (lubs):

$$\text{cpo } D = \text{po } D \wedge \forall X. \text{ chain } D \, X \Rightarrow \exists x. \text{ is\_lub } D \, X \, x,$$

where

$$\text{chain } D \, X = (\forall n. \, X \, n \in \text{set } D) \wedge (\forall n. \text{ rel } D(X \, n)(X(n+1)))$$
$$\text{is\_ub } D \, X \, x = x \in \text{set } D \wedge \forall n. \text{ rel } D(X \, n)x$$
$$\text{is\_lub } D \, X \, x = \text{is\_ub } D \, X \, x \wedge \forall y. \text{ is\_ub } D \, X \, y \Rightarrow \text{rel } D \, x \, y.$$

Hence, a chain of elements of a partial order is a non-decreasing sequence of type $num \to V$.

The set of continuous functions is the subset of the set of monotonic functions that preserve lubs of chains:

$$\text{mono}(D, E) =$$
$$\{f \in \text{set } D \to \text{set } E \mid \forall xy \in \text{set } D. \text{ rel } D \, x \, y \Rightarrow \text{rel } E(f \diamond x)(f \diamond y)\}$$
$$\text{cont}(D, E) =$$
$$\{f \in \text{mono}(D, E) \mid$$
$$\forall X. \text{ chain } D \, X \Rightarrow f \diamond (\text{lub } D \, X) = \text{lub } E(\lambda n. f \diamond (X \, n))\}.$$

Note that continuous (and monotonic) functions are set functions and therefore have type $V$.

The continuous function space construction on cpos is defined as the pair consisting of the set of continuous functions between two cpos and the pointwise ordering relation on functions:

$$\text{cf}(D, E) = (\text{cont}(D, E), \lambda fg. \forall x \in \text{set } D. \text{ rel } E(f \diamond x)(g \diamond x)).$$

The construction always yields a cpo if its arguments are cpos.

## 3.2 The Inverse Limit Construction

Just as chains of elements of cpos have least upper bounds, there are "chains" of cpos that have "least upper bounds", called *inverse limits*. The ordering relation on elements is generalised to the notion of embedding morphisms between cpos. A certain constant Dinf, parametrised by a chain of cpos, can be proven once and for all to yield the inverse limit of the chain. In this section, we give a brief overview of a formalisation of the inverse limit construction.

---

[3] Our notion of completeness is sometimes called $\omega$-completeness. Similarly, the chains are sometimes called $\omega$-chains.

Embedding morphisms come in pairs with projections, forming the so-called embedding-projection pairs:

$\mathsf{projpair}(D, E)(e, p) =$
$\quad e \in \mathsf{cont}(D, E) \wedge p \in \mathsf{cont}(E, D) \wedge$
$\quad p \circ e = \mathsf{Id}(\mathsf{set}\, D) \wedge \mathsf{rel}(\mathsf{cf}(E, E))(e \circ p)(\mathsf{Id}(\mathsf{set}\, E)).$

The conditions make sure that the structure of $E$ is richer than that of $D$ (and can contain it). $D$ is embedded into $E$ by $e$ (one-one) which in turn is projected onto $D$ by $p$.

Embeddings uniquely determine projections (and vice versa). Hence, it is enough to consider embeddings

$\mathsf{emb}(D, E)e = \exists p.\ \mathsf{projpair}(D, E)(e, p)$

and define the associated projections, or *retracts* as they are often called, using the choice operator:

$R(D, E)e = \varepsilon p.\ \mathsf{projpair}(D, E)(e, p).$

For readability, $\mathsf{emb}(D, E)e$ will sometimes be written using the standard mathematical notation $e : D \lhd E$. Similarly, $R(D, E)e$ is sometimes written as $e^R$.

Embeddings are used to form chains of cpos in a similar way that the ordering on elements of cpos is used to form chains. A chain of cpos is a pair $(\mathbf{D}, \mathbf{e})$ consisting of a sequence of cpos $D_n$ and a sequence of embeddings $e_n$ where $e_n : D_n \lhd D_{n+1}$ for all $n : num$:

$$D_0 \overset{e_0}{\lhd} D_1 \overset{e_1}{\lhd} \cdots \overset{e_{n-1}}{\lhd} D_n \overset{e_n}{\lhd} \cdots .$$

The notion of (embedding-projection) chain of cpos is formalised as follows:

$\mathsf{emb\_chain}\, \mathbf{D}\, \mathbf{e} = (\forall n.\ \mathsf{cpo}(\mathbf{D}\, n)) \wedge (\forall n.\ \mathsf{emb}(\mathbf{D}\, n, \mathbf{D}(\mathsf{SUC}\, n))(\mathbf{e}\, n)).$

In HOL terms, we write $\mathbf{D}\, n$ and $\mathbf{e}\, n$ instead of $D_n$ and $e_n$.

The inverse limit Dinf can now be defined as follows:

$\mathsf{Dinf}\, \mathbf{D}\, \mathbf{e} =$
$\quad (\{x \in \prod_{n \in \mathsf{Num}} \mathsf{set}(\mathbf{D}(\mathsf{Num2num}\, n)) \mid \cdots \},$
$\quad \lambda xy.\ \forall n.\ \mathsf{rel}(\mathbf{D}\, n)(x \diamond (\mathsf{num2Num}\, n))(y \diamond (\mathsf{num2Num}\, n))),$

where the dependent product construction on sets is used. Informally, the underlying set of Dinf is defined as the subset of all infinite tuples $x$ on which the $n$-th projection $e_n^R$ maps the $(n + 1)$-st index to the $n$-th index for all $n$: $e_n^R(x_{n+1}) = x_n$. The underlying relation is defined componentwise. The annoying num2Num conversions, which also appear in the omitted part, could be avoided by using the set of numbers Num instead of the type of numbers $num$ to represent chains of cpos. However, the present choice makes proofs simpler. This issue is discussed further in Section 4.

The details of proving the property that Dinf yields the inverse limit of a given chain of cpos will not be given here (see [1]). Informally speaking, it must satisfy a certain commutivity condition, which says that there is a family of embeddings from the elements of the chain to Dinf such that the resulting diagrams commute, and it must be universal with this property in the sense that there is a unique embedding of Dinf into any other cpo with this property.

## 3.3 Construction of $D_\infty$

In general, a recursive domain (isomorphism) equation can have the form

$$D \cong \mathcal{F}(D)$$

where $\mathcal{F}$ is a continuous covariant functor (a term of category theory). In particular, this means that $\mathcal{F}$ preserves inverse limits of chains of cpos and consists of both a construction on cpos and a construction on embeddings. Note that a domain equation is stated using only the cpo construction part of a functor.

The inverse limit construction provides solutions of any domain equation of the above form. We shall sketch very roughly how to obtain a solution of the equation

$$D \cong [D \to D],$$

which is stated using the standard notation for the continuous function space construction (defined as cf in HOL-ST). It can be proved once and for all that the continuous function space construction with an appropriate embedding is a continuous covariant functor.

First, a specific (but still parametrised) chain is constructed by iterating the continuous function space functor, starting at any cpo $D$ with an embedding $e :$ $D \lhd \mathrm{cf}(D, D)$. From this chain of cpos we obtain a partially concrete instantiation of Dinf, called Dinf_cf, which, due to the fact that the functor for the function space preserves inverse limits, yields a parametrised model of the untyped $\lambda$-calculus:

$\forall De.$
  $\mathrm{cpo}\, D \Rightarrow \mathrm{emb}(D, \mathrm{cf}(D, D))e \Rightarrow$
  $\mathrm{Dinf\_cf}\, D\, e \cong \mathrm{cf}(\mathrm{Dinf\_cf}\, D\, e, \mathrm{Dinf\_cf}\, D\, e).$

Choosing one concrete starting point (with for instance one element), and defining $D_\infty$ to abbreviate the corresponding instantiation of Dinf_cf, a concrete nontrivial model of the untyped $\lambda$-calculus is obtained:

$$D_\infty \cong \mathrm{cf}(D_\infty, D_\infty).$$

(See [1] for further details.)

# 4 Set Theory in Isabelle

Isabelle/ZF supports set theory in first order logic [21, 22]; it is an extension of a first order logic instantiation of the generic theorem prover Isabelle [23] with axioms of Zermelo-Fraenkel set theory. In contrast, HOL-ST supports set theory in higher order logic. Formalising the inverse limit construction in Isabelle/ZF revealed various differences between the two systems. This section summarises the conclusions of the comparison presented in [3].

## 4.1 First-order versus Higher-order Set Theory

The formalisation presented above exploited higher order logic as much as possible. Hence, cpos were HOL pairs, ordering relations were HOL functions and chains were HOL functions from the HOL type of natural numbers *num* to *V*. Alternatively, we could have chosen to do more work in set theory. For instance, the natural number argument of chains could have been in the set Num and cpos could have been represented by (non-reflexive) relations of set theory.

It makes a difference whether sets of ST or types of HOL are used. Using the latter, set membership conditions are avoided and furthermore, type checking is done automatically in ML. Using sets, type checking, i.e. ensuring terms are in the right sets, is done by theorem proving, and it is done late.

Obviously, exploiting the additional power of set theory will require leaving higher order logic and paying a price. It is an interesting and difficult question which parts of the formalisation should be done in set theory and which should be done in higher order logic. As noted in Section 3.2, the definition of the inverse limit constructor Dinf could have been simplified if chains of cpos had been represented using the set of natural numbers instead of the HOL type. However, experiments showed that it was worth paying the inconvenience of the translation functions at this stage since many theorems and proofs became quite horrible later, with the set approach. This in turn is related to the choice of representing ordinary chains in higher order logic; had they been represented using set theory numbers it would have been more natural to stick to this.

In Isabelle/ZF, there is much less choice. Most of the development must be done in set theory since the first order logic is so weak; for instance, it does not provide natural numbers, pairs, or functions (the individuals of the logic are sets). Therefore, chains must be represented as set functions and cpos must be represented as set theory pairs where both the set and the relation components are sets. In principle, the function type of Isabelle's polymorphic higher-order meta logic could be used to represent chains, but the definition of cpos must quantify over chains and in first order logic it is only possible to quantify over individuals (sets), not meta level functions. The meta logic is meant for expressing and reasoning in logic instantiations of Isabelle, the so-called object logics, not for formalising concepts in object logics.

The consequence of doing more work in set theory is that terms and proofs become more complicated, since set membership conditions appear more often. For instance, to prove that a term is a chain it must be shown that it is a set function, i.e. that it is a relation on the right sets that defines a function (though, usually the λ-abstraction is used and then it is only necessary to type check the body of this, due to a pre-proved theorem). Similarly, constructions on cpos have more complicated definitions and proofs: before applying the relation of a cpo to its arguments, these must be shown to be in the right sets.

## 4.2 Proof Support

Whilst there are benefits from the more powerful higher order logic of HOL-ST, Isabelle/ZF has the advantage of providing better proof support for set theory. Though proofs in principle are longer in terms of number of proof steps (of applying theorems), due to the additional type conditions, they are in fact much shorter in terms of lines, and easier to write. Note that the disadvantage of Isabelle/ZF mentioned above is logic dependent and could be avoided by using Isabelle/HOL, which supports a HOL-like higher order logic, rather than Isabelle/FOL as a basis for ZF set theory (though this would introduce translation functions as in HOL-ST). The better proof support of Isabelle/ZF is not logic dependent and thus would be preserved if we moved to Isabelle/HOL.

Isabelle provides an elegant proof infrastructure. Meta logic theorems can express both object logic theorems, inference rules and tactics. It does not implement a separate inference rule and tactic for each operation as in HOL; instead, the same theorem is applied either in a forward or backward fashion. Hence, there are very few tools for forward and backward proof. In fact, the main way to prove theorems is by the principle of resolution, which supports both styles.

The notion of resolution in Isabelle supports 'real' backward proofs better than they are supported in HOL. In Isabelle/ZF, one almost always works from the conclusion of a goal backward towards the assumptions, due to the design of Isabelle resolution tactics. In HOL-ST, one often ends up doing a lot of sometimes ugly assumption hacking working forward from the assumptions towards the conclusion, due to the design of HOL resolution tactics. More natural backward strategies like conditional rewriting and a matching modus ponens style strategy like MATCH_MP_TAC, which may be viewed as a simplified form of Isabelle resolution, are not well supported in HOL. These strategies are particularly useful in HOL-ST, compared to HOL, due the fact that many theorems have set membership assumptions.

The backward proofs of the formalization were usually more than 50% shorter in Isabelle/ZF (we used the default subgoal package of HOL). Its subgoal module provides a kind of flat structure on proof states where all subgoals can be accessed directly by an index specified by tactics; proving subgoal $i$ results in all subgoals with a higher index to be shifted such that the indexing contains no holes. This makes it possible to access all subgoals, possibly at the same time, and to prove several subgoals by just repeating a tactic supplied with the right list of theorems—no matter where they would appear in a HOL proof tree.

Another reason why proofs are shorter is that usually the instantiation of both existentially and universally quantified variables is handled automatically in Isabelle. It provides a notion of unknown variables which can be instantiated in proofs. This means that witnesses and proper instantiations are constructed behind the scenes, possibly in stages. Unification is essential for allowing this. Existential quantifiers are often introduced by backward strategies, for instance, when employing the transitivity of a cpo relation or the facts that function composition preserves the function set, continuity or embeddings: such theorems have free variables in the antecedents that do not appear in the consequent.

HOL-ST proofs could probably be simplified if there was a way of handling existential quantifiers. At the moment, witnesses must be provided on the spot and manually (some user-contributed tools for solving existential goals automatically are available, but we have not tried them with HOL-ST). Furthermore, it is often necessary to instantiate universally quantified variables manually.

## 5 Conclusions

The case study on formalising a model of the $\lambda$-calculus via the inverse limit construction shows that combining set theory and higher order logic in the same theorem prover provides a useful system for doing mathematics. The simplicity and convenience of higher order logic can be exploited as well as the expressive power of set theory. Rather than working in set theory only, it was shown that set and type theoretic reasoning can be mixed to advantage by exploiting set theory only when it is necessary and working in higher order logic the rest of the time. The logic of Isabelle's set theory, being first order, is weaker and therefore larger parts of the formalisation had to be done in set theory.

One of the main disadvantages of set theory is the presence of explicit type (set membership) conditions. This means that type checking is done late by theorem proving whereas in higher order logic type checking is done early in ML. Furthermore, type checking is automatic in HOL but cannot be fully automated in set theory. Thus, the comparison with Isabelle/ZF showed that proofs required more theorems since certain concepts were formalised in logic in HOL which had to be represented in set theory in Isabelle/ZF (since its first order logic is weaker). On the other hand, the comparison also showed that Isabelle/ZF supports proofs in set theory much better than HOL-ST, which generally speaking lacks ways of handling conditional theorems conveniently, and also does not provide any support for unknown variables for quantifier reasoning like in Isabelle/ZF. Since the better proof support in Isabelle/ZF is not logic dependent, extending Isabelle/HOL with ZF set theory might be a way of combining the benefits of HOL-ST and Isabelle/ZF. However, this would introduce the disadvantage of translation functions between types and sets as in HOL-ST.

In summary: it is not yet clear to us whether set theory in higher order logic is right, or just more support for set theory in first order logic is needed. Points requiring further consideration include the following.

1. ST in HOL yields a need for ugly translation functions (num2Num etc). Can these be hidden by suitable parsing and printing functions together with specialised theorem proving support?

2. ST in FOL yields a lot of set membership conditions. Can these be better automated, perhaps with theorem proving tools that mimic 'type checking'?

3. The construction of $D_\infty$ might be possible in a type theory with dependent types. Are there natural examples that really require set theory? Embedding Z semantics is a candidate.

# Acknowledgements

The research described here is partly supported by EPSRC grant GR/G23654 and partly by an HCMP fellowship under the EuroForm network. We are grateful to Larry Paulson, who read a preliminary draft of this paper and suggested some improvements. Francisco Corella, whose PhD thesis [6] contains many insights into mechanizing set theory (including the use of simple type theory as the underlying logic) has had a significant influence on the second author's thinking.

# References

1. S. Agerholm. Formalising a model of the $\lambda$-calculus in HOL-ST. Technical Report 354, University of Cambridge Computer Laboratory, November 1994.

2. S. Agerholm. *A HOL Basis for Reasoning about Functional Programs*. PhD thesis, BRICS, Department of Computer Science, University of Aarhus, December 1994. Available as Technical Report RS-94-44.

3. S. Agerholm. A comparison of HOL-ST and Isabelle/ZF. Technical Report 369, University of Cambridge Computer Laboratory, 1995.

4. D. Cantone, A. Ferro, and E. Omodeo, editors. *Computable Set Theory*, volume 1. Clarendon Press, Oxford, 1989.

5. R. L. Constable et al. *Implementing Mathematics with the Nuprl Proof Development System*. Prentice-Hall, 1986.

6. F. Corella. Mechanizing set theory. Technical Report 232, University of Cambridge Computer Laboratory, 1991.

7. G. Dowek, A. Felty, H. Herbelin, G. Huet, C. Murthy, C. Parent, C. Paulin-Mohring, and B. Werner. The Coq proof assistant user's guide - version 5.8. Technical Report 154, INRIA-Rocquencourt, 1993.

8. Roman Matuszewski (ed). *Formalized Mathematics*. Université Catholique de Louvain, 1990 –. Subscription is \$10 per issue or \$50 per year (including postage). Subscriptions and orders should be addressed to: Fondation Philippe le Hodey, MIZAR, Av.F.Roosevelt 35, 1050 Brussels, Belgium (fax: +32 (2) 640.89.68).

9. W. M. Farmer, J. D. Guttman, and F. Javier Thayer. IMPS: An interactive mathematical proof system. *Journal of Automated Reasoning*, 11(2):213–248, 1993.

10. S. Finn and M. P. Fourman. *L2 – The LAMBDA Logic*. Abstract Hardware Limited, September 1993. In LAMBDA 4.3 Reference Manuals.

11. M. J. C. Gordon. Merging HOL with set theory: preliminary experiments. Technical Report 353, University of Cambridge Computer Laboratory, 1994.

12. M. J. C. Gordon and T. F. Melham, editors. *Introduction to HOL: A Theorem-proving Environment for Higher-Order Logic*. Cambridge University Press, 1993.

13. F. K. Hanna, N. Daeche, and M. Longley. Veritas+: a specification language based on type theory. In M. Leeser and G. Brown, editors, *Hardware specification, verification and synthesis: mathematical aspects*, volume 408 of *Lecture Notes in Computer Science*, pages 358–379. Springer-Verlag, 1989.

14. C. B. Jones. *Systematic Software Development using VDM*. Prentice Hall International, 1990.

15. L. Lamport. TLA+. Available on the World Wide Web at the URL: http://www.research.digital.com/SRC/tla/tla.html.

16. Z. Luo and R. Pollack. LEGO proof development system: User's manual. Technical Report ECS-LFCS-92-211, University of Edinburgh, LFCS, Computer Science Department, University of Edinburgh, The King's Buildings, Edinburgh, EH9 3JZ, May 1992.

17. L. Magnusson and B. Nordström. The ALF proof editor and its proof engine. In *Types for Proofs and Programs: International Workshop TYPES '93*, number 806 in Lecture Notes in Computer Science, pages 213–237. Springer-Verlag, 1994.

18. D. A. McAllester. *ONTIC: A Knowledge Representation System for Mathematics*. MIT Press, 1989.

19. P. M. Melliar-Smith and John Rushby. The enhanced HDM system for specification and verification. In *Proc. Verkshop III*, volume 10 of *ACM Software Engineering Notes*, pages 41–43. Springer-Verlag, 1985.

20. L. C. Paulson. *Logic and Computation: Interactive Proof with Cambridge LCF*. Cambridge Tracts in Theoretical Computing 2, Cambridge University Press, 1987.

21. L. C. Paulson. Set theory for verification: I. From foundations to functions. *Journal of Automated Reasoning*, 11(3):353–389, 1993.

22. L. C. Paulson. Set theory for verification: II. Induction and Recursion. Technical Report 312, University of Cambridge Computer Laboratory, 1993.

23. L. C. Paulson. *Isabelle: A Generic Theorem Prover*, volume 828 of *Lecture Notes in Computer Science*. Springer-Verlag, 1994.

24. K. D. Petersen. Graph model of lambda in higher order logic. In J. J. Joyce and C. H. Seger, editors, *Proceedings of the 6th International Workshop on Higher Order Logic Theorem Proving and its Applications*, volume 780 of *Lecture Notes in Computer Science*. Springer-Verlag, 1994.

25. G. Plotkin. *Domains*. Course notes, Department of Computer Science, University of Edinburgh, 1983.

26. PVS World Wide Web page. http://www.csl.sri.com/pvs/overview.html.

27. Piotr Rudnicki. *An Overview of the MIZAR Project*. Unpublished; but available by anonymous FTP from menaik.cs.ualberta.ca in the directory pub/Mizar/Mizar_Over.tar.Z, 1992.

28. M. Saaltink. Z and EVES. Technical Report TR-91-5449-02, Odyssey Research Associates, 265 Carling Avenue, Suite 506, Ottawa, Ontario K1S 2E1, Canada, October 1991.

29. M. Smyth and G. D. Plotkin. The category-theoretic solution of recursive domain equations. *SIAM Journal of Computing*, 11, 1982.

30. J. M. Spivey. *The Z Notation: A Reference Manual*. Prentice Hall International Series in Computer Science, 2nd edition, 1992.

# Automatically Synthesized Term Denotation Predicates: A Proof Aid

Paul E. Black* and Phillip J. Windley

Computer Science Department, Brigham Young University, Provo UT 84602, USA

**Abstract.** In goal-directed proofs, such as those used in HOL, tactics often must operate on one or more specific assumptions. But goals often have many assumptions. Currently there is no good way to select or denote assumptions in HOL88. Most mechanisms are sensitive to inconsequential changes or are difficult to use.

Denoting assumptions by filters (matching) makes it easier to maintain large proofs and reuse pieces. But writing the filter predicate can be time-consuming and distracting.

We describe an aid to proof building which synthesizes filter functions from terms. Given examples of terms which should and should not be matched, the function creates an ML predicate which can be used, for example, with `filter` or `FILTER_ASM_REWRITE_TAC`.

This paper reviews past discussions on denotation methods, the design and implementation of the filter synthesizer, applicable AI classification techniques, and possible application to more general term handling and recognition.

## 1 Introduction

Proofs take a lot of time to create. To make proofs more widely applicable, the time and expertise necessary to create a proof must be reduced. One approach is to reuse large portions of proof scripts (tactics) in similar proofs. To be able to reuse portions of proof scripts, minor differences in goals should require minimal changes to the proof script.

Changes arise because the theorem to be proved changes and because the HOL implementation changes. One area which has caused lots of possibly avoidable work is in denoting or choosing assumptions. Currently there is no good way to denote or choose assumptions in HOL88.

There are many ways to denote assumptions, but all have drawbacks.

- Denoting assumptions by their position in the assumption list is simple, but if other tactics are used or tactic implementation changes, the position of assumptions may change.
- Denoting assumptions by quotation works regardless of position, but fails if the goal changes even slightly.

---

\* This work was sponsored by the National Science Foundation under NSF grant MIP–9412581

- Naming assumptions, then referring to them by name is robust. However it is not available in HOL88 and is just becoming available in HOL90.
- Denoting assumptions by filtering is insensitive to the order of assumptions and can be insensitive to changes in variable names and details of the theorem to be proved. However it is time-consuming and tedious to come up with appropriate filter predicates.

In Sect. 2 we review these approaches and others which have been suggested over the years in more detail. We compare their advantages and disadvantages. In Sect. 3 we present the design of the proof aid and its implementation. Section 4 presents alternative classification techniques and algorithms for constructing predicates. Finally we give some examples in Sect. 5, and Sect. 6 is our conclusions and ideas for future work.

## 2 Approaches to Denoting or Selecting Assumptions

Over the years the HOL community has suggested and discussed many approaches to denoting or selecting assumptions. Here we enumerate them and compare their strengths and weaknesses.

### 2.1 Denoting by Quotation

The HOL "handbook" [7] mentions the "general problem of denoting assumptions" and says

> The only straightforward way to denote them in the existing system is to supply their quoted text. Though adequate, this method may result in bulky ML expressions; and it may take some effort to present the text correctly (with necessary type information, etc.).

### 2.2 Denoting by Position

The book goes on to describe two other approaches: treating the assumption set as a stack and intercepting and manipulating results without them being added as assumptions.

The first approach uses *pop* operations to minimize the number of assumptions and denote the top assumption without explicit quotation. This is workable if the number of assumptions is small. The generalization of this approach, denoting assumptions by their position in the list, is sensitive to changes in tactic implementation or proofs.

### 2.3 Immediate Use

The second approach employs tacticals such as DISJ_CASES_THEN and DISCH_THEN to use results immediately without ever making them assumptions.

Ching-Tsun Chou [5] gave a detailed explanation of theorem continuations and their use. These techniques reduces the problem of denoting assumptions by reducing the number of assumptions and using some results directly. However proofs still have many assumptions at some points.

## 2.4 Reference by Position, Denoting by Quotation

David Shepherd [13] suggests a rather radical approach. The user refers to assumptions by their position in the assumption list, and a tactic recording system replaces the reference with a quotation of the assumption. The user then saves a "tactic script" for future use. The HOL system would require changes. Worse yet, when the proof is rerun after small changes, such as variable names, the quoted assumption would not match, the tactic would fail, and the script would require human maintenance.

Even more radical would be to have a system recording the tactics which are used and which tactic added which assumption. One refers to assumptions by position, and the tactic recording system modifies the step which added the assumption so it will save the assumption for use at this point. This would require even more extensive changes, and could be accomplished by naming (explained in Sect. 2.6).

## 2.5 Saving Terms and Denoting by Quotation

Paul Curzon [6] suggested saving assumptions in ML variables during the proof, then using them, e.g. via ASSUME, when needed. This is insensitive to changes of position or nature of term. The variable name could serve as documentation, too. The drawback is that some tactics don't return the results, although Chou's [5] theorem continuations could be adapted to the purpose.

## 2.6 Denoting by Name

Sara Kalvala [10] developed an extension of HOL90 in which terms and other structures can have labels (among other things). She also presented some specialized tactics which denote assumptions by label. Nuprl 4 allows users to label assumptions [12], too. Labeling assumptions is probably the best approach in the long run: proof scripts are insensitive to most changes and names can act as documentation or hints.

## 2.7 Denoting by Filter Predicate

John Harrison [9] suggested using a filter predicate, such as
```
FIRST_ASSUM(SUBST1_TAC o assert (curry $= "x:num" o lhs o concl))
```
This can be done in HOL88 and is less sensitive to changes in proof scripts than denotation by quotation or position. Jim Alves-Foss [2] pointed out that although proofs written with filters are insensitive to many changes, they may

not be very readable. Denoting assumptions by quotation documents what assumption is used at a step.

Another drawback is that it is difficult to write filter functions. If the assumptions to be denoted are very different from all the others, a few selectors, such as is_abs or \t.is_comb t & rator t = $+, are sufficient. But a predicate of any significant depth will be a confusing jumble of selectors little better than the car's, cdr's, and cadddr's sometimes seen in LISP. For example, here is a predicate used in Uinta verification:

```
let machine_pred thm =
    let tm = concl thm in
    (((rand o fst o dest_eq o snd o dest_forall) tm)
                            = "lmdr(t + 1):*wordn") ? false;;
```

Clearly the predicate is checking for lmdr(t + 1) at some specific place in the term, but it is not clear where. When the term changes, it will take some work to rewrite the predicate.

# 3  The Predicate Synthesizer

In this section we describe a new tool which synthesizes a predicate from examples of terms. We detail the design requirements and explain the implementation.

## 3.1  Requirements

The tool, find_filter, takes two lists: a list of terms and a list of indices of those terms which *must* be matched. Terms whose indices are not listed must *not* be matched. Terms which must match are called **positive examples** in Artificial Intelligence literature [14]. Terms which must *not* match are called **negative examples**. The synthesized function is returned as a string with full types and can be incorporated in the proof script for later use. The type of the tool is then find_filter: term list -> int list -> string, and the synthesized function has the type : term -> bool.

The synthesized function should be insensitive to minor changes in the proof. Therefore it should recognize general differences and similarities instead of specifics which are peripheral to the proof. As an example of what we *don't* want, the function \t. t = pos1 \/ t = pos2 \/ . . .\/ t = posN (where pos1 through posN are the positive examples) always filters exactly (as long as no term is both a positive and negative example), but it is verbose and sensitive to any change in the positive terms.

The synthesized function must be fast since some proofs may filter assumptions hundreds or thousands of times. Therefore the function should be a series of predicates, such as

```
\t.is_forall t & vname (bndvar (rand t))='t'
                & is_imp (body (rand t)) ? false
```

The tool is acceptable even if it takes some minutes to synthesize the filter function since the user only runs it occasionally, and then at human interaction speeds. The synthesized function should be somewhat readable, so the user can gauge how general the function actually is. This is not crucial since even if the proof changes enough to invalidate the function, the new assumptions can be added to the "training set" and the tool rerun to yield a more robust function.

## 3.2  Implementation Philosophy

The tool should compare the positive and negative example and synthesize not just *some* function which discriminates between the two sets, but a function which covers the most possibilities. That is, the function should continue to discriminate correctly in the future as far as possible. We would have to be clairvoyant to write a program which always chooses correctly, but how good can we do in practice?

Carbonell, et. al. [3] considers this problem as learning a classification from examples. The examples came from the external environment, that is, the examples are not specifically chosen or designed to teach the concept. Both positive and negative examples without "noise" are available, and the resultant classification must be correct for all examples. In learning the classification, the synthesizer should find features of terms which make up the concept "desired terms." The general approach we chose is to search a rule-version space [14].

### Abstraction Level: One Possible Feature Set

Terms often have a general form in common, for instance, all positive example terms may be implications while none of the negative examples are. This suggests extracting features based on parts closest to the top, or root, of the abstract syntax tree or parse tree of a term. Consider the term `"!x y. x < 3 /\ y > 0"`. The most general recognition function of the term is `\(t:term).true`; all detail of the term has been abstracted away. Since there is no detail, we call it a level 0 abstraction; none of the parse tree of the term is used. The function which only includes one level of detail from the top, that is the level 1 abstraction from the top, is `\t.is_forall t`. It checks that the first level of the term matches, but nothing more. The level 2 top abstraction is

`\t.is_forall t & is_conj (body (rand (body (rand t)))) ? false`

A rigorous description more clearer phrased as testing for a match with a "term pattern." A term pattern is a term which may have specially named variables that match any (sub)term. (The special variables are named `STARn` reminiscent of the Kleene Star.) The level 0 abstraction of any term is just a `STAR` variable, i.e., a pattern which matches anything (the type is copied from the term). A level n abstraction, where $n > 0$, is as follows. The level n abstraction of a constant or variable is that constant or variable: an exact match. The level n abstraction of a $\lambda$-term is `mk_abs` of the level $n - 1$ abstraction of the bound variable and the body.

The level n abstraction of a combination term is more complex. If the rator is a binder, such as ∀ or ∃, it is the level $n - 1$ abstraction of the bound variable and body of the binder's abstraction:

```
mk_comb(
    rator t,
    mk_abs(
        abstraction (n-1) (bndvar (rand t)),
        abstraction (n-1) (body (rand t))
    )
)
```

If the rator is not a binder, it is the combination of the level $n$ abstraction of the rator and the level $n - 1$ abstraction of the rand. That is, if an operator is at level n, *all* its operands are at level $n + 1$.

## 3.3 Implementation

The filter function is synthesized in several steps.

1. Find one or more term patterns such that all positive terms are matched by at least one pattern and the pattern do not match any negative terms.
   (a) Arbitrarily choose a positive example. Find the greatest abstraction (least level n, from Sect. 3.2 above) of it which matches all the positive examples and none of the negative examples.
   (b) Slightly generalize the abstraction if possible.
   (c) If no abstraction matches all the positives and none of the negatives, find a pattern for the first positive and a pattern for the rest of the positives separately.
2. Convert the term pattern(s) into an ML predicate which matches the same terms. The ML predicate is constructed as a string so it can be easily printed. The output can then be used as a filter function in a proof.

Obviously this limited exploration of possible term patterns may fail to find a set which works. If none of these work, find_filter ends with the message, no filter found.

Abstractions are slightly generalized if possible. A new pattern, with some penultimate node of the pattern tree replaced by a STAR, is formed and checked against the negatives. If it does not match any of the negatives, it is used. For example, suppose a term pattern is STAR2 /\ STAR3 ==> STAR3 \/ STAR4. The new patterns formed by slightly generalizing it are STAR2 /\ STAR3 ==> STAR5 and STAR6 ==> STAR3 \/ STAR4. Both of these are slightly more general than the original pattern.

In some cases no single pattern matches all of the positives and none of the negatives. To handle these, the positives are split into two sets, the first positive and the rest. The code attempts to find patterns which match the first, but not the negative, and the rest, but not the negatives. The final predicate then checks for a match with either of these patterns. Clearly this could be enhanced to create as many patterns as needed to match all of the positives, but none of the negatives.

The final term pattern (or patterns) is converted to a string of appropriate ML code. For instance, a conjunction in the pattern becomes `is_conj` of the appropriate selection *and* (`&`) predicates to test both conjuncts, if needed. STARs match anything, so no predicate is needed for them. Special characters are quoted so the text can be incorporated directly into proof scripts. We give examples of the synthesized functions in Sect. 5.

# 4 Alternative Algorithms and Classification Techniques

This section discusses some alternative algorithms and approaches we considered, but haven't explored. For proofs in other subjects, such as mathematics, software, or protocols, or other styles of proofs, different selection functions may be better. Additionally term recognition may be generally useful in developing tactics. So we mention some alternate approaches here.

## 4.1 Pattern Terms and General Match Function

Recognition functions can get verbose. Using pattern terms with a general match function yields more succinct functions, but much slower execution. For example using a general match function and a pattern the level 2 top abstraction of `"!x y. x < 3 /\ y > 0"` is `tmatch "!STAR1 STAR2. STAR3 /\ STAR4"`. The function `tmatch` checks each part of the pattern against the corresponding part of the term. This is used in the implementation for flexibility, but patterns are converted to functions for speed.

If the terms to be selected contain some particular function somewhere in them, a search-and-match function could be used. If `subtmatch` searches for a matching subterm, `subtmatch "lsim(STAR):num -> bool"` t searches t checking for a combination with `lsim` as the operator.

## 4.2 Other AI Techniques

The way we apply rule-version space search misses common subexpressions in positive examples. Common subexpression location with DAGs, as in compilers [1], is probably not useful since compilers look for *exact* matches. Discrimination networks [4] may be a means of finding common subexpressions. The common subexpressions could then be found with a search-and-match function.

The work of Feng and Muggleston [8] may be applicable. They are concerned with finding selectors which classify positive and negative examples of higher order terms. The statistical feature selection of Kira and Rendell [11] is another way of extracting concepts from large numbers of features. This is especially applicable since the set of potential features is the power set of term nodes.

## 4.3 Methods of Synthesizing Selection Predicates

The current method uses abstractions from the top down as features for building the selection function. Other kinds of features may lead to better selection functions.

**Bottom-Up and General Abstraction** Rather than starting at the top and allowing detail downward, details from the bottom of the term's parse tree upward might be used. A reasonable, informal definition of the level n abstraction from the bottom may be a parse tree of height n above any leaf node. Since constants and variables names are rarely good identifiers, all leaf nodes may be replaced with STAR's. In contrast to top abstraction in Sect. 3.2 the level 1 abstractions from the bottom for the term "!x y.x < 3 /\ y > 0" are "STAR1 < STAR2" and "STAR3 > STAR4". The level 2 bottom abstraction is "STAR1 < STAR2 /\ STAR3 > STAR4".

Clearly one could search for and check abstractions found in the middle, too. For example, another pattern for the above term is "STAR1 /\ STAR2". It is not clear how features could be chosen efficiently when terms are large. Since there are $O(2^n)$ nodes in a term of depth $n$, the total number of features grows as $O(2^{2^n})$.

**Maximum Distance** An entirely different approach is to define a metric of distance between terms in some multi-dimensional space. Denotation predicates would then specify hyperrectangles or separating hyperplanes. The synthesizing function would choose those predicates which maximize separation between the positive and negative examples.

## 5 Examples

### 5.1 Manufactured Examples

In order to illustrate the operation of **find_filter**, I present some small, contrived examples.

**Simplest** Suppose the assumption list at some point is

```
["!x. x /\ y ==> y /\ x";
 "!a. a /\ b"]
```

The following finds a filter for the first assumption:

```
find_filter (fst (top_goal())) [1];;
'let f = \(t:term).is_forall t & is_imp (body (rand t)) ? false;;'
: string
```

If one of the negative examples also has an implication, the match must be more exact.

```
find_filter ["!x. x /\ y ==> y /\ x";
             "!a. a /\ b";
             "!x. x ==> d"] [1];;
'let f = \(t:term).is_forall t & is_imp (body (rand t))
           & is_conj (rand (rator (body (rand t)))) ? false;;'
: string
```

A filter for two very different terms is a conjunction.

```
find_filter ["!x. x /\ y ==> y /\ x";
             "!x. a /\ b";
             "!x. x ==> d"] [2;3];;
'let f = \(t:term).is_forall t & is_conj (body (rand t))
         or is_forall t & is_imp (body (rand t))
           & vname (rand (rator (body (rand t))))='x'
 ? false;;': string
```

Clearly the duplicated test of is_forall t could be done once to shorten and
speed the predicate.

## 5.2   Examples from Uinta

In this section we show find_filter used with one of biggest, most complex assumption list we could find in a real proof. In the proof of the general interpreter
in Uinta [15], there is a point where a goal has 21 fairly complex assumptions.
To save space, only a few of the assumptions are used and shown here. The
assumptions shown are typical. These assumptions are *not* shown with enough
type information to be reentered: that would make them even bigger. To get
enough type information, print terms with set_flag('show_types', true).

**First Example**  We begin by synthesizing a selector for one of the most complex
assumptions. The assumption is

```
"!t.
  (select
   gi2
   (substate
    gi2
    (substate gi1(s'(Temp_Abs(\t'. sync gi1(s' t')(e' t'))t))))
    (subenv gi2(subenv gi1(e'(Temp_Abs(\t'.
                                  sync gi1(s' t')(e' t'))t))))) =
   k) /\
  sync
  gi2
  (substate gi1(s'(Temp_Abs(\t'. sync gi1(s' t')(e' t'))t)))
  (subenv gi1(e'(Temp_Abs(\t'. sync gi1(s' t')(e' t'))t))) ==>
  (subout gi2(subout gi1(p'(Temp_Abs(\t'.
```

```
                                    sync gi1(s' t')(e' t'))t))) =
output
gi2
k
(substate
 gi2
 (substate gi1(s'(Temp_Abs(\t'. sync gi1(s' t')(e' t'))t))))
 (subenv gi2(subenv gi1(e'(Temp_Abs(\t'.
                                    sync gi1(s' t')(e' t'))t)))))"
```

When we run **find_filter**, we get

```
\t.is_forall t & is_imp (body (rand t)) &
    is_conj (rand (rator (body (rand t))))
? false
```

The selection function may be informative by itself: it shows that at an abstract level, the structure of the assumption is `!STAR1. STAR2 /\ STAR3 => STAR4`.

**Second Example** In this example we use **find_filter** to find a selector for two structurally similar assumptions. The assumptions are

```
"(!s' e' p' k. INST_CORRECT gi2 s' e' p' k) ==>
 (!s' e' p' k. OUTPUT_CORRECT gi2 s' e' p' k) ==>
 (!s' e' p'.
   implementation gi2 s' e' p' ==>
   (!t.
     sync gi2(s' t)(e' t) ==>
     (?n. Next(\t'. sync gi2(s' t')(e' t'))(t,t + n))))"

"(!s' e' p' k. INST_CORRECT gi1 s' e' p' k) ==>
 (!s' e' p' k. OUTPUT_CORRECT gi1 s' e' p' k) ==>
 (!s' e' p'.
   implementation gi1 s' e' p' ==>
   (!t.
     sync gi1(s' t)(e' t) ==>
     (?n. Next(\t'. sync gi1(s' t')(e' t'))(t,t + n))))"
```

The selector is simply `\t.is_imp t`. No other assumptions are implications at the top.

**Third Example** The last example finds a filter for one of the simplest assumptions:

```
"implementation gi1 s' e' p'"
```

There is another similar assumption (shown below), so the filter must be very specific.

```
"implementation
 gi2
 (\x. substate gi1(s'(Temp_Abs(\t. sync gi1(s' t)(e' t))x)))
 (\x. subenv gi1(e'(Temp_Abs(\t. sync gi1(s' t)(e' t))x)))
 (\x. subout gi1(p'(Temp_Abs(\t. sync gi1(s' t)(e' t))x)))"

\t.cname (rator (rator (rator (rator t)))) = 'implementation' &
    vname (rand (rator (rator (rator t)))) = 'gi1'
    ? false
```

# 6  Conclusions

Until assumptions can be labeled, filtering the assumption list is the most robust way to write proof scripts. Having an aid to synthesize filter functions will encourage people to use filters instead of denoting assumptions by position. With a large user group, more experience, and a wider range of styles, more hueristics can be captured and the filter synthesizer improved. In addition, machine learning applied to terms may be the basis for other operations or proof functions.

## Acknowledgements

We thank Tony R. Martinez for his suggestions about AI sources. We are indebted to the reviewers whose extensive comments pointed out vague and confusing parts of the paper.

## References

1. Alfred V. Aho, Ravi Sethi, and Jeffrey D. Ullman. *Compilers, Principles, Techniques, and Tools.* Addison-Wesley Publishing Co., 1986.
2. Jim Alves-Foss. message of 1 November 1991 to info-hol mailing list. (Available electronically at http://lal.cs.byu.edu/lal/hol-documentation.html)
3. Jaime G. Carbonell, Ryszard S. Michalski, and Tom M. Mitchell. (Eds.) *Machine Learning. An Artificial Intelligence Approach.* Tioga Publishing Company, Palo Alto, California, 94302, 1983, page 9.
4. Eugene Charniak, Christopher K. Riesbeck, and Drew V. McDermott. *Artificial Intelligence Programming.* Lawrence Erlbaum Associates, 1980, chapters 11 and 14, pp. 121–130 and 162–176.
5. Ching-Tsun Chou. "A Note on Interactive Theorem Proving with Theorem Continuation Functions," in *Higher Order Logic Theorem Proving and Its Applications (HOL '92),* edited by Luc Claesen and Michael Gordon, Elsevier Science Publishers B.V., 1992, pp. 59–69.
6. Paul Curzon. *Re: Accessing Assumptions.* message of 25 May 1994 to hol2000 mailing list. (Available electronically at http://lal.cs.byu.edu/lal/hol-documentation.html)

7. M. J. C. Gordon and T. F. Melham. *Introduction to HOL. A theorem proving environment for higher order logic.* Cambridge University Press, 1993, Section 24.5, pp. 384–396.

8. Cao Feng and Stephen Muggleston. "Towards Inductive Generalisation in Higher Order Logic," in *Machine Learning: Proceedings of the Ninth International Workshop (ML92)* edited by Derek Sleeman and Peter Edwards, Morgan Kaufmann Publishers, 1992, pp. 154–162.

9. John Harrison. *Selecting Assumptions.* message of 31 October 1991 to info-hol mailing list. (Available electronically at http://lal.cs.byu.edu/lal/hol-documentation.html)

10. Sara Kalvala, Myla Archer, Karl Levitt. "Implementation and Use of Annotations in HOL," in *Higher Order Logic Theorem Proving and Its Applications (HOL '92)*, edited by Luc Claesen and Michael Gordon, Elsevier Science Publishers B.V., 1992, pp. 407–426.

11. Kenji Kira and Larry A. Rendell. "A Practical Approach to Feature Selection," in *Machine Learning: Proceedings of the Ninth International Workshop (ML92)* edited by Derek Sleeman and Peter Edwards, Morgan Kaufmann Publishers, 1992, pp. 249–256.

12. Miriam Leeser. *assumption numbering, user interfaces ....* message of 26 May 1994 to hol2000 mailing list, archived at http://lal.cs.byu.edu/lal/hol-documentation.html.

13. David Shepherd. *Re: Accessing Assumptions.* message of 26 May 1994 to hol2000 mailing list. (Available electronically at http://lal.cs.byu.edu/lal/hol-documentation.html)

14. Steven L. Tanimoto. *The Elements of Artificial Intelligence.* Computer Science Press, 1987.

15. Phillip J. Windley and Michael Coe. "A Correctness Model for Pipelined Microprocessors," in *Proceedings of the 1994 Conference on Theorem Provers in Circuit Design*, edited by Thomas Kropf and Ramayya Kumar, 1994.

# On the refinement of symmetric memory protocols

J.-P. Bodeveix, M. Filali

IRIT - Université Paul Sabatier
118 Route de Narbonne
F-31062 Toulouse cédex France
email: {bodeveix, filali}@irit.fr

**Abstract.** In this paper, we present the formalization and the validation of memory protocols for multiprocessor architectures. Memory access is supposed to be atomic and can be observed by all the processors of the architecture. We first introduce a framework for the specification and the validation by refinements of parallel programs. For this purpose, we propose a new state representation allowing strong typing. After introducing refinements in the context of transition systems, we validate some refinement properties of sequential and parallel statements. At last, we define a representation of multiprocessor cache protocols exploiting the symmetry of such algorithms and show its correctness.

## 1   Introduction

Our research is concerned with the formalization and the validation of memory protocols for multiprocessor architectures. In this paper, we consider protocols where memory access is atomic and can be observed by all the processors of the architecture: the so called snooping protocols.

The first part of this paper introduces a framework for the specification and the validation by refinements of parallel programs. For this purpose, we propose a new state representation allowing strong typing. Then, we adapt the semantics of usual sequential and parallel statements to our representation and show that classical or intuitive properties are still valid. After introducing refinements in the context of transition systems, we propose and validate some refinement properties concerning sequential and parallel statements.

The second part of this paper deals with a representation of multiprocessor cache protocols exploiting the symmetry of such algorithms. After presenting the basic ideas of this approach, we propose a write-invalidate protocol, give a symmetric version of the algorithm and show the correctness of the latter.

## 2   Program state logic

In this section, we first review some existing approaches for representing program states. Then, we present a "genuine" state representation, its motivation being to allow the use of well typed notations of Pascal like programming languages. In some way, we try to reduce the gap between the syntactic domain where we are used to reason and the semantic domain where we can rigorously prove program properties.

## 2.1 Overview of some state representations

We consider two usual state representations. The first one is "untyped" since all variables are supposed to be of a fixed type (integer). The second one allows for the use of typed variables. However, some casting remains necessary and semantically invalid statements can be encoded.

**States as mono valued functions.** In this approach, the space of program variables **Var** and a space of values **Val** are assumed. A program state is a mapping from variables to values. Its type is **Var** $\rightarrow$ **Val**. For instance, in their programing logics theory[Tre92, vW90], **Var** is the type **string** and **Val** is the set of natural numbers. Then in order to reuse such logics, one has to encode all its variables as values of type **Val**.

**States as type union valued functions.** This representation has been introduced by [APP93] to describe UNITY[CM88] logics. A state is represented by a function from an enumeration of variable representatives to the union of variable values types. Then a variable is defined as a function from states to its type domain.

Suppose we have the following declaration in a PASCAL like language:

   **VAR i, j: INTEGER; b: BOOLEAN;**

Then, we define the abstract data type **Rep** of variable representatives and the union of variable types **Types** as follows:

   **type Rep = I | J | B**
   **type Types = INT integer | BOOL boolean**

Destructors are associated to this type in order to down-cast union-typed expressions to their effective type. Here, two partial functions are defined:

   **dest_int: Type $\rightarrow$ integer     dest_int (INT n) = n**
   **dest_bool: Type $\rightarrow$ boolean  .. dest_bool (BOOL b) = b**

A state $\sigma$ is represented by a function from **Rep** to **Types**.

Now, a variable is defined as a function from states to its type. For instance, the variable i is defined as follows:

   $i : (\textbf{Rep} \rightarrow \textbf{Types}) \rightarrow \textbf{integer}$     $i(\sigma) = \textbf{dest\_int } (\sigma(I))$

This representation raises a problem concerning the definition of assignment. An expression must be converted to the union type before being assigned. Thus, the type checker cannot detect that an expression is assigned to a variable of a different type. Type checking will indeed occur at proof time: an access to a badly assigned variable cannot be reduced to its value as destructor functions are partial.

## 2.2 The proposed state representation

A state is represented by a boolean valued function over *variable- value* associations. The space of well typed variable–value associations is introduced as an abstract data type of which constructors are the variables of the program. Consequently, a variable can also be considered as a function from well–typed values to the previously defined abstract data type. In a given state, the variable values are such that the association variable–value is mapped to true through the state function. Then, such a state function can also be considered as a set of associations variable–value.

In order to give a concrete view of such a representation we consider two examples concerning scalar and array variables.

**Scalar variables.** Suppose we have the following declaration in a PASCAL like language:

    VAR i, j: INTEGER; b: BOOLEAN;

Then, we define the abstract data type **declarations** of well typed variable – value associations by its constructors:

    type declarations = i INTEGER | j INTEGER | b BOOLEAN

This type declaration introduces three functions (the constructors of the data type):

    i,j : INTEGER $\rightarrow$ declarations

    b  : BOOLEAN $\rightarrow$ declarations

The state where the variables i, j and b have respectively the values $i_0, j_0$ and $b_0$ is represented by the function **state$_0$** : **declarations** $\rightarrow$ **bool** such that[1]:

$$\textbf{state}_0(i(x)) = (x = i_0) \wedge \textbf{state}_0(j(x)) = (x = j_0) \wedge \textbf{state}_0(b(x)) = (x = b_0)$$

In our approach, the operation which consists in the creation of a "pair" (variable,value) defined by the application of the variable to the value is not injective if the variable is not supposed to be a type constructor. More precisely, this application can give the same association for different pairs. The definition of injectivity would require here quantification over types and a generalized equality ($\doteq$) between elements of different types. The expression of such a property would state the existence of a set of variables $\mathcal{V}$ such that[2]:

$$\forall(x_1 : *x_1 \rightarrow *s)\, (v_1 : *x_1)\, (x_2 : *x_2 \rightarrow *s)\, (v_2 : *x_2)$$
$$x_1 \in \mathcal{V} \wedge x_2 \in \mathcal{V} \Rightarrow x_1(v_1) = x_2(v_2) \Rightarrow (x_1 \doteq x_2) \wedge (v_1 \doteq v_2)$$

This formula raises two typing problems:

– the set $\mathcal{V}$ contains elements of different types $(x_1, x_2)$.
– equality is applied between objects of different types $(x_1 \doteq x_2)$.

In fact, we would like to specify that $\mathcal{V}$ is the set of constructors of some type. In order to overcome this problem, we define weaker properties on individual variables by introducing the two polymorphic predicates **IS_VAR** and **D_VAR**.

---

[1] We note that such a representation allows for multivalued variables. However in this paper we do not use such a feature.

[2] type variables are prefixed by *

$$\text{IS\_VAR } (v : *v \rightarrow *s) = \forall\, a\, b.\ (v(a) = v(b)) = (a = b)$$
$$\text{D\_VAR } (x : *x \rightarrow *s, y : *y \rightarrow *s) = \forall\, vx\ vy.\ (x(vx) \neq y(vy))$$

**Array variables.** The same formalism can be used to represent arrays. For the following declaration:

    VAR t: ARRAY [INDEX] of INTEGER;

we associate the data type:

    type declarations = t INDEX INTEGER

In the same way, this type declaration introduces the function:

    t : INDEX → INTEGER → declarations

The state $\text{state}_0$ where all the elements of the array t are 0 except the one at index $i_0$ where the value is $v_0$ is encoded as follows.

$$\forall i\ x.\ \text{state}_0(t(i, x)) = \text{if } (i = i_0) \text{ then } x = v_0 \text{ else } x = 0$$

**Expressions and variable access.** Although in our representation a variable may be multivalued, we define expressions as *functions* over states:

    expression : *state → *exp_type

Then the standard operators (boolean and arithmetic operators) can be lifted to such expressions. For this purpose, we define a unary and a binary polymorphic lifting functions. For instance, we define the **Bop_Lift** function as follows:

$$\text{Bop\_Lift } op = \lambda\, e_1\ e_2\ st.\ op\ (e_1(st), e_2(st))$$

We use the same convention as [APP93] where the name of lifted operator is the name of the original operator suffixed by *. For instance, **Bop_Lift (<) = <\***.

Note that the multivaluation of variables could have been extended to expressions. However, in this framework, the definition of boolean algebra operators becomes tricky and some classical properties are lost.

As specific expressions, we define the variable access function **val**. This function must choose *one* of the values associated to the variable by the state. Moreover, in a given state, this choice must be the same for every access in the same expression. For this purpose, we use the Hilbert choice function[GM94] denoted $\epsilon$. The term $\epsilon\, x : \sigma.\ P(x)$ denotes an arbitrary but fixed variable of type $\sigma$. This term verifies the predicate $P$ if a term verifying $P$ exists. Then, the polymorphic function **val** is defined as follows:

    val : (*v → *decl) → (*decl → bool) →        *v
          v          ,        st         ↦ ε x. st(v(x))

Note that **val(v)** is a function from states of type *decl → bool to variable values of type *v.

# 3   Statements logic

In order to allow for non determinism, we define statements as binary relations over states[Tre92]. If we consider states as sets of associations variable-value, a statement can also be interpreted as a function which consumes some elements of a set and produces new ones. However, we have not pursued further in this direction. An interesting study would be to explore the links with linear logic [Gir87].

In the following, we give the semantics of some basic statements like assignments, conditional, alternative and parallel statements. We remark that the semantics of the conditional and the alternative statements is the usual one. However the assignment and parallel statements need to be revisited. Note that our purpose is not to define the semantics of some programming language but to define some generic transitions over the state space.

In the last section, we establish some well known results and give some examples.

## 3.1 Assignments

The assignment statement $x := e$ establishes a relation between two states $st_1$ and $st_2$ such that the value of the variable $x$ in $st_2$ is the value of the expression $e$ in $st_1$, any other variable binding remaining unchanged. The assignment is defined as follows:

$$x := e = \lambda st_1 \, st_2. \, \forall y. \, st_2(x(y)) = (y = e(e1)) \, \wedge$$
$$\forall s. \, (\forall y. \, s \neq x(y)) \Rightarrow st_2(s) = st_1(s)$$

From this definition, we can prove the following theorems which state explicitly the usual properties of the assignment statement:

$$(x := e)(st_1, st_2) \Rightarrow \text{val } (x)(st_2) = e(st_1)$$
$$\text{D\_VAR } (x, y) \Rightarrow (x := e)(st_1, st_2) \Rightarrow \text{val } (y)(st_2) = \text{val } (y)(st_1)$$

We have only defined the single assignment statement. The multiple assignment statement will be defined in section 3.4 through the parallel constructor $\|$.

## 3.2 Alternatives

The definition of statements as binary relations over states yields a straightforward definition of non deterministic statements as in CSP[Hoa85]. We consider the binary operator $\|$ and the generalized one indexed by a set of alternatives $A$: $\|_A$.

$$i_1 \| i_2 = \lambda st_1 st_2. \, i_1(st_1, st_2) \, \vee \, i_2(st_1, st_2)$$
$$\|_A i = \lambda st_1 st_2. \, \bigvee_{a \in A} i_a(st_1, st_2)$$

The associativity and commutativity of the binary alternative operator are easily proved.

## 3.3 Conditional Statements

We introduce successively two conditional statements. The first one ($\longrightarrow$) is similar to a guarded statement[Dij76], while the second one (**IF**) is similar to the usual **if-then** statement.

$i \longrightarrow c = \lambda\ st_1\ st_2.\ c(st_1) \wedge i(st_1, st_2)$
$i\ \textbf{IF}\ c\ = \lambda\ st_1\ st_2.\ \textbf{if}\ c(st_1)\ \textbf{then}\ i(st_1, st_2)\ \textbf{else}\ (st_1 = st_2)$

## 3.4 Parallel statements

The basic idea of the parallel construct is to *superpose* the changes performed by several statements. A variable binding is changed by a parallel statement if it is changed by one of the composing statements; otherwise it remains unchanged.

**The binary PAR constructor.** From the intuitive definition of parallelism, we first define a binary parallel operator $\|$ between two statements $i_1$ and $i_2$. The parallel statement updates a state so that a binding is present in the final state if it is present in the final state of one of the statements modifying it. This is encoded as follows:

$$i_1\ \|\ i_2 = \lambda\ st\ st'.\ \exists\ st_1\ st_2.\ i1(st, st_1) \wedge i2(st, st_2)\ \wedge$$
$$\forall s.\ st'(s) =\ \textbf{if}\ st(s)\ \textbf{then}\ st_1(s) \wedge st_2(s)$$
$$\textbf{else}\ st_1(s) \vee st_2(s)$$

We should remark that the $\|$ constructor requires a point by point interpretation of states contrary to other constructors such as conditional alternatives.

**The generalized PAR constructor.** The previous definition can be easily extended to an indexed set of statements $ins_i$ in the following way:

$$\textbf{PAR}\ ins = \lambda\ st\ st'.\ \exists\ e'.\ \forall\ i.\ ins_i(st, e'_i)\ \wedge$$
$$\forall\ s.\ st'(s) =\ \textbf{if}\ st(s)\ \textbf{then}\ \forall\ i.\ e'_i(s)\ \textbf{else}\ \exists\ i.\ e'_i(s)$$

**Multi-assignments.** The parallel operator lets us introduce multi-assignment statements as a parallel combination of single assignments. Then, single assignment theorems (3.1) can be extended, for instance, to two-assignment statements. It should be noted that we do not introduce multi-assignments as such, i.e., an assignment of a list of expressions to a list of variables, since the elements of the respective lists do not have a unique type.

$$\textbf{D\_VAR}\ (x, y) \wedge (x := ex\ \|\ y := ey)(st_1, st_2) \Rightarrow$$
$$\text{val}\ (x)(st_2) = ex(st_1) \wedge \text{val}\ (y)(st_2) = ey(st_1)$$

$$\textbf{D\_VAR}\ (x, z) \wedge \textbf{D\_VAR}\ (y, z) \wedge (x := ex\ \|\ y := ey)(st_1, st_2) \Rightarrow$$
$$\text{val}\ (z)(st_2) = \text{val}\ (z)(st_1)$$

The previous two rules illustrate the idea of the superposition of changes made by the components of a parallel statement. However, the behaviour of a multi-assignment to a single variable (which is usually syntactically forbidden) is counter intuitive; for instance, let us consider the multi-assignment $x := 1 \parallel x := 2$. If in the initial state the value of $x$ is 0, then in the final state $x$ is multivalued. But, if the initial value of $x$ is 1, then the final value of $x$ is 2! Another semantics could be to allow a non-deterministic behaviour in such cases, or to state explicitly how values are combined[BC84]. Nevertheless, we have done with this definition for its conciseness and since we do not consider parallel updates of the same variable.

At last, general multi-assignment rules cannot be stated because of typing problems: each variable has its own type and a set of such variables cannot be defined. However, we will see in section 5.2 how to overcome this problem by introducing a kind of meta theorem.

## 3.5 A relation between parallel and sequential statements

Parallel constructs may be transformed into sequential constructs[3] given some independence hypothesis expressed using the following e_indep_x predicate:

An expression is said to be independent of a variable $x$ if any assignment to $x$ does not modify its value.

$$\text{e\_indep\_x}(\text{exp, } x) = \forall st_1 \; st_2 \; e. \; (x := e)(st_1, st_2) \Rightarrow \text{exp}(st_1) = \text{exp}(st_2)$$

The following theorem states the equivalence between a sequential and a parallel assignment.

$$\forall x \; y \; ex \; ey. \; \text{IS\_VAR} \; (x) \wedge \text{IS\_VAR} \; (y) \wedge \text{D\_VAR} \; (x, y) \wedge \text{e\_indep\_x} \; (ey, x) \Rightarrow$$
$$(x := ex \parallel y := ey) = \text{Seq} \; (x := ex, y := ey)$$

# 4 Program refinements

First we just recall the refinement definition, similar to the one given in [LT87].

**Definition 1 (Refinement relation)** *Given two transition systems s and s',
s simulates s' iff there exits a simulation relation such that:*

- *to each initial state of s at least one initial state of s' corresponds,*
- *if from a state e1, s can move to e2, and e1 corresponds to the state e1' of
  s' then there exists a state e2' such that e2' corresponds to e2 and s' can
  move from e1' to e2'.*

We formalize such a definition as follows:

$$\forall \; s \; s'. \; s \; \text{Sim} \; s' = \exists \; R.$$
$$\forall \; e. \; \text{Init} \; s \; e \Longrightarrow \exists \; e'. \; R \; e \; e' \wedge \text{Init} \; s' \; e' \wedge$$
$$\forall \; e1 \; e1' \; e2. \; \text{Next}_s \; e1 \; e2 \wedge R \; e1 \; e1' \Longrightarrow \exists \; e2'. \; R \; e2 \; e2' \wedge \text{Next}_{s'} \; e1' \; e2'$$

---

[3] Seq $i_1 i_2 = \lambda st \; st'. \; \exists st''. \; i_1 \; st \; st'' \wedge i_2 \; st'' \; st'$

In a transition system derived from a program, we can label each transition by its corresponding statement. Then, the refinement between labeled transition systems can be established independently for each kind of label. More precisely, we have the following result:

**Theorem 1 (Refinement through operators)** *Given two transition systems $S^1(\Sigma^1, Q_0^1, T^1)$ and $S^2(\Sigma^2, Q_0^2, T^2)$ with the same alphabet $A$, a sufficient condition for $S^2$ to simulate $S^1$ is that there exists a relation $\varphi$ such that:*

- *for each label $l \in A$, $\xrightarrow{l}_{S^2} \sqsubseteq_\varphi \xrightarrow{l}_{S^1}$*
- *$\forall e \in \Sigma_{S^2}$ , $\exists e' \in \Sigma_{S^1}$ such that $\varphi\, e\, e'$*

## 4.1 Refinement of sequential statements

In [BFR94], we have already presented refinement rules for the conditional and alternative constructs. In this paper, we investigate two new refinement rules for the sequential construct.

- The first one allows the independent refinement of the elements of a sequence.
- The second one allows the refinement of a sequence by a single statement that refines independently each element of the sequence.

**Independent refinement.** A sequence of statements can be refined by refining each of its components separately:

$$\frac{i_1 \sqsubseteq_\varphi i_1' \quad i_2 \sqsubseteq_\varphi i_2'}{\text{Seq } (i_1, i_2) \sqsubseteq_\varphi \text{Seq } (i_1', i_2')}$$

**Sketch of the proof:**

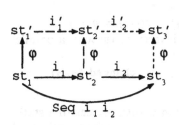

Fig. 1. SEQ refinement

The figure 1 represents the states connected by the sequential construct and its components as well as the states connected by the refinement relation $\varphi$. Given $st_1, st_3, st_1'$ such that $Seq(i_1, i_2)(st_1, st_3)$ and $\varphi(st_1, st_1')$, we look for a state $st_3'$ such that $st_1' \xrightarrow{Seq(i_1', i_2')} st_3'$ and $\varphi(st_3, st_3')$.
By the first refinement hypothesis: $i_1 \sqsubseteq_\varphi i_1'$, there exists a state $st_2'$ such that $st_1' \xrightarrow{i_1'} st_2'$ and $\varphi(st_2, st_2')$ and finally, by the second refinement hypothesis: $i_2 \sqsubseteq_\varphi i_2'$, there exists a state $st_3'$ such that $st_2' \xrightarrow{i_2'} st_3'$ and $\varphi(st_3, st_3')$.

**Joint refinement.** We first introduce the predicate **stable** $(\varphi, i)$:

**Definition 2 (Stability)** *The refinement relation $\varphi$ is invariant through the statement $i$.*

$$stable\ (\varphi, i) = \forall st\ st'_1\ st'_2.\ \varphi(st, st'_1) \wedge i(st'_1, st'_2) \Rightarrow \varphi(st, st'_2)$$

The idea of the joint refinement rule is to split the refinement of a sequence of statements by a single statement into the refinement of each component through a *projection* of the initial refinement operator.

$$\frac{i \sqsubseteq_{\varphi_1} i'_1 \quad i \sqsubseteq_{\varphi_2} i'_2 \quad stable\ (\varphi_1, i'_2) \quad stable\ (\varphi_2, i'_1)}{i \sqsubseteq_{\varphi_1 \wedge \varphi_2} Seq\ (i'_1, i'_2)}$$

**Sketch of the proof:**

The figure 2 represents the states connected by the sequential construct and its components as well as the states connected by the refinement relation $\varphi$.

Given $st_1, st_2, st'_1$ such that $st_1 \xrightarrow{i} st_2$, $\varphi_1(st_1, st'_1)$ and $\varphi_2(st_1, st'_1)$, we look for a state $st'_3$ such that $st'_1 \xrightarrow{Seq(i'_1, i'_2)} st'_3$ and $\varphi_i(st_2, st_3)$.

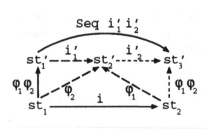

**Fig. 2.** Joint SEQ refinement

– By the refinement hypothesis $i \sqsubseteq_{\varphi_1} i'_1$, there exists a state $st'_2$ such that $st'_1 \xrightarrow{i'_1} st'_2$ and $\varphi_1(st_2, st'_2)$.
– By the stability hypothesis **stable** $(\varphi_2, i'_1)$, we have $\varphi_2(st_1, st'_2)$.
– By the refinement hypothesis $i \sqsubseteq_{\varphi_2} i'_2$, there exists a state $st'_3$ such that $st'_2 \xrightarrow{i'_2} st'_3$ and $\varphi_2(st_2, st'_3)$.
– By the stability hypothesis **stable** $(\varphi_1, i'_2)$, we have $\varphi_1(st_2, st'_3)$.
– Thus there exists a state $st'_3$ satisfying the refinement requirements.

## 4.2 Refinement of parallel constructs

**Binding invariance.** A binding is invariant or unchanged through a statement $i$ if it has the same status in any states connected by $i$.

$$\textbf{unchanged}\ (s, i) = \forall st_1\ st_2.\ i(st_1, st_2) \Rightarrow st_1(s) = st_2(s)$$

We can prove the intuitive result concerning invariance through a parallel statement: if $i_1$ and $i_2$ are functional, we have:

$$\textbf{unchanged}\ (s, i_1 \parallel i_2) = \textbf{unchanged}\ (s, i_1) \wedge \textbf{unchanged}\ (s, i_2)$$

**Projections.** We define the projection of a relation $\varphi$ with respect to a statement $i$ (denoted by $\varphi_i$) as a relation which only depends on bindings unchanged by $i$.

$$\varphi_i = \lambda\ st\ st'.\ \exists\ st''.\ \varphi(st, st'') \wedge \forall s.\ \textbf{unchanged}\ (s, i) \Rightarrow st'(s) = st''(s)$$

Then, the projected relation is easily shown to be stable by the statement $i$:
**stable** $(\varphi_i, i)$.

Furthermore, if $\varphi$ is functional ($\mathtt{IsF}\ (\varphi)$) and if the statements $i_1$ and $i_2$ are independent (they cannot change a binding together[4]), then $\varphi$ is the conjunction[5] of its projections over $i_1$ and $i_2$.

$$\mathtt{IsF}\ (\varphi) \wedge \mathtt{indep}\ (i_1, i_2) \Rightarrow \varphi = \mathtt{And}\ (\varphi_{i_1}, \varphi_{i_2})$$

**Refinement theorem.** The following refinement rule states that the refinement of a parallel construct can be split into the refinement of its components given some hypothesis.

$$\frac{\mathtt{IsF}\ (\varphi)\quad \mathtt{indep}\ (i'_1, i'_2)\quad i \sqsubseteq_{\varphi_{i'_2}} i'_1\quad i \sqsubseteq_{\varphi_{i'_1}} i'_2}{i \sqsubseteq_\varphi i'_1 \parallel i'_2}$$

# 5 Refinements rules and tactics

## 5.1 HOL refinement theories

All the refinement theory we have developed is definitional. Starting with the initial theory, we have elaborated the presented framework without introducing any axiom. Consequently, since the initial HOL theory is consistent, the developed one is consistent as well.

## 5.2 Meta theorems in HOL

The theorems stated in this paragraph assume that we have a data type defined by an indexed set of constructors. Such a data type cannot be characterized in HOL since this would require at least sets with elements of different types. Consequently generals theorems about these data types cannot be stated rigorously in HOL. However, HOL *conversions* provide a way to generate specific instances of these theorems for a given data type.

In the following, we suppose that we have a data type $\mathtt{Decl}$ and an indexed family of typed constructors $(C_i : T_i \rightarrow \mathtt{Decl})_{i \in I}$ such that:

- $\forall i \in I.\ \mathtt{IS\_VAR}\ (C_i)$
- $\forall i\, j \in I.\ i \neq j \Rightarrow \mathtt{D\_VAR}(C_i, C_j)$

**Assignments.** The single assignment meta-rule applied to an assignment statement expresses the value of each variable in the new state in terms of expressions evaluated in the current state.

$$\bigwedge_{i \in I} \left\{ \begin{array}{l} \forall e : T_i.\ (C_i := e)(st_1, st_2) = \quad \forall x : T_i.\ st_2(C_i(x)) = (x = e(st_1))\ \wedge \\ \qquad\qquad \bigwedge_{j \neq i} \forall x : T_j.\ st_2(C_j(x)) = st_1(C_j(x)) \end{array} \right.$$

We illustrate the application of this meta rule on a data type coming from the study of the refinement of multiprocessor memory models. This generic data type describes an array of cache and status registers, and a global memory:

---

[4] $\mathtt{indep}\ (i_1, i_2) = \forall s.\ \mathtt{unchanged}\ (s, i_1) \vee \mathtt{unchanged}\ (s, i_2)$
[5] $\mathtt{And}\ (\varphi_1, \varphi_2) = \lambda\, x\ y.\ \varphi_1(x, y) \wedge \varphi_2(x, y)$

```
C_decls = VAL *ind *val | STATE *ind *state | MEM *val
```

The HOL conversion associated to the assignment meta rule proves the following theorem, given the previous data type:

```
|- (∀ x0 ex e1 e2. ((VAL x0) := ex)e1 e2 =
   (∀ x0' x1.
     e2(VAL x0' x1) = ((x0' = x0) => (x1 = ex e1) | e1(VAL x0' x1))) ∧
   (∀ x0' x1. e2(STATE x0' x1) = e1(STATE x0' x1)) ∧
   (∀ x. e2(MEM x) = e1(MEM x))) ∧
   ...
```

**Multi-Assignments.** The multi-assignment meta rule is similar to the single assignment one except that the effects of the different assignments are superposed. The meta rule is formulated as follows:

$$\forall e_1 : T_1 \cdots e_n : T_n \; (C_{i_1} := e_1 \parallel \cdots \parallel C_{i_n} := e_n)(st_1, st_2) =$$
$$\forall x : T_{i_1}. \; st_2(C_{i_1}(x)) = (x = e_1(st_1)) \wedge$$

$$\cdots$$

$$\forall x : T_{i_n}. \; st_2(C_{i_n}(x)) = (x = e_n(st_1)) \wedge$$
$$\bigwedge_{i \notin \{i_1, \cdots, i_n\}} \forall x : T_i. \; st_2(C_i(x)) = st_1(C_i(x))$$

We illustrate the application of the multi-assignment rule on the same data type. Here, the HOL conversion needs the variables assigned to. The generation of all combinations of assignments is not realistic. Here, the conversion proves the following theorem defining the multi-assignment of the three variables MEM, STATE p and VAL q:

```
|- (MEM := m) || (STATE p := s) || (VAL q := v) =
   (λ e1 e2. (∀ x0 x1.
     e2 (VAL x0 x1) = ((x0 = q) => (x1 = v e1) | e1(VAL x0 x1))) ∧
     (∀ x0 x1. e2 (STATE x0 x1) =
       ((x0 = p) => (x1 = s e1) | e1(STATE x0 x1))) ∧
     (∀ x. e2 (MEM x) = (x = m e1)))
```

**Refinements tactics.** Refinement theorems concerning the sequential and parallel constructs have been also implemented in HOL as proof tactics. Thus, these tactics can be used by the backward proof engine to reduce a goal into subgoals. To give the flavor of the application of a tactic, we have peeked a fragment from the proofs developed for the study of refinements between multiprocessor memory models.

Suppose we have to prove the following refinement property:

```
∀ p st ost states. st IN states ∧ ost IN states ⇒
  REF_Op
  (((((STATE p) := (CST st)) || ((VAL p) := (val MEM))) IF
    ((val(STATE p)) =* (CST ost)))
  ((A_STATE := (REPLACE_B_P(val A_STATE)(CST ost)(CST st))) ||
    (A_VAL := ((val A_MEM) INSERT_P (val A_VAL))))
  (λ e1 e2. e2 = C2Af states e1)
```

The application of the parallel-sequential transformation of independent assignments (section 3.5) yields the following subgoal:

```
REF_Op
 (((((STATE p) := (CST st)) || ((VAL p) := (val MEM))) IF
  ((val(STATE p)) =* (CST ost)))
 (Seq (A_STATE := (REPLACE_B_P(val A_STATE)(CST ost)(CST st)))
      (A_VAL := ((val A_MEM) INSERT_P (val A_VAL))))
 (And (Pr2_Phi_X(λ x y. y = C2Af states x)A_VAL)
      (Pr2_Phi_X(λ x y. y = C2Af states x)A_STATE))"
 2  ["st IN states" ]
 1  [" ost IN states" ]
```

Then, the application of the tactic associated to the joint sequential refinement theorem of section 4.1 splits the current goal into four subgoals.

# 6 Atomic memory models

In this section, we are especially interested in shared memory multiprocessors [Lit90]. In such an environment, a value is implemented in central memory and for efficiency reasons can be replicated in processor caches. Values are generally grouped into blocks, the block being the unit of transfer between memory and caches. However, in our study, we make a practical simplifying assumption: values are the unit of transfer (we do not consider the problem of mapping a value in a block). Three operations are generally considered: the usual read and write and the *flush* operation due to the limited capacity of a cache.

Several memory models have been proposed in the literature. A memory coherency defines the sequences of values, one is allowed to observe after read and write operations issued by the different processors. In [BFR94], we have characterized the usual coherency models and established a refinement ordering between them. We have adopted a state based approach since the abstract algorithms studied are expressed within such a formalism[PD92].

Now, we only study the strongest memory model, qualified as atomic, and one of its implementations.

We first present the abstract view of an atomic memory then a multiprocessor implementation of such a memory model.

## 6.1 The abstract atomic memory

An atomic memory is similar to a single register. It is defined as a transition system over the generic states typed by *val. Its initial state is characterized by the predicate M_INIT v where v is the initial value of the memory cell. We consider the read (M_READ v) and write (M_WRITE v) transitions. The transition M_READ v maps a state with value v to itself. The transition M_WRITE v maps any state to the state with value v. Thus, the value v is read after the M_WRITE v transition. In order to represent the abstract view of cache flushes, we also consider the M_FLUSH transition, defined as a SKIP statement.

```
M_decls = M_VAL *val
M_INIT  |- M_INIT v = (val M_VAL) =* (CST v)
M_READ  |- M_READ v = SKIP IF ((val M_VAL) =* (CST v))
M_WRITE |- M_WRITE v = M_VAL := (CST v)
M_FLUSH |- M_FLUSH = SKIP
```

## 6.2 A multiprocessor implementation

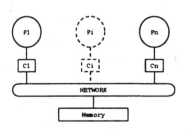

Fig. 3. A multiprocessor architecture

We study cache coherence protocols for shared memory multiprocessors as illustrated by the figure 3. Data used by coherence protocols are specified by the generic declaration

```
C_decls = VAL *ind *val
       | STATE *ind *state
       | MEM *val
```

representing an indexed collection of caches with two fields: a value and a state respectively of type *val and *state, and a centralized memory of type *val.

With respect to a set **states** of valid states, e.g., *Dirty, Shared, Valid, Exclusive*, the following invariant links the concrete and abstract representations:

```
C2Mf |- C2Mf states e1 (M_VAL v) =
  ∀* i. val(STATE i) ∈* (CST states) ⟹* val(VAL i) =* (CST v) ∧*
  ∀* i. ¬* (val(STATE i) ∈* (CST states)) ⟹* val MEM =* (CST v)
  e1
```

The value of the single register is either the value stored in any of the valid caches, or else the value stored in memory.

Two types of atomic coherence algorithms are generally considered: the *write invalidate* and the *write update* protocols[Ste90]. Informally, the principles of these protocols are to read a value in the cache if it is locally available. They can be distinguished by their behavior on write requests:

- Write invalidate protocols invalidate distant valid caches and update the local cache.
- Write update protocols updates the local cache as well as the distant valid caches.

The algorithm considered in this paper belongs to the write invalidate family. Each cache can be in one of the two states: C_Valid or C_Invalid. The algorithm is defined by the transition system of which the initialization predicate and the transitions are:

```
C_INIT |- C_INIT v = (M_INIT v) o (C2Mf C_Valid)
C_READ |- C_READ i v =
    (((((STATE i) := (CST C_Valid)) || ((VAL i) := (val MEM))) IF
     ((∀* j. (val (STATE j)) =* (CST C_Invalid)) ∧*
      ((val MEM) =* (CST v))))
   [] (ALT j.
       (((STATE i) := (CST C_Valid)) || ((VAL i) := (val (VAL j)))) IF
       (((val (STATE j)) =* (CST C_Valid)) ∧* ((val (VAL j)) =* (CST v))))
C_WRITE |- C_WRITE i v =
     (PAR j. ((STATE j) := (CST C_Invalid)) IF (CST ¬ (i = j))) ||
     (((STATE i) := (CST C_Valid)) || ((VAL i) := (CST v)))
```

A valid initial state in the concrete model is a state that can be linked to a state of the abstract model through the representation invariant.

A value v can be read by the processor i if either no processor owns a valid copy and v is the current memory value or a processor j owns a valid copy and its value is v.

When a value is written by processor i all the other processor copies are invalidated and the local copy is set to v and stated as valid.

Starting from the previous description, we could directly establish the refinement proof with the abstract model given in section 6.1. However, in such a proof, it would not be possible to turn profit from the symmetries of the problem. In the next section, we first review one approach that has tried to do so then we outline how we have proceeded in order to establish the refinements between the abstract, symmetric and concrete models.

# 7 Exploiting symmetries in cache coherency protocols

Validation algorithms are usually based on exhaustive state space generation methods. Although efficient technics have been proposed to reduce the state space, it remains huge for simple protocols. Thus, they are not well suited for algorithms dedicated to scalable architectures since the complexity of the validation increases with the size of the architecture and we cannot use the same validation process for any instance of the algorithm. A first idea would be to reason by induction over the size of the architecture. However, cache coherency algorithms cannot easily be stated as primitive recursive functions. Thus, the usual induction rules cannot be applied in this context.

Another way of dealing with such problems is to reduce the size of the state space. For this purpose, we exploit the symmetry of the problem to define an equivalence relation over states. Thus the exploration is limited to the quotient space. In the field of cache coherency protocols, this idea has been studied by Pong and Dubois[PD92].

## 7.1 The approach of Pong and Dubois

Pong and Dubois use the symmetry of cache coherence algorithms to reduce the state space and allow an exhaustive exploration in order to prove the validity of the algorithm. This symmetry is used to define an equivalence relation over the states defining a bounded quotient space. The basic idea of Pong and Dubois is to not distinguish individual elements of the architecture being in the same state, but to count them. Furthermore, the counting process does not use natural numbers: naturals greater than one are identified. They introduce four counting symbols: 0, 1, + for at least one, and * for unknown. For instance, $Shared^+$, $Invalid^*$, $Dirty^0$ represents the set of states where at least one cache is shared, some are invalid and no one is dirty. In their paper, some cache coherency protocols[AB86] are developed using such a formalism. For each statement of a given algorithm, they derive a transition over a system of which states are defined as t-uples of counting symbols. Then, they generate all minimal reachable states from a given initial state and check that these states verify the specified properties. However, the paper is not concerned with the validation of the transformation of the multiprocessor algorithm to the transition system over the quotient space. More precisely, they do not show that the multiprocessor algorithm is indeed a refinement of the proposed equivalent one.

## 7.2 A generic symmetric memory model

Thanks to transitivity of refinements, the validity of a memory protocol can be established through an intermediate symmetric model. Thus, two refinement steps must be proved. Here, we have defined a symmetric model for write invalidate protocols. In the following, we only report on the refinement between the symmetric model and the atomic one.

**The transition system.** In the considered symmetric model, we also do not distinguish individual elements being in the same state. The symmetric representation we adopt is that of a bag. Moreover, since we do not presume of the correctness of the studied algorithms, we consider a set of cached values. Data used by the symmetric model are as follows:

```
A_decls = A_VAL (*val)set | A_STATE (*state)bag | A_MEM *val
```

The representation invariant which links the representation of the symmetric model representation and the abstract representation is given by the A2Mf relation[6]:

```
A2Mf |- A2Mf e1(M_VAL v) =
```
$C\_Valid \in_B(val\ A\_STATE\ e1) \Rightarrow (val\ A\_VAL\ e1 = \{v\})\ |\ (val\ A\_MEM\ e1 = v)$

We define the three cache access operations A_READ, A_WRITE and A_FLUSH on the symmetric model introduced above:

- the A_READ v operation allows reading a value v if no current cache state is valid and the memory value is v, or if a cache is in the valid state and the set of cached values is {v}. After a read access, an invalid cache becomes valid. This is simulated by either a SKIP statement if the local cache is valid, or by a REPLACE statement if not.

---

[6] $\in_B$ denotes the bag membership predicate

- After a **A_WRITE** v operation, the current cache is the only one in the valid state, and the set of cached values is {v}.
- A **A_FLUSH** operation may invalidate a cache and copy a cached value to memory.

For instance, the **A_READ** v operation is defined in HOL as follows:

```
A_READ v =
   (A_STATE := REPLACE_B_P (val A_STATE) (CST C_Invalid) (CST C_Valid)
   ∥ A_VAL := (CST {v})
   ) IF ¬*((CST C_Valid) ∈_B* (val A_STATE)) ∧* (val A_MEM) =* (CST v)
 ∥ SKIP IF ((CST C_Valid) ∈_B* (val A_STATE)) ∧* (val A_VAL) =* (CST {v})
 ∥ A_STATE := REPLACE_B_P (val A_STATE) (CST C_Invalid) (CST C_Valid) IF
      (CST C_Valid) ∈_B* (val A_STATE) ∧*
      (CST C_Invalid) ∈_B* (val A_STATE) ∧*
      (val A_VAL) =* (CST {v})
```

**Correctness of the symmetric model.** The correctness of the symmetric model is established by proving the refinement between the operations of the atomic memory model and the corresponding operations of the symmetric one. We have proved the following theorems:

$$A\_READ_v \sqsubseteq_{A2Mf} M\_READ_v$$
$$A\_WRITE_v \sqsubseteq_{A2Mf} M\_WRITE_v$$
$$A\_FLUSH \sqsubseteq_{A2Mf} M\_FLUSH$$

The proof of these results mainly use refinement theorems and conversions establishing the relation between variable values after a multiple assignment.

## 8 Conclusion

In this paper, we have been concerned with the refinement of snooping memory protocols. To achieve this goal, we have introduced a new state representation allowing for strong typing and defined the semantics of sequential and parallel statements. Some refinement results have been established. This framework has been used to specify and validate a symmetric view of a write invalidate cache protocol. The next step of this study will be to validate the transformation between the concrete and the symmetric versions of the model by the proof of a refinement relation between them. This step could be made easier by the introduction of symmetric statements and expressions and related theorems. Such tools could then be used to validate classical symmetric (write invalidate or write update) protocols such as Illinois and Firefly.

This paper was mainly concerned with local properties of programs. A further study would be to consider state sequences for reasoning about behavioral program properties[CT94].

# References

[AB86]   J. Archibald and J.-L. Baer. Cache coherence protocols: Evaluation using a multiprocessor simulation model. *ACM Transactions on Computer Systems*, 4(4):273–298, nov 1986.

[APP93]  F. Andersen, K. D. Petersen, and J.S. Pettersson. Program verification using HOL-UNITY. In *Higher Order Logic Theorem Proving and its Applications*, volume 780 of *Lecture Notes in Computer Science*. Springer-Verlag, 1993.

[BC84]   G. Berry and L. Cosserat. The ESTEREL synchronous programming language and its mathematical semantics. volume 197 of *Lecture Notes in Computer Science*, pages 389–448, Berlin, Germany, 1984. Springer-Verlag.

[BFR94]  J.-P. Bodeveix, M. Filali, and P. Roché. Towards a HOL theory of memory. In *Higher Order Logic Theorem Proving and its Applications*, volume 859 of *Lecture Notes in Computer Science*, pages 49–64. Springer-Verlag, sep 1994.

[CM88]   K.M. Chandy and J. Misra. *Parallel Program Design, A Foundation*. Addison-Wesley, 1988.

[CT94]   C. Ching-Tsun. Mechanical verification of distributed algorithms in higher order logic. In *Higher Order Logic Theorem Proving and its Applications*, volume 859 of *Lecture Notes in Computer Science*, pages 158–176. Springer-Verlag, 1994.

[Dij76]  E.W. Dijkstra. *A Discipline of Programming*. Englewood Cliffs New Jersey: Prentice Hall, 1976.

[Gir87]  J.-Y. Girard. Linear logic. *Theoretical Comp. Science*, 50:1–102, 1987.

[GM94]   M.J.C. Gordon and T.F. Melham. *Introduction to HOL*. Cambridge University Press, 1994.

[Hoa85]  C.A.R. Hoare. *Communicating Sequential Processes*. Prentice Hall, 1985.

[Lit90]  D. Litaize. Architectures multiprocesseurs à mémoire commune. In *Deuxième symposium architectures nouvelles de machines*, pages 1–40, sep 1990.

[LT87]   N.A. Lynch and M.R. Tuttle. Hierarchical correctness proofs for distributed algorithms. In *Proceedings of the sixth annual ACM symposium on principles of distributed computing*, pages 137–151, aug 1987.

[PD92]   F. Pong and M. Dubois. The verification of cache coherence protocols. Technical Report CENG-92-20, USC, nov 1992.

[Ste90]  P. Stenstrom. A survey of cache coherence schemes for mutliprocessors. *Computer*, 23(6):11–25, jun 1990.

[Tre92]  G. Tredoux. Mechanizing execution sequence semantics in HOL. *South African Computer Journal*, (7), July 1992.

[vW90]   J. von Wright. *A lattice-theoretical basis fro program refinement*. PhD thesis, Abo Akademi Finland, 1990.

# Combining Decision Procedures
# in the HOL System[*]

Richard J. Boulton

University of Cambridge Computer Laboratory, New Museums Site, Pembroke Street,
Cambridge CB2 3QG, United Kingdom

**Abstract.** A HOL implementation of Nelson and Oppen's technique for combining decision procedures is described. The principal advantage of this technique is that the procedures for the component theories (e.g. linear arithmetic, lists, uninterpreted function symbols) remain separate. Equations between two variables are the only information that need be communicated between them. Thus, code for deciding the component theories can be reused in a combined procedure and the latter can easily be extended. In addition, efficiency techniques used in the component procedures can be retained in the combined procedure.

## 1  Introduction

The construction of program verification tools in the late 1970's and early 1980's drove the development of decision procedures for quantifier-free formulas over various theories and combinations of theories. Much of this work was in two streams, one by Shostak [13, 14, 15, 16, 17] and the other by Nelson and Oppen [10, 11, 12, 9]. Shostak's work is the basis for the decision procedures in PVS [19], and the Eves verification environment [4] makes use of some of Nelson and Oppen's techniques.

Decision procedures are an important tool in theorem provers and verification systems. They allow much low-level reasoning to be performed automatically. Lemmas and simplifications that appear trivial may take many minutes or even hours to prove 'by hand', especially for less experienced users. Decision procedures can relieve users of some of this burden. The HOL system [6] suffers from a relative lack of decision procedures. There are procedures for propositional tautologies, a subset of linear arithmetic formulas (over the natural numbers) [3, Chapter 5] and a Knuth-Bendix completion procedure [18]. Procedures have also been written for elementary real algebra [7] and for a special case of solving linear systems of equations [1].

A limitation of decision procedures for individual theories such as linear arithmetic is that formulas to be proved rarely conform to the pure theory. They usually involve symbols from other theories. Sometimes replacement of nonconforming subterms with new variables does not affect the truth of the formula

---

[*] Research supported by the Engineering and Physical Sciences Research Council of Great Britain under grant GR/J42236.

and in such cases proof of the generalised formula followed by re-instantiation is successful. Often, however, the decision procedure cannot be used. For this reason there has been much interest in decision procedures for combinations of theories, both by Shostak, Nelson and Oppen, and by others [8, 5].

In this paper, an implementation of Nelson and Oppen's algorithm for combining decision procedures for disjoint theories is described. This is limited to quantifier-free formulas (or equivalently, formulas that when placed in prenex normal form contain only universal quantifiers). However, many of the formulas arising in verification proofs conform to this restriction. The principal feature of the Nelson-Oppen technique is that decision procedures for the component theories remain separate. Equations between variables are the only information that need be communicated between them. Thus, code for deciding the component theories can be reused in a combined procedure and the latter can easily be extended.

As with all proof procedures in the HOL system there is the added difficulty that all reasoning must ultimately be performed by primitive inference rules of the logic. It is not sufficient just to compute whether a formula is true or false since this fact cannot be used for further reasoning. Having to justify all steps by primitive inferences has implications for both the ease of coding and for efficiency. On the latter issue the Nelson-Oppen technique scores highly because efficiency techniques used in the component procedures can be retained in the combined procedure. For example, the linear arithmetic procedure in the HOL arith library uses a highly simplified representation of linear inequalities when performing operations that may not lead to a successful proof. This representation could not be used if symbols from other theories had to be included. Also, some efficiency would be lost in communicating between component procedures if the information passed was complex. However, very little computation is required in transforming one procedure's representation of an equation between variables to the representation used by some other procedure.

An outline of the Nelson-Oppen technique is given in the next section. Then Sections 3 and 4 describe the implementation of the technique in HOL. Section 5 describes one way of obtaining a component procedure suitable for use in combination, and Sect. 6 gives some example procedures. Sect. 7 is a brief illustration of how a new component procedure may be added. The issue of maintaining efficient representations in component procedures is considered in Sect. 8, and some results are presented in Sect. 9.

## 2 The Nelson-Oppen Technique

Nelson and Oppen [10] describe a method for combining decision procedures for disjoint quantifier-free theories into a single decision procedure for the combined theory. A decision procedure for a theory is an algorithm for determining whether a logical formula is valid in the theory. In the case of quantifier-free theories the variables of the formula are implicitly universally quantified. So, the algorithm decides whether the formula is true for all possible instantiations of the variables. A theory consists of a number of function, predicate and constant symbols together with axioms (actually definitions or theorems in HOL) about them. Two theories are disjoint if they have none of these symbols in common. The combination of two such theories contains the symbols and axioms from both.

The formulas considered involve the usual logical connectives ($\neg$, $\land$, $\lor$, $\Longrightarrow$, $\Leftrightarrow$), conditional expressions (written $\ldots \Rightarrow \ldots | \ldots$ in HOL) and atoms. Atoms are formed by applying some predicate of the theory. The component theories may have the equality predicate in common.

The first step in deciding a formula is to convert its negation to disjunctive normal form: implication, if-and-only-if (equality on boolean values) and boolean-valued conditional expressions are eliminated and the result is stratified into a disjunction in which each disjunct is a conjunction of literals. Literals are atoms or negated atoms. To validate the original formula it is sufficient to prove that each disjunct of the normalised negation is unsatisfiable, i.e.:

$$\vdash ((\forall x_1 \ldots x_n.\ f[x_1, \ldots, x_n]) = \text{T}) = ((\exists x_1 \ldots x_n.\ f'[x_1, \ldots, x_n]) = \text{F})$$

where $f'[x_1, \ldots, x_n]$ is the normalised $\neg f[x_1, \ldots, x_n]$. So, the basic procedure must be able to show that a conjunction of literals is unsatisfiable (false for all possible instantiations of the variables).

Consider an example conjunction taken from Nelson and Oppen's paper and modified slightly to suit HOL:

$$(x \leq y) \land (y \leq x + \text{HD(CONS 0 (CONS } x \text{ NIL)))} \land P(h(x) - h(y)) \land \neg P(0)\ .$$

(The term CONS 0 (CONS $x$ NIL) will also be written as $[0; x]$.) The formula involves symbols from three theories: linear arithmetic over natural numbers, lists under equality, and uninterpreted function symbols under equality. Here, the uninterpreted function symbols ($P$ and $h$) are HOL variables but they may also be constants whose properties are not known by the decision procedure.

The next step is to make each literal homogeneous, that is to contain symbols from only one theory. This is achieved by replacing each heterogeneous subterm with a new variable and adding a new literal to the conjunction asserting the equality of the variable and the subterm. The first conjunct in the example is already homogeneous (in the theory of linear arithmetic). The top-level operator of the second conjunct is $\leq$, so its arguments should also belong to the theory of linear arithmetic. The variables $y$ and $x$ and the operator $+$ are fine, but HD belongs to the theory of lists. So, its application is replaced by a new variable $v_1$:

$$(x \leq y) \land (y \leq x + v_1) \land P(h(x) - h(y)) \land \neg P(0) \land (v_1 = \text{HD } [0; x])\ .$$

This process is repeated on the other conjuncts and recursively on the new conjuncts. Each new variable is existentially quantified, so the original conjunction is logically equivalent to:

$\exists v_1 \ldots v_6.$

$(x \leq y) \wedge (y \leq x + v_1) \wedge P(v_3) \wedge \neg P(v_6) \wedge$
$(v_1 = \text{HD } [v_2; x]) \wedge (v_2 = 0) \wedge$
$(v_3 = v_4 - v_5) \wedge (v_4 = h(x)) \wedge (v_5 = h(y)) \wedge (v_6 = 0) \ .$

Showing that the body of this formula is unsatisfiable is sufficient to show that the original conjunction is unsatisfiable. The new conjuncts are separated to give to the component decision procedures:

| | | |
|---|---|---|
| $x \leq y$ | $P(v_3)$ | $v_1 = \text{HD } [v_2; x]$ |
| $y \leq x + v_1$ | $\neg P(v_6)$ | |
| $v_2 = 0$ | $v_4 = h(x)$ | |
| $v_3 = v_4 - v_5$ | $v_5 = h(y)$ | |
| $v_6 = 0$ | | |

Each component procedure must be capable of deducing equalities between variables from its conjunction. For example, the list procedure should deduce $v_1 = v_2$. This new equation is given to the other procedures that already know about both variables. Thus, the arithmetic procedure receives the equation and is able to deduce that $v_1 = 0$ and hence that both $x \leq y$ and $y \leq x$. This implies $x = y$. The full propagation is given below:

| | | |
|---|---|---|
| $x \leq y$ | $P(v_3)$ | $v_1 = \text{HD } [v_2; x]$ |
| $y \leq x + v_1$ | $\neg P(v_6)$ | |
| $v_2 = 0$ | $v_4 = h(x)$ | |
| $v_3 = v_4 - v_5$ | $v_5 = h(y)$ | |
| $v_6 = 0$ | | |
| | | $v_1 = v_2$ |
| $x = y$ | | |
| | $v_4 = v_5$ | |
| $v_3 = v_6$ | | |
| | unsatisfiable | |

Unfortunately, propagation of equalities between variables is not sufficient for all theories. In some theories a disjunction of equalities between variables can be deduced without the individual equalities being true. Case splitting is required for such theories. For example, in natural and integer arithmetic the conjunction $(x \leq y) \wedge (y \leq x + 1) \wedge (z = x + 1)$ entails the disjunction $(y = x) \vee (y = z)$ but not either of the equations alone. Nelson and Oppen refer to theories that do not produce splits as *convex*. Rational and real arithmetic under addition are convex, as are the theories of lists and uninterpreted function symbols. The theory of sets is non-convex.

The full equality propagation procedure is given below. Nelson and Oppen prove its correctness.

1. Assign the separated conjunctions $C_i$ to the $n$ component procedures $P_i$.
2. If any $C_i$ is unsatisfiable so is the original conjunction (Halt).

3. If any $C_i$ entails an equality between variables, then add the equality to all the other conjunctions that already involve both variables but which do not themselves entail the equality, and go to step 2.

4. If any $C_i$ entails a disjunction $u_1 = v_1 \vee \ldots \vee u_k = v_k$ of equalities between variables, then apply the procedure recursively to the $k$ formulas

$$(u_i = v_i) \wedge C_1 \wedge \ldots \wedge C_n \quad (1 \leq i \leq k) \ .$$

If all these formulas are unsatisfiable, so is the original conjunction.

In HOL, this procedure must apply inference rules to justify the final result as a theorem. How this can be achieved is described in subsequent sections.

## 3 Normalisation

A formula to be proved is converted to prenex normal form (all quantifiers outermost). If the result contains existential quantifiers then the combined decision procedure is not applicable. Otherwise, the formula is universally quantified or has no quantifiers. It is sufficient to prove that the body of this formula is true. This is achieved by putting its negation into disjunctive normal form and proving that each disjunct is false. The latter step requires that the conjunctions be made homogeneous. The implementations of these normalisation procedures are described in the following subsections.

### 3.1 Disjunctive Normal Form

Conversion to disjunctive normal form takes place in two steps. First, negations are pushed down through the formula, with implications, boolean equalities and boolean-valued conditionals being expanded. The second step distributes conjunctions over disjunctions to obtain disjunctive normal form.

The rules for moving a negation through conjunctions, disjunctions and other negations are:

$$\neg(x \wedge y) \longrightarrow \neg x \vee \neg y$$
$$\neg(x \vee y) \longrightarrow \neg x \wedge \neg y$$
$$\neg\neg x \longrightarrow x$$

The rules are applied before processing the subterms (i.e. $\neg x$ and $\neg y$ for the first and second rules) so that the negations are swept down in one pass. The other three operators are expanded in different ways according to the required normal form and whether or not they are negated:

| Source | Disjunctive Expansion | Conjunctive Expansion |
|---|---|---|
| $x \Longrightarrow y$ | $\neg x \vee y$ | $\neg x \vee y$ |
| $x \Leftrightarrow y$ | $(x \wedge y) \vee (\neg x \wedge \neg y)$ | $(\neg x \vee y) \wedge (x \vee \neg y)$ |
| $b \Rightarrow x \mid y$ | $(b \wedge x) \vee (\neg b \wedge y)$ | $(\neg b \vee x) \wedge (b \vee y)$ |
| $\neg(x \Longrightarrow y)$ | $x \wedge \neg y$ | $x \wedge \neg y$ |
| $\neg(x \Leftrightarrow y)$ | $(x \wedge \neg y) \vee (\neg x \wedge y)$ | $(x \vee y) \wedge (\neg x \vee \neg y)$ |
| $\neg(b \Rightarrow x \mid y)$ | $(b \wedge \neg x) \vee (\neg b \wedge \neg y)$ | $(\neg b \vee \neg x) \wedge (b \vee \neg y)$ |

Observe that boolean equalities are expanded in one step rather than first expanding to a conjunction of implications and then expanding the implications. This saves several costly inferences and is an example of using pre-proved lemmas to improve efficiency. The form of the rules has also been chosen so that the negations are as deep down in the term as they can be.

To see why it is inefficient to expand to a conjunction of disjunctions when aiming for disjunctive normal form, consider the expansion of boolean equality:

$$x \Leftrightarrow y \longrightarrow (\neg x \vee y) \wedge (x \vee \neg y) .$$

In obtaining disjunctive normal form, the conjunction will be distributed over the disjunctions to produce a term of the form

$$(\neg x \wedge x) \vee (\neg x \wedge \neg y) \vee (y \wedge x) \vee (y \wedge \neg y) .$$

The first and last disjuncts can be discarded, but by this time $x$, $y$, $\neg x$, and $\neg y$ will have been normalised so that it is difficult for the procedure to see this.

When negations reach the level of atomic formulas they stop. These negated and non-negated formulas must themselves be normalised. This could be done in another pass over the term but it is more efficient to do it while dealing with the negations as this avoids costly repeated traversals of the term. The normaliser for literals is an argument of the disjunctive-normal-form function. It may introduce conjunctions and disjunctions, so it is important to delay the stratification of the whole formula until the atomic formulas have been normalised.

## 3.2 Homogenisation

The vital step in making a conjunction of literals homogeneous is the replacement of a subterm by a new variable. The HOL theorem that achieves this is

$$\vdash \forall P\ t.\ P(t) = \exists v.\ P(v) \wedge (v = t) .$$

Consider the term $y \leq x + \text{HD}\ [0; x]$. The heterogeneous subterm is replaced by a new variable and the result (called the *template*) becomes the body of an abstraction: $\lambda v_1.\ y \leq x + v_1$. This abstraction is used to instantiate $P$ in the theorem, while $t$ is instantiated with $\text{HD}\ [0; x]$. The result is

$$\vdash (\lambda v_1.\ y \leq x + v_1)(\text{HD}\ [0; x]) = \exists v.\ (\lambda v_1.\ y \leq x + v_1)(v) \wedge (v = \text{HD}\ [0; x]) .$$

After $\beta$-reduction in the appropriate places and renaming of the existentially quantified variable, the theorem becomes

$$\vdash (y \leq x + \text{HD}\ [0; x]) = \exists v_1.\ (y \leq x + v_1) \wedge (v_1 = \text{HD}\ [0; x]) .$$

The procedure is applied recursively to the equation for the new variable so that $\text{HD}\ [0; x]$ can be made homogeneous.

Terms in which more than one subterm have to be replaced (e.g. $v_3 = h(x) - h(y)$, the new equation obtained when $P(h(x) - h(y))$ is homogenised), require a little more work. First, $\lambda v_5.\ v_3 = v_4 - v_5$ and $h(y)$ are used to instantiate the general theorem:

$$\vdash (v_3 = v_4 - h(y)) = \exists v_5.\ (v_3 = v_4 - v_5) \wedge (v_5 = h(y)) \qquad (1)$$

Then $\lambda v_4.\ v_3 = v_4 - h(y)$ and $h(x)$ are used:

$$\vdash (v_3 = h(x) - h(y)) = \exists v_4.\ (v_3 = v_4 - h(y)) \wedge (v_4 = h(x)) \qquad (2)$$

Substituting (1) into (2) yields:

$$\vdash (v_3 = h(x) - h(y)) = \exists v_4.\ (\exists v_5.\ (v_3 = v_4 - v_5) \wedge (v_5 = h(y))) \wedge (v_4 = h(x))\ .$$

Finally, the $\exists v_5$ is pulled out through the conjunction:

$$\vdash (v_3 = h(x) - h(y)) = \exists v_4\ v_5.\ ((v_3 = v_4 - v_5) \wedge (v_5 = h(y))) \wedge (v_4 = h(x))\ .$$

This process generalises to three or more subterms but note that multiple existential quantifiers may then have to pulled through the conjunction.

It remains to say how heterogeneous subterms are identified. The top-level operator of the literal is tested using discriminator functions until one is found that recognises it. There is one discriminator for each component decision procedure. Negation and equality are ignored in obtaining the top-level operator since they are shared by all the theories. This makes it possible for the term to have two top-level operators, e.g. $v_3 = h(x) - h(y)$ has $v_3$ and $-$. (The word 'operator' is used for convenience; $v_3$ is not really an operator.) In this case some arbitration occurs (see below). Subterms are then examined top-down until the matching discriminator differs from the one for the top-level operator.

The discriminators take a term and either raise an exception or return a list of subterms together with a function to rebuild a term from the subterms (or modified versions of them). The function is used to build the template around the new variables. Thus, the type of a discriminator is

```
term -> ((term list -> term) * (term list)) .
```

The discriminators for the component theories are placed in a list which gives them a priority (the earlier in the list, the higher the priority). Instead of looking for the first successful discriminator, the procedure filters the list retaining the successful ones. Normally the head of the filtered list is taken as the discriminator to use, but when arbitrating between the two sides of an equality the discriminator used is the first to appear in both the list for the left-hand side and the list for the right-hand side. If there is no such discriminator but both sides of the equation are recognised by at least one discriminator individually and are not variables, the first discriminator for the left-hand side is used. The requirement that the terms are not variables prevents non-terminating repeated replacement of a variable by a new variable.

If no discriminator is successful for the top-level operator then it belongs to none of the component theories. The literal is ignored in the hope that the remaining literals entail false without it. If no discriminator is successful for a subterm then it is considered to be a heterogeneity.

Before returning the homogeneous formula, all the newly introduced existential quantifiers are pulled to the outside.

# 4 The Decision Process

The result of homogenisation is an existentially quantified conjunction of homogeneous literals. It is sufficient to prove that the conjunction (call it $C$) is false. The discriminator functions are used again to separate the conjuncts of $C$ into groups, one for each component decision procedure. Denoting these groups by new conjunctions $C_1, \ldots, C_n$, it is easy to obtain theorems $C \vdash C_1, \ldots, C \vdash C_n$ by assuming $C$, breaking the result into theorems for each conjunct, and then regrouping them.

Any procedure for which there are no conjuncts is dropped. So, for a formula that only involves symbols from one theory, the combined decision procedure immediately degenerates into a single-theory procedure. There may also be conjuncts that do not belong to any of the component theories. These are ignored. So, when a formula involves only symbols from one theory plus unrecognised symbols, the combined decision procedure is equivalent to replacing unrecognised subterms by new variables before applying the single-theory procedure, a technique that is already used in the linear arithmetic procedure of the HOL arith library.

The procedures that remain are paired with their theorem $C \vdash C_i$ and sorted for priority according to a user-supplied weighting function. For example, the procedures might be weighted by increasing number of variables in $C_i$. Fewer variables implies fewer potential new equalities between variables.

Processes are constructed from each pair of a theorem and a decision procedure. The processes interact by passing theorems of the form $C \vdash u = v$. Such a theorem is generated by decision procedure $i$ inferring $u = v$ from its conjunction $C_i$. A process $j$ may use the theorem by adding it to its conjunction. The new theorem is $C \vdash (u = v) \wedge C_j$. The key point here is that because both the original theorem and the variable equality theorem have $C$ as hypothesis, no new hypotheses are introduced into the theorem held by the process.

A process is like an object with the theorem $C \vdash C_i$ as its data ($C_i$ may change) and methods based around the decision procedure for the theory. The input and output messages for a process are specified by the ML data types `request` and `response`, respectively:

```
datatype request =
    ProveFalse | GetEquality | GetEqualities | AddEquality of thm;

datatype response = Thm of thm | None;
```

A `ProveFalse` request either yields a response Thm ($C \vdash F$), or None if the decision procedure cannot derive false from the conjunction. Similarly, `GetEquality` and `GetEqualities` either yield theorems of the form $C \vdash u = v$ and $C \vdash u_1 = v_1 \vee \ldots \vee u_k = v_k$, respectively, or None. The conjunction held by a process may entail more than one equality (or disjunction of equalities) at once. However, the process should only provide one at a time. The result of an `AddEquality` request is always None. A process should be able to receive the equality it generated itself and know to ignore it. This allows new equalities to be broadcast to *all* the component processes.

The interface to processes is deliberately abstract to give the implementor as much freedom as possible. The full information required for each component theory is given by the following ML record:

```
type decision_procedure =
   {Name : string,
    Description : string,
    Author : string,
    Discriminator : term -> (term list -> term) * term list,
    Normalizer : term -> thm,
    Procedure : thm -> incremental_procedure};
```

The type `incremental_procedure` is specified by:

```
datatype incremental_procedure =
   Increment of request -> response * incremental_procedure;
```

An `incremental_procedure` accepts a request and returns a response and a modified version of itself. It is initially generated from one of the $C \vdash C_i$ theorems. The `Normalizer` is a conversion which may be used to normalise literals prior to the decision process.

The co-operation between processes closely follows the description in Sect. 2 but is in a functional style. The only feature worthy of further comment is the handling of disjunctions of equalities. When a process provides a theorem of the form $C \vdash u_1 = v_1 \vee \ldots \vee u_k = v_k$, it is not possible to send a theorem $C \vdash u_i = v_i$ to the other processes because $C$ does not entail one of the equalities alone. Instead, the theorem $u_i = v_i \vdash u_i = v_i$ is passed to the other processes. There are $k$ recursive invocations of the co-operation function, one for each equation in the disjunction. If they are all successful, $k$ theorems of the form $C, u_i = v_i \vdash F$ are obtained. The equation is discharged to give $C \vdash (u_i = v_i) \implies F$ from which $C \vdash (u_i = v_i) = F$ follows. Together these entail $C \vdash (u_1 = v_1 \vee \ldots \vee u_k = v_k) = F$ which, along with the original theorem for the disjunction, gives $C \vdash F$.

## 5 Generating a Component Procedure

Ideally, component decision procedures should be able to deduce new equalities directly and hence efficiently. However, suppose that the only procedure available is one that can prove that an unsatisfiable conjunction of literals equals false. The procedure must allow equalities as literals. Then, new equalities between variables ($u = v$) can be obtained by testing all the possibilities until one is found for which $\vdash (\neg(u = v) \wedge C_i) = F$, i.e. $\vdash C_i \implies (u = v)$. It follows from $C \vdash C_i$ that $C \vdash u = v$.

The implementation of an `incremental_procedure` from a basic procedure with the above properties and a theorem $C \vdash C_i$ is best described imperatively. However, an imperative coding (using ML reference values) does not work well in practice because of case splits. A case split causes multiple copies of the process to be invoked with slightly different data. This is easy in a functional style but requires copies of the process to be made in an imperative style.

The state consists of four things: the theorem $C \vdash C_i$ (called th), a list eqs of known equalities between variables in $C_i$, a list possible_eqs of possible new equalities, and a list get_eqs of the possible equalities that have not been tried since $C_i$ was last changed. The free variables vars in $C_i$ must also be known but do not change because new equalities are only added if they are between two variables already in $C_i$. The variable eqs is initialised to the equalities between variables in $C_i$, and possible_eqs and get_eqs are initialised to all the possible pairings of vars with vars, less symmetric duplicates, reflexive equations, and the contents of eqs.

Using the basic procedure in attempting to prove false from $C_i$ is straight-forward — simply apply the procedure to $C_i$. If successful, a little manipulation is required to obtain $C \vdash \mathsf{F}$.

To get an equality, remove the first equation $u = v$ from get_eqs or return None if it is empty. Apply the basic procedure to $\neg(u = v) \wedge C_i$. If this fails, repeat with the new head of get_eqs. Otherwise, manipulate the theorem into the required form as indicated above and add $u = v$ to eqs and remove it from possible_eqs. If $u = v$ were not added to eqs, the process would not know to ignore the subsequent AddEquality broadcast by the co-operation function.

When given an equality $C \vdash u = v$ to add, only do so if both $u$ and $v$ are in vars and both $u = v$ and $v = u$ are not in eqs. Then, conjoin the equation onto $C_i$ to form a new th, add $u = v$ to eqs, and remove it from possible_eqs. Set get_eqs to the new value of possible_eqs because since $C_i$ has changed all the possible equalities should be tried again.

One optimisation is to maintain a flag indicating whether an attempt has been made to prove false since $C_i$ was last changed. If it has then there is no point in trying again. Another is to look for new equalities as transitive conse-quences of known ones before applying the basic procedure. The overall technique seems reasonable for convex theories because no case splitting is then required and GetEqualities can always return None. Testing all possible disjunctions of equalities between variables would require more complex data structures and would probably be very costly. Even the above algorithm is quadratic in the number of variables.

# 6    Example Procedures

In order to test the combination procedure on Nelson and Oppen's example, three component decision procedures are required, for linear arithmetic, lists, and uninterpreted function symbols. The HOL arith library has a linear arith-metic decision procedure. Unfortunately, though, its core procedure for falsifying conjunctions does not allow equalities as literals but only $\leq$ inequalities. It is therefore necessary to expand equalities into two inequalities during normalisa-tion, and to perform the same operation whenever a new equality is added or has to be proved. The other two theories can be decided by a congruence closure procedure as described by Nelson and Oppen [11] though the theory of lists can be decided more efficiently [12]. An experimental version of congruence closure has been implemented in HOL by the current author.

The method described in Sect. 5 can be used to generate suitable functions to use with the combined procedure since the three theories are convex. (Actually, linear natural number arithmetic is non-convex but the procedure in the arith library is not capable of deducing the properties that lead to case splitting.)

## 7 Adding a Procedure

Adding an existing decision procedure to the combined procedure is straightforward if the theory it decides is convex. (The theory must also be disjoint with respect to the theories already handled by the combined procedure.) A function make_incremental_procedure is provided which implements the algorithm described in Sect. 5.

As an example, consider the theory of equality over pairs. This consists of three symbols: the infix ',' which constructs a pair from two arguments, and 'FST' and 'SND', the projection functions. Suppose that the ML function PAIR_CONV can prove the unsatisfiability of a conjunction of literals from the theory of pairs, e.g. $\vdash ((x, y) = p) \land \neg(\text{FST } p = x) = F$. Then a suitable procedure for inclusion in the combined procedure is obtained by applying make_incremental_procedure to CONJ and PAIR_CONV. CONJ is the HOL inference rule that conjoins the conclusions of two theorems. It is used for adding an equation between variables to a conjunction of literals, and is taken as a parameter for flexibility. For example, an arithmetic procedure might require that the equality $m = n$ be added as $(m \leq n) \land (n \leq m)$.

As stated in Sect. 4, a component procedure must also have a discriminator and a normaliser. For pairs the normaliser need do nothing, so the identity conversion ALL_CONV is used. The discriminator looks something like this:

```
exception PairDiscrim;
fun pair_discrim tm =
    if (is_var tm)
    then (fn _ => tm,[])
    else let val (f,args) = strip_comb tm
             fun reconstruct args' = list_mk_comb (f,args')
         in  if (is_const f) andalso
                (member (#Name (dest_const f),length args)
                      [(",",2),("FST",1),("SND",1)])
             then (reconstruct,args)
             else raise PairDiscrim
         end;
```

It recognises variables and applications of ',', 'FST' and 'SND' as being valid terms of the theory of pairs. (The function member returns true if its first argument is equal to one or more of the elements of its second argument.) If the term is not recognised, the exception `PairDiscrim` is raised. Otherwise a reconstruction function and the subterms are returned. (See Sect. 3.2.)

The full information for the component decision procedure for pairs is thus:

```
val pair_proc =
  {Name = "pairs",
   Description = "Theory of equality on pairs",
   Author = "...",
   Discriminator = pair_discrim,
   Normalizer = ALL_CONV,
   Procedure = make_incremental_procedure CONJ PAIR_CONV};
```

Finally, `pair_proc` is added to the list of component procedures.

The function `make_incremental_procedure` can also be used for non-convex theories but, because it does not generate disjunctions of equations between variables, the resulting combined procedure will not be a full decision procedure.

## 8 Laziness

When a procedure is being generated as described in Sect. 5, the basic procedure being used must fail quickly if the technique is to be in any way efficient. Lazy theorems [2] can be of great help in this respect since they delay the application of inference rules until a theorem is actually required. Thus, an unsuccessful proof attempt involves only direct computation on the underlying data structures; no inference rules are applied. The Standard ML type underlying a lazy theorem is

```
(term list * term) * (unit -> thm) .
```

There is a list of terms for hypotheses and a term for the conclusion. The function with argument type `unit` delays the inferences.

In the specialised realm of a decision procedure the data structures used need not be full HOL terms. For example, the linear arithmetic procedure in the HOL `arith` library uses an `int * (string * int) list` to represent normalised inequalities; `(~1, [("m",~3), ("n",1)])` represents the inequality $1 + 3m \leq n$. Computation on these structures is faster than on terms. To accommodate this, lazy theorems can be generalised to *polymorphic lazy theorems* [3, Sect. 5.4] in which a type variable (`'a`) is used in place of terms: `'a * (unit -> thm)`. With the Nelson-Oppen technique these lazy..theorems can be passed between component procedures efficiently because the representation for equations between variables will always be simple. If more complex information had to be passed, considerable computation could be required to transform one procedure's representation to another's.

An important feature of lazy theorems is that the proof function component is assignable. Once it has been applied to `()`:`unit` to obtain a real theorem it is replaced by that theorem. So, the proof function is applied at most once. Hence, the inferences are performed at most once. This is significant when passing a lazy theorem between co-operating procedures because it may be used by more than one of them.

## 9   Results

Table 1 gives run times and garbage collection times in seconds, and the number of primitive inferences, for applications of the linear arithmetic decision procedure in the HOL `arith` library and of the combined procedure described in this paper to the following examples:

1. $m \leq n \wedge \neg(m = n) \Longrightarrow \text{SUC } m \leq n$
2. $p + 3 \leq n \Longrightarrow (\forall m. \, ((m \text{ EXP } 2 = 0) \Rightarrow (n - 1) \mid (n - 2)) > p)$
3. $\neg(x \leq y \wedge y \leq x + \text{HD } [0; x] \wedge P(h \, x - h \, y) \wedge \neg(P \, 0))$
4. $(\text{HD } x = \text{HD } y) \wedge (\text{TL } x = \text{TL } y) \wedge \neg(\text{NULL } x) \wedge \neg(\text{NULL } y) \Longrightarrow (f \, x = f \, y)$

The tests were done using HOL90.7 on a Sun SparcStation 2 with 64 Mbytes of real memory. Lazy theorems were not used. The combined procedure incorporated component procedures for linear arithmetic, lists, and uninterpreted function symbols.

**Table 1.** Execution times and number of inferences for the examples

| Example | Arithmetic Procedure | | | Combined Procedure | | |
|---|---|---|---|---|---|---|
| | Run (s) | GC (s) | Inferences | Run (s) | GC (s) | Inferences |
| 1 | 0.26 | 0.00 | 177 | 0.29 | 0.00 | 181 |
| 2 | 0.57 | 0.02 | 381 | 0.42 | 0.00 | 298 |
| 3 | | failed | | 8.30 | 0.03 | 5799 |
| 4 | | failed | | 0.27 | 0.00 | 247 |

## 10   Conclusions and Future Work

Decision procedures for combinations of theories are important as can be seen from the level of automation they provide for the PVS proof checker. Having to use inference rules to justify the procedures makes them difficult to write in HOL, especially to be fast. Nelson and Oppen's approach to combining decision procedures alleviates the problems somewhat by allowing the code for individual theories to remain separate and efficiency techniques used in them to be retained. It also allows a combined procedure to be extended easily for new (disjoint) theories.

Possible future work includes developing a basic procedure for linear natural arithmetic that accepts equalities rather than only $\leq$ inequalities. Procedures for decidable properties of user-defined concrete types might be automatically generated and integrated into the combined procedure. A procedure for deciding certain properties in set theory would also be useful. Finally, the linear arithmetic procedure for natural numbers might be adapted to handle integers, rationals, and reals. Procedures for all of these number systems should be able to cohabit in a combined procedure because they use different HOL constants for the arithmetic operators and so are disjoint.

## Acknowledgements

I have had useful discussions on decision procedures with members of the Computer Science Laboratory of SRI International (Menlo Park), Paul Jackson of Cornell University, and Mike Gordon, John Harrison and John Herbert in Cambridge.

## References

1. C. M. Angelo, L. Claesen, and H. De Man. Reasoning about a class of linear systems of equations in HOL. In T. F. Melham and J. Camilleri, editors, *Proceedings of the 7th International Workshop on Higher Order Logic Theorem Proving and Its Applications*, volume 859 of *Lecture Notes in Computer Science*, pages 33–48, Valletta, Malta, September 1994. Springer-Verlag.

2. R. J. Boulton. Lazy techniques for fully expansive theorem proving. *Formal Methods in System Design*, 3(1/2):25–47, August 1993.

3. R. J. Boulton. *Efficiency in a Fully-Expansive Theorem Prover*. PhD thesis, University of Cambridge Computer Laboratory, New Museums Site, Pembroke Street, Cambridge CB2 3QG, U.K., May 1994. Technical Report 337.

4. D. Craigen, S. Kromodimoeljo, I. Meisels, W. Pase, and M. Saaltink. EVES: An overview. In S. Prehn and W. J. Toetenel, editors, *VDM'91 Formal Software Development Methods*, volume 551 of *Lecture Notes in Computer Science*, pages 389–405. Springer-Verlag, 1991.

5. L. Fribourg. A decision procedure for a subtheory of linear arithmetic with lists. Research Report LiTH-IDA-R-91-33, Department of Computer and Information Science, Linköping University, Linköping, Sweden, October 1991.

6. M. J. C. Gordon and T. F. Melham, editors. *Introduction to HOL: A theorem proving environment for higher order logic*. Cambridge University Press, 1993.

7. J. Harrison. A HOL decision procedure for elementary real algebra. In J. J. Joyce and C.-J. H. Seger, editors, *Proceedings of the 6th International Workshop on Higher Order Logic Theorem Proving and its Applications (HUG'93)*, volume 780 of *Lecture Notes in Computer Science*, pages 426–436, Vancouver, B.C., Canada, August 1993. Springer-Verlag, 1994.

8. T. Käufl. Cooperation of decision procedures in a tableau-based theorem prover. Technical Report 19/89, University of Karlsruhe, Institut für Logik, Komplexität und Deduktionssysteme, 1989.

9. G. Nelson. *Techniques for Program Verification*. PhD thesis, Stanford University, 1980. Revised version: Technical Report CSL-81-10, Xerox PARC, June 1981.

10. G. Nelson and D. C. Oppen. Simplification by cooperating decision procedures. *ACM Transactions on Programming Languages and Systems*, 1(2):245–257, October 1979.

11. G. Nelson and D. C. Oppen. Fast decision procedures based on congruence closure. *Journal of the ACM*, 27(2):356–364, April 1980.

12. D. C. Oppen. Reasoning about recursively defined data structures. *Journal of the ACM*, 27(3):403–411, July 1980.

13. R. E. Shostak. On the SUP-INF method for proving Presburger formulae. *Journal of the ACM*, 24(4):529–543, October 1977.

14. R. Shostak. Deciding linear inequalities by computing loop residues. Technical report, Computer Science Laboratory, SRI International, Menlo Park, California, 1978.

15. R. E. Shostak. An algorithm for reasoning about equality. *Communications of the ACM*, 21(7):583–585, July 1978.

16. R. E. Shostak. A practical decision procedure for arithmetic with function symbols. *Journal of the ACM*, 26(2):351–360, April 1979.

17. R. E. Shostak. Deciding combinations of theories. *Journal of the ACM*, 31(1):1–12, January 1984.

18. K. Slind. Completion as a derived rule of inference. Research Report 90/409/33, Department of Computer Science, The University of Calgary, 2500 University Drive N.W., Calgary, Alberta, Canada T2N 1N4, 1990.

19. N. Shankar, S. Owre, and J. M. Rushby. *The PVS Proof Checker: A Reference Manual.* Computer Science Laboratory, SRI International, Menlo Park CA 94025, March 1993. Beta Release.

# Deciding Cryptographic Protocol Adequacy with HOL

Arca Systems, Inc.
ESC/ENS
Hanscom AFB, MA 01731-2116

**Abstract.** A *cryptographic protocol* is an algorithm involving exchanges of encrypted information carried out by principals in a distributed environment. It is intended to produce secure communications, even if every message can be read by, or originate with, every principal. This paper gives a definitional HOL formalization of a "belief logic" based on the full Gong, Needham, and Yahalom [2] logic for analyzing whether protocols achieve desired communication conditions. This gives the "belief logic" a sound formal basis. The paper also sketches the algorithm for a possible HOL tactic automatically constructing proofs that protocols achieve desired communication conditions if they do achieve them.

## 1 Introduction

It is often necessary to communicate securely across a network, even if this network is vulnerable to attack. Possible attacks include the following:

- Wiretapping, so messages can be read by other than their intended recipients.
- Interference, so messages can be prevented from reaching their intended recipients.
- Modification, so parts of messages, particularly source and destination labels, can be falsified.
- Insertion, so new, recorded, or modified messages can be placed on the network. Inserting messages containing parts of previously recorded messages is called a *playback* attack.
- Code breaking, so even encrypted information can be read and modified if the attacker has enough time to break the code.

In a network vulnerable to attack, one never knows who, if anyone, will actually receive any message sent, or who actually sent any message received. All communication might be impossible.

* The author wishes to thank Key Software, particularly Doug Weber, for providing the computing facilities used for this work. Geoffrey Hird answered questions on the Gong, Needham, Yahalom logic. Shiu-Kai Chin provided a copy of HOL90.7. This work was partially supported by Air Force Materiel Command's Electronic Systems Center/Software Center (ESC/ENS), Hanscom AFB, through the Portable, Reusable, Integrated Software Modules (PRISM) contract.

There are general *authentication* principles, though, that allow people to trust that their communications are with whom they think they are, even if these communications occur over a network vulnerable to attack. Informal statements of two such principles follow:

- If Jack is confident that only he and Jill know an encryption key $K$, and Jack receives a message that decodes to something meaningful when decrypted with $K$, then Jack can be confident that Jill originally sent this message — though not necessarily recently.
- If Jack sends Jill a number that he has never used for this purpose before, subsequently receives a message that he can be confident originally came from Jill, and this message contains something depending on the number Jack sent Jill, then Jack can be confident that Jill sent this message after he sent his number.

Cryptographic protocols are intended to exploit principles like these to produce secure communication over a network even when the network is vulnerable to attack. Different protocols serve different purposes, make different assumptions about which network users can be trusted, and operate under different timing and computational feasibility restrictions.

Determining whether a protocol actually gives the form of secure communication desired is difficult, though. Even when one treats the protocol abstractly, and ignores possible codebreaking or software failures, it is easy to miss potential attacks. In a famous example, a protocol published by Needham and Schroeder [6] was later found to have a serious vulnerability to a playback attack [7]. Even at the abstract level, a formal, machine-checked analysis is needed.

For this purpose, Burrows, Abadi, and Needham [1] developed a modal logic (called the *BAN logic*), for reasoning directly about the *beliefs* held by protocol principals. This logic formalizes authentication principles such as those described above and deductions made from these principles.

Gong, Needham, and Yahalom [2] extended the BAN logic by distinguishing belief from *possessing* information, by introducing *recognizability* to express that only some data items can be recognized as meaningful information, and by introducing new rules of inference letting principals in a protocol reason about other principals' beliefs. They also introduced a "not originated here" notion to make explicit protections against replays of messages sent earlier in the current execution of a protocol. Gong [3] further extended this work with rules to eliminate infeasible protocols. This paper will call the belief logic developed by Gong, Needham, and Yahalom [2], with the extensions developed by Gong [3], the *GNY logic*.

Odyssey Research Associates [4, 5] expressed part of the GNY logic in HOL, but did so using axioms, giving no assurances that the logic is consistent or that the objects and relations dealt with actually exist. Burrows, Abadi, and Needham [1] gave an English description of a formal semantics for the BAN logic, and Gong, Needham, and Yahalom [2] gave a similar English description of a formal semantics for the GNY logic, but neither of these descriptions was formalized in HOL or another system requiring existence proofs.

This paper gives a definitional formalization, in HOL, of a protocol inference system based on the full GNY logic. It gives a concrete recursive type `:Term` for the pieces of information exchanged during protocols, and a concrete recursive type `:Assertion` for making assertions about the states of protocol principals. It gives inductive definitions of the assertions true at a particular stage during the execution of a protocol, allowing deductions about arbitrarily highly nested levels of belief. The paper also sketches the algorithm for a possible tactic automatically proving that assertions hold at a stage in a protocol if they do indeed hold, and briefly describes related results from using the belief logic developed here to analyze a variant of the Kerberos protocol [8].

The major difference between this paper's approach and the GNY logic is that this paper interprets trustworthiness in terms of the accuracy of transmissions' claimed properties rather than in the accuracy of beliefs. Principals acquire other principals' beliefs only when these beliefs are expressed by transmissions. This paper also defines the relationship between having adequate reason to believe that a term is fresh or that a principal recognizes it, and possessing this term, more carefully, and explicitly treats recognition as a *relationship* between a principal and a term, not simply as a property of the term. This paper allows the analysis of intermediate stages of protocol executions, which the GNY logic does not, but it uses a simpler, single-variable class of reversible-modification functions. Other differences, briefly noted, seem to be errors in the GNY logic.

The rest of the paper is organized as follows: Section 2 gives basic assumptions about and terminology for the network model, protocols, protocol principals, and principals' possessions and beliefs. Section 3 gives the formal definitions of terms and assertions. Section 4 gives the inference rules for making deductions about assertions, including assertions about arbitrarily nested levels of belief. Finally, Section 5 sketches the algorithm for a possible HOL tactic automatically proving that assertions hold when they do, and briefly describes related results.

The paper uses the notational conventions of Slind's HOL90.7.

## 2 Computational Model

This section gives assumptions and terminology for the rest of the paper.

A *network*, or distributed environment, consists of *principals* connected by communication links. Messages on these links constitute the only communications between principals. Any principal can place a message on any link and can see or modify any message on any link.

A *protocol* is a distributed algorithm, carried out by principals acting as state machines; the protocol determines which messages the principals send as functions of their internal states. A protocol is divided into *stages* by message transmissions; the number of stages is always finite. A run, or *session*, is a particular execution of a protocol. This paper will only consider sessions that seem to end successfully, not considering possible alarm, adaptation, or retry provisions for other sessions.

At each stage of a protocol session, each principal has a set of *possessions*, pieces of information either in the principal's possession or computable from other pieces of information in the principal's possession. An actual set of possessions is always finite, but a potential set of possessions is typically infinite.

At each stage of a protocol session, each principal also has a set of *beliefs*. Principals believe a proposition if they can be confident, even though this confidence is not absolute, that this proposition is true. Belief can be based, for instance, on the near impossibility of quickly decrypting encrypted information without having the needed key.

For the remainder of this paper, "believe" will be interpreted as "believe with confidence". Although the consequences of false beliefs in conjunction with a protocol could be investigated, this paper will not do so. Rather, it will investigate whether the beliefs that can be (validly) attained include desired authentication conditions.

Every principal starts a session with initial sets of possessions and beliefs, and expands these sets by receiving messages using the inference rules in Section 4. Every principal's possession and belief sets increase monotonically during a session, but a principal need not still have a possession or belief that this principal had in an earlier session.

This paper assumes that the principals involved in a session do not send or receive messages other than the ones in the session; in particular, it assumes that the principals do not simply give away their secrets.

## 3 Terms and Assertions

This section gives the formal definitions of the types `:Term` and `:Assertion`, and discusses their intended interpretations. For the remainder of this paper, let *term* and *assertion* denote elements of types `:Term` and `:Assertion`, respectively. The formal, inductive, definition of truth for assertions is given in Section 4.

Types `:Term` and `:Assertion` are mutually recursive, defined using Gunter's `mutrec` library in HOL90.7's `contrib` directory. The definition of these types uses a call to `MutRecTypeFunc` with an argument of type `:MutRecTypeInputSig`, and uses additional definitions and rewrites to get the effect of infix constructors for these types. For the sake of simplicity, though, this paper will describe these types as if they were ordinary, independent, concrete recursive types.

Types `:Term` and `:Assertion` are both polymorphic in each of the following four type variables, whose intended instantiations are noted:

- `'principal` — type of protocol principals;
- `'key` — type of encryption/decryption keys;
- `'message` — type of plaintext messages;
- `'data` — type of arbitrary other data, e.g., time stamps.

Objects of type `:Term` are complete *descriptions* of the corresponding pieces of data. The recipient of such a piece of data, particularly if it is encrypted, need not be able to identify it or distinguish it from another piece of data, even though HOL could prove that the two corresponding :Term elements are distinct.

The :Term constructors Pr, Ky, Ms, and Da denote principal names (or equivalently, addresses), keys, plaintext, and other data of an arbitrary type. Constructors Sencrypt, Pencrypt, Sdecrypt, and Pdecrypt denote secret- and public-key encryption and decryption. Constructor Hash denotes an effectively one-to-one, one-way function — i.e., a function such that it is practically impossible to find an X and Y such that (Hash X) and (Hash Y) are equal if X and Y are not equal, but it is nevertheless impossible to compute X from (Hash X). Constructor Revmod denotes a non-identity function whose inverse is readily computable. Constructor Att denotes an *attributed* term, one with an associated assertion. (The assertion corresponds to an *extension* in the GNY logic [2]). A protocol with sending an attributed term requires that the sender have adequate reason to believe the assertion associated with this term. The infix operator ;; is the pairing operator for :Term elements.

```
'Term =
  Pr of 'principal |
  Ky of 'key |
  Ms of 'message |
  Da of 'data |
  Hash of Term |
  Revmod of Term |
  Sencrypt of 'key => Term |
  Pencrypt of 'key => Term |
  Sdecrypt of 'key => Term |
  Pdecrypt of 'key => Term |
  Att of Term => Assertion |
  ;; of Term => Term'
```

The :Assertion constructors and informal descriptions of their intended meanings, follow:

- Fresh: The data item was created for this run of the protocol, not recorded from an earlier run. Principals believe a data item is fresh if they believe that, whenever the item comes into their possession, they can confidently identify it as having been created for this run of the protocol.
- KnowsFromSelf: The principal can always identify any message or part of a message that the principal conveyed at any earlier time.
- Trustworthy: If the principal was the source of a data item with an associated assertion, this is adequate reason for believing the associated assertion.
- PrivateKey: The key is one of the principal's private keys.
- PublicKey: The key is one of the principal's public keys.
- Conveyed: The principal released the data item at some earlier time. Principals believe a data item was conveyed by a particular principal if, whenever the item comes into their possession, they can confidently identify it as released by that other principal.

- **Possesses**: The principal possesses the data item or is capable of computing it from other data items the principal possesses. If the data item has an associated assertion, the principal either received the data item as having this property or had independent, adequate reason for believing it.
- **Receives**: The principal receives the data item, and the data item's sender was actually capable of sending it.
- **Recognizes**: The principal can identify the data item as meaningful information — e.g., it has an expected form. Principal **p1** believes principal **p2** recognizes the data if **p1** believes **p2** can identify it as meaningful if it comes into **p2**'s possession.
- **Sends**: The first principal sends the second principal the data item. As a uniform prohibition against impossible protocols, the second principal only receives the data item if the first principal possesses the data item and was either told or had adequate reason to believe that it had any associated properties, and the principals are distinct.
- **SharedSecret**: The two principals are the only ones, other than principals they both trust, who possess the data item; they can safely use it to identify each other or to encrypt data they send each other.
- **Believes**: The principal has confidence in the accuracy of the assertion.
- **&&**: Conjunction operator for :**Assertion** elements.

```
'Assertion =
  Fresh of Term |
  KnowsFromSelf of 'principal |
  Trustworthy of 'principal |
  PrivateKey of 'principal => 'key |
  PublicKey of 'principal => 'key |
  Conveyed of 'principal => Term |
  Possesses of 'principal => Term |
  Receives of 'principal => Term |
  Recognizes of 'principal => Term |
  Sends of 'principal => 'principal => Term |
  SharedSecret of 'principal => 'principal => Term |
  Believes of 'principal => Assertion |
  && of Assertion => Assertion'
```

## 4   Inference Rules

This section describes the inference rules inductively defining a function **GNY** that maps a protocol, session stage, and assertion to a truth value. For this function and its subfunctions, a **protocol** is a list of assertions, and a **stage** is an index into this list. The 0th element in the list gives each session's initial conditions. The other elements of the list, thought of as assertions becoming true at later stages in the session, are intended to be **Sends** assertions.

All except the first two of the 63 rules defining **GNY** use the inductively defined relation **Relativizes**, which holds for two lists of assertions if the second is a belief-relativized analog of the first. (As an example, for assertions A1 and A2, [(Believes P A1);(Believes P A2)] is a belief-relativized analog of [A1;A2], as is [(Believes Q (Believes P A1));(Believes Q (Believes P A2))]. By using Relativizes, these rules capture the GNY "Rationality" rule, which makes each rule in the GNY logic an infinite schema, each member of the schema being a belief-relativized analog of the schema's simplest, basic rule [2].

Each of the 61 rules defining GNY that uses Relativizes thus has a simplest, and most important, special case as a corollary. For the GNY rule

GAN1
```
|- !protocol stage A3.
    (?A1 A2 a1 a2.
       GNY protocol stage A1 /\
       GNY protocol stage A2 /\
       Relativizes [a1; a2; a1 && a2] [A1; A2; A3]) ==>
    GNY protocol stage A3
```

for example, that most important special case is the theorem

AN1
```
|- !protocol stage a1 a2.
     GNY protocol stage a1 /\ GNY protocol stage a2 ==>
     GNY protocol stage (a1 && a2)
```

All these special cases can be generated automatically by an appropriate special-purpose rule.

For the sake of avoiding clutter, the remainder of this section will describe only the first two of the rules defining GNY, and the "most important special case" corollaries of the other 61 rules. It will call these corollaries rules.

GNY is also defined in terms of the function NotFromHere, which captures the idea that a term cannot be a replay of something extracted from one of a principal's own transmissions earlier in the current session, and the unspecified constant KeyPair, which identifies the relationship between the public and private parts of a public/private key pair. This paper consistently treats k1 as the public member of such a pair and k2 as the corresponding private member.

Function NotFromHere, not given in this paper, expresses a more stringent "not originated here" condition than that given in the GNY logic [2]. It treats anything in a term or subterm as having been made available to hostile principals, regardless of whether a hostile principal could actually obtain this data. Such a function is only possible because a term is a complete description of a data item. NotFromHere is also defined so that it always has value "false" when it is evaluated on protocols with non-zero elements that are not Sends assertions.

The first 23 of the 63 rules in the inductive definition of GNY do not have explicit corresponding rules in the original GNY logic [2]. The additional rules give new functionality (e.g., evaluating assertions at intermediate stages in protocols, and having the conjunction operator &&), impose Gong's protocol-feasibility restrictions [3], define a somewhat different notion of "belief", or were implicit in the GNY logic.

The first two rules, "Base Case" and "Monotonicity", describe how new assertions become and remain true as the protocol progresses.

```
BC1
|- !protocol stage. GNY protocol stage (EL stage protocol)
MN1
|- !protocol stage assertion.
     GNY protocol stage assertion ==>
     (!n. GNY protocol (stage + n) assertion)
```

The next three rules, not given here, define the conjunction operator **&&**; one of them, **AN1**, was discussed earlier.

The next four "Self Belief" rules state that principals believe that they possess what they possess, possess what they believe that they possess, believe that they believe what they believe, and believe what they believe that they believe. These rules do not have analogs in the GNY logic because belief and possession are more restricted here, and terms are consistently interpreted as representing their values here.

```
SB1
|- !protocol stage p t.
     GNY protocol stage (Possesses p t) ==>
     GNY protocol stage (Believes p (Possesses p t))
SB2
|- !protocol stage p t.
     GNY protocol stage (Believes p (Possesses p t)) ==>
     GNY protocol stage (Possesses p t)
SB3
|- !protocol stage p a.
     GNY protocol stage (Believes p a) ==>
     GNY protocol stage (Believes p (Believes p a))
SB4
|- !protocol stage p a.
     GNY protocol stage (Believes p (Believes p a)) ==>
     GNY protocol stage (Believes p a)
```

The next four rules, not given here, were implicit in the GNY logic. The first states that the **SharedSecret** relation is symmetrical. The others state that if one possesses the decryption of the encryption of a term, or the encryption of the decryption of a term, whether with symmetric- or public-key encryption, then one possesses the term.

The next two rules express Gong's [3] "Eligibility" restrictions for eliminating infeasible protocols: one can only send what one possesses, cannot create a possession by sending it to oneself, and can only make sending something express a property if one has adequate reason to believe this property. The rules impose these conditions by having the **Receive** for a **Send** only happen if the **Send** meets these conditions.

**EL1**
```
|- !protocol stage p1 p2 t.
   (GNY protocol stage (Sends p1 p2 t) /\
    GNY protocol stage (Possesses p1 t)) /\
   ~(p1 = p2) ==>
   GNY protocol stage (Receives p2 t)
```
**EL2**
```
|- !protocol stage p t a.
   GNY protocol stage (Possesses p t) /\
   GNY protocol stage (Believes p a) ==>
   GNY protocol stage (Possesses p (Att t a))
```

Note that these rules do not eliminate all infeasible protocols; they allow the sending of a term to convey more information than the actual value of the term could convey.

The next eight rules, not given here, were implicit in the GNY logic. They state that (Att t a) can be replaced by t, or vice versa, in deciding whether a term is fresh or is recognized or conveyed by a particular principal, and that if one receives or possesses (Att t a), then one receives or possesses t.

The next eight rules correspond to GNY "Possession" rules P1 through P8. These are simple, intuitive, and set the pattern for many of the remaining rules. The idea behind them is that things received are possessed, and anything computable from what is possessed is considered to be possessed.

**P1**
```
|- !protocol stage p t.
   GNY protocol stage (Receives p t) ==>
   GNY protocol stage (Possesses p t)
```
**P2**
```
|- !protocol stage p t1 t2.
   GNY protocol stage (Possesses p t1) /\
   GNY protocol stage (Possesses p t2) ==>
   GNY protocol stage (Possesses p (t1 ;; t2))
```
**P3**
```
|- !protocol stage p t1 t2.
   GNY protocol stage (Possesses p (t1 ;; t2)) ==>
   GNY protocol stage (Possesses p t1) /\
   GNY protocol stage (Possesses p t2)
```
**P4**
```
|- !protocol stage p t.
   GNY protocol stage (Possesses p t) ==>
   GNY protocol stage (Possesses p (Revmod t)) /\
   GNY protocol stage (Possesses p (Hash t))
```
**P5**
```
|- !protocol stage p t.
   GNY protocol stage (Possesses p (Revmod t)) ==>
   GNY protocol stage (Possesses p t)
```

**P6**

```
|- !protocol stage p k t.
    GNY protocol stage (Possesses p (Ky k)) /\
    GNY protocol stage (Possesses p t) ==>
    GNY protocol stage (Possesses p (Sencrypt k t)) /\
    GNY protocol stage (Possesses p (Sdecrypt k t))
```

**P7**

```
|- !protocol stage p k1 t.
    GNY protocol stage (Possesses p (Ky k1)) /\
    GNY protocol stage (Possesses p t) ==>
    GNY protocol stage (Possesses p (Pencrypt k1 t))
```

**P8**

```
|- !protocol stage p k2 t.
    GNY protocol stage (Possesses p (Ky k2)) /\
    GNY protocol stage (Possesses p t) ==>
    GNY protocol stage (Possesses p (Pdecrypt k2 t))
```

P2 and P5 here are less general than the corresponding GNY rules, since those rules also consider functions of more than one variable that are reversible in each variable. P3 here corrects an error by explicitly mentioning possessing the second element of the pair. Here, possessing (Revmod t} is in P4 rather than P2.

The next eleven rules correspond to GNY "Freshness" rules F1 through F11. The idea behind these rules is that something computable only from something created for the current session must have been created for the current session. The first four of these rules say that pairs with identifiably fresh members, and reversible modifications of identifiably fresh terms, are identifiably fresh, and identifiable encryptions of identifiably fresh terms, or encryptions made with identifiably fresh keys, are identifiably fresh.

**F1**

```
|- !protocol stage p t t2 t1.
    GNY protocol stage (Believes p (Fresh t)) ==>
    GNY protocol stage (Believes p (Fresh (t ;; t2))) /\
    GNY protocol stage (Believes p (Fresh (t1 ;; t))) /\
    GNY protocol stage (Believes p (Fresh (Revmod t)))
```

**F2**

```
|- !protocol stage p t k.
    GNY protocol stage (Believes p (Fresh t)) /\
    GNY protocol stage (Possesses p (Ky k)) ==>
    GNY protocol stage (Believes p (Fresh (Sencrypt k t))) /\
    GNY protocol stage (Believes p (Fresh (Sdecrypt k t)))
```

**F3**

```
|- !protocol stage p t k1.
    GNY protocol stage (Believes p (Fresh t)) /\
    GNY protocol stage (Possesses p (Ky k1)) /\
    GNY protocol stage (Possesses p t) ==>
    GNY protocol stage (Believes p (Fresh (Pencrypt k1 t)))
```

**F4**

```
|- !protocol stage p t k2.
    GNY protocol stage (Believes p (Fresh t)) /\
    GNY protocol stage (Possesses p (Ky k2)) /\
    GNY protocol stage (Possesses p t) ==>
    GNY protocol stage (Believes p (Fresh (Pdecrypt k2 t)))
```

F3 and **F4** here differ from the corresponding GNY rules in explicitly requiring
(**Possesses p t**); otherwise, one could obtain the term in question without
being able to identify it, thus having no reason to believe that it is fresh.

The next two rules, not given here, say that if one member of a public/private
key pair is fresh, the other must be also. These rules differ from the corresponding
GNY rules in explicitly requiring possession of the key believed to be fresh.

In the next three rules, if it is possible to decrypt or encrypt something with
a fresh key, and get something meaningful, the thing decrypted or encrypted
must have been produced with the fresh key, and thus be fresh itself.

**F7**

```
|- !protocol stage p t k.
    GNY protocol stage (Believes p (Recognizes p t)) /\
    GNY protocol stage (Believes p (Fresh (Ky k))) /\
    GNY protocol stage (Possesses p (Ky k)) ==>
    GNY protocol stage (Believes p (Fresh (Sencrypt k t))) /\
    GNY protocol stage (Believes p (Fresh (Sdecrypt k t)))
```

**F8**

```
|- !protocol stage p t k1 k2.
    (GNY protocol stage (Believes p (Recognizes p t)) /\
     GNY protocol stage (Believes p (Fresh (Ky k1))) /\
     GNY protocol stage (Possesses p (Ky k2))) /\
    KeyPair k1 k2 ==>
    GNY protocol stage (Believes p (Fresh (Pencrypt k1 t)))
```

**F9**

```
|- !protocol stage p t k2 k1.
    (GNY protocol stage (Believes p (Recognizes p t)) /\
     GNY protocol stage (Believes p (Fresh (Ky k2))) /\
     GNY protocol stage (Possesses p (Ky k1))) /\
    KeyPair k1 k2 ==>
    GNY protocol stage (Believes p (Fresh (Pdecrypt k2 t)))
```

**F8** and **F9** here differ from the corresponding GNY rules by requiring possession
of **k2** and **k1**, respectively; this was the apparent intent of the GNY rules.

The next two rules make inferences from or about the values of one-way
functions. If it is possible to confirm that something is the value of a one-way
function applied to something fresh, that value can be believed fresh. Similarly,
if the one-way function value is believed fresh, the thing this value was created
from can be believed fresh.

```
F10
|- !protocol stage p t.
    GNY protocol stage (Believes p (Fresh t)) /\
    GNY protocol stage (Possesses p t) ==>
    GNY protocol stage (Believes p (Fresh (Hash t)))
F11
|- !protocol stage p t.
    GNY protocol stage (Believes p (Fresh (Hash t))) /\
    GNY protocol stage (Possesses p (Hash t)) ==>
    GNY protocol stage (Believes p (Fresh t))
```

The next six rules, not given here, correspond to GNY "Recognizability" rules R1 through R6. These rules are similar to the "Possession" rules; the idea behind all of them is that anything that a principal can compute something recognizable as meaningful information from is itself recognizable as meaningful information. R3 and R4 here differ from GNY rules R3 and R4 by requiring possession of **k2** and **k1**, respectively, rather than **k1** and **k2**. R6 here differs from GNY rule R6 in imposing the requirement (Believes p (Recognizes p (Hash t))); otherwise everything received by every principal would be regarded as being recognized by the principal receiving it.

The next seven rules correspond to GNY "Interpretation" rules I1 through I7. All allow conclusions about which principal initially conveyed (i.e., released) a particular piece of information.

If one can decrypt a message with a key one believes is known to at most one other person, if the result is meaningful information, if one believes that the key or message decrypted were created in the current session, and if the message cannot be a replay of something one sent earlier in the session, then one can believe that the other person actually knows the key and initially conveyed the message.

```
I1
|- !protocol stage p k t q.
     (GNY protocol stage (Receives p (Sencrypt k t)) /\
      GNY protocol stage (Possesses p (Ky k)) /\
      GNY protocol stage (Believes p (SharedSecret p q (Ky k))) /\
      GNY protocol stage (Believes p (Recognizes p t)) /\
      GNY protocol stage (Believes p (Fresh (t ;; Ky k)))) /\
     NotFromHere protocol p stage (Sencrypt k t) ==>
     GNY protocol stage (Believes p (Conveyed q t)) /\
     GNY protocol stage (Believes p (Conveyed q (Sencrypt k t)))/\
     GNY protocol stage (Believes p (Possesses q (Ky k)))
```

If one receives a message encrypted with one's own public key, has the corresponding private key and a piece of information one believes is known to at most one other person, decrypts the message to find something meaningful containing this information, and believes one's public key or something in the message was created in the current session, and if the encrypted message cannot be a replay

of something one sent earlier in the session, then one can believe that the other
person knows one's public key and initially conveyed the encrypted message.

I2
```
|- !protocol stage p k1 t x k2 q.
    (GNY protocol stage (Receives p (Pencrypt k1 (t ;; x))) /\
     GNY protocol stage (Possesses p (x ;; Ky k2)) /\
     GNY protocol stage (Believes p (PublicKey p k1)) /\
     GNY protocol stage (Believes p (SharedSecret p q x)) /\
     GNY protocol stage (Believes p (Recognizes p (t ;; x))) /\
     GNY protocol stage (Believes p (Fresh (t ;; x ;; Ky k1))))/\
    KeyPair k1 k2 /\
    NotFromHere protocol p stage (Pencrypt k1 (t ;; x)) ==>
    GNY protocol stage (Believes p (Conveyed q (t ;; x))) /\
    GNY protocol stage
     (Believes p (Conveyed q (Pencrypt k1 (t ;; x)))) /\
    GNY protocol stage (Believes p (Possesses q (Ky k1)))
```

This rule differs from GNY rule I2 in requiring that p possess x, as suggested by
GNY rule I2'; this apparently corrects an error.

If one receives a value that is a one-way function of something containing
what one believes is a secret known to at most one other person, if one believes
something the one-way function was applied to in creating the value was created
in the current session, and if the value cannot be a replay of something one sent
out earlier in the session, then one can believe that the other person initially
conveyed the value and thereby indicated conveying the data from which the
value was computed.

I3
```
|- !protocol stage p t x q.
    (GNY protocol stage (Receives p (Hash (t ;; x))) /\
     GNY protocol stage (Possesses p (t ;; x)) /\
     GNY protocol stage (Believes p (SharedSecret p q x)) /\
     GNY protocol stage (Believes p (Fresh (t ;; x)))) /\
    NotFromHere protocol p stage (Hash t ;; x) ==>
    GNY protocol stage (Believes p (Conveyed q (t ;; x))) /\
    GNY protocol stage (Believes p (Conveyed q (Hash (t ;; x))))
```

If one receives something decrypted with someone else's private key, uses
that person's public key to recover the original thing decrypted, and recognizes
this thing as meaningful information, then one can believe that the other person
initially conveyed the message. If, in addition, one believes that the thing sent
or the public key used was created in the current session, one can believe that
the person who initially conveyed the message possesses the message and the
private key used.

I4
```
|- !protocol stage p k2 t k1 q.
```

```
      (GNY protocol stage (Receives p (Pdecrypt k2 t)) /\
       GNY protocol stage (Possesses p (Ky k1)) /\
       GNY protocol stage (Believes p (PublicKey q k1)) /\
       GNY protocol stage (Believes p (Recognizes p t))) /\
      KeyPair k1 k2 ==>
      GNY protocol stage (Believes p (Conveyed q t)) /\
      GNY protocol stage (Believes p (Conveyed q (Pdecrypt k2 t)))
I5
|- !protocol stage p k2 t k1 q.
      (GNY protocol stage (Receives p (Pdecrypt k2 t)) /\
       GNY protocol stage (Possesses p (Ky k1)) /\
       GNY protocol stage (Believes p (PublicKey q k1)) /\
       GNY protocol stage (Believes p (Recognizes p t)) /\
       GNY protocol stage (Believes p (Fresh (t ;; Ky k1)))) /\
      KeyPair k1 k2 ==>
      GNY protocol stage (Believes p (Possesses q (Ky k2 ;; t)))
```

Anyone who conveys something created for the current session can be presumed to still possess it, and anyone who conveys a list of things can be presumed to convey each item in the list.

```
I6
|- !protocol stage p q t.
      GNY protocol stage (Believes p (Conveyed q t)) /\
      GNY protocol stage (Believes p (Fresh t)) ==>
      GNY protocol stage (Believes p (Possesses q t))
I7
|- !protocol stage p q t1 t2.
      GNY protocol stage (Believes p (Conveyed q (t1 ;; t2))) ==>
      GNY protocol stage (Believes p (Conveyed q t1)) /\
      GNY protocol stage (Believes p (Conveyed q t2))
```

The next five rules, not given here, correspond to the GNY logic's "Being Told" rules T2 through T6. They state that a principal who has received a data item can be considered to have received everything computable from that data item. These rules, which are similar to "Possession" rules, allow the "Interpretation" rules to be applied to messages, parts of messages, and things computable from parts of messages, in being able to believe that a particular principal conveyed a data item. No analog of GNY rule T1 is needed here because "not originated here" restrictions are handled with the NotFromHere function rather than a preliminary syntactic analysis. T2 here corrects an apparent error in GNY rule T2 by explicitly mentioning receiving the second element of a pair, and T5 here is less general than GNY rule T5, considering only single-variable reversible-modification functions.

There are no analogs here of GNY "Jurisdiction" rules J1 and J3. For J1, here principals only acquire other principals' beliefs when these beliefs are expressed by actions. Even if a principal believes another principal to be "honest and competent", as in the GNY logic [2], this belief would not convey the information needed to know what the other principal's beliefs *are*. GNY rule J3 follows from a belief-relativized form of **SB4**.

The next rule corresponds to GNY "Jurisdiction" rule J2, but it differs from rule J2 in two ways. It has the simpler conclusion **(Believes p a)**, as well as **(Believes p (Believes q a))**. The simpler conclusion follows from the apparent intent of GNY rule J1.

```
J2
|- !protocol stage p q t a.
    GNY protocol stage (Believes p (Conveyed q (Att t a))) /\
    GNY protocol stage (Believes p (Fresh t)) /\
    GNY protocol stage (Believes p (Trustworthy q)) ==>
    GNY protocol stage (Believes p a) /\
    GNY protocol stage (Believes p (Believes q a))
```

The last three rules, not given here, correspond to GNY rules I1', I2', and I3'. In these rules, the **KnowsFromSelf** constructor essentially replaces the **Fresh** constructor and a **NotFromHere** side condition in rules I1, I2, and I3. The third rule here does *not* impose the condition **(Believes p (Recognizes p (t ;; x)))**, which is useless when looking at **Hash** values.

## 5   Decision Procedure Algorithm

This section sketches an algorithm for a potential HOL tactic showing that the belief logic described here is *decidable* — i.e., that for any protocol and session stage, the characteristic function of the set of assertions identified as true by **GNY** for this protocol and stage is Turing computable. It then lists several problems that must be solved before a true decision procedure for this belief logic can be implemented, and briefly describes related results.

The algorithm proceeds by induction on protocol stage. By induction hypothesis, the algorithm can decide the validity of all the conditions that must be satisfied before **Sends** assertions give rise to corresponding **Receives** assertions. For each principal, it is thus possible to determine exactly which terms, including exactly which attributed terms, that principle has received.

All assertions, other than **Sends** and **Receives** assertions, are assertions about principals' possessions and beliefs. The set of all terms that a principal possesses is decidable, since the only question for a given term is whether the principal possesses either the term or all of its subterms. Since the model treats it as impossible to decrypt anything without having the necessary key, and impossible to undo the effect of a one-way function, it is only necessary to examine the subterms of the finite set of terms initially possessed or received by the principal; it is not necessary to consider possible combinations of arbitrarily many terms, leading to potentially infinite searchs.

For similar reasons, if a principal's initial beliefs and all the beliefs the principle acquires through transmissions are given by a feasible protocol, all the principal's "believes fresh", "believes self recognizes", and so on, sets are decidable. It is only necessary to look at finite sets of terms and their subterms, including attributed terms, and term/principle pairs initially assumed to have these properties, to identify all beliefs of these forms that can possibly be derived.

For deciding beliefs about other principals' beliefs, the algorithm simply relativizes the process it uses for other types of beliefs. (Yes, "simply" is a joke.)

The catch is the "given by a feasible protocol". What if initial or acquired beliefs are highly iterated, or inconsistent with other conditions — e.g., that $A$ believes $B$ believes $C$ believes $A$ recognizes $X$, but $A$ has no means of recognizing $X$? All these troublesome possibilities arise from protocols that are impossible or invalid in some way, but a decision-procedure tactic needs to detect all these conditions, as well as solve all the practical problems in carrying out the complicated algorithm just sketched. This is an ample area for future research.

The author has used the HOL implementation of the belief logic described here, along with simple tactics and utilities, to prove 29 desired properties of the "ticket granting service" variant of the Kerberos protocol [8]. These properties include assertions that principals obtain session keys forwarded from trusted servers, believe these keys are secrets shared with particular other principals, and believe that these other principals possess these same keys and believe these same beliefs. The proofs are short, being about 9 lines long on average. Further, they follow regular patterns that could easily be mechanized in a tactic of the form sketched. The author will provide details in a later paper.

# References

1. M. Burrows and M. Abadi and R. Needham. A Logic of Authentication. In *Proceedings of the 12th Symposium on Operating Systems Principles*, Litchfield Park, AZ, December 1989. ACM.
2. L. Gong and R. Needham and R. Yahalom. Reasoning about Belief in Cryptographic Protocols. In *Proceedings of the Symposium on Security and Privacy*, pages 234–248, Oakland, CA, May 1990. IEEE.
3. L. Gong. Handling Infeasible Specifications of Cryptographic Protocols. In *Proceedings of Computer Security Foundations Workshop IV*, pages 99–102, Franconia NH, June 1991. IEEE Computer Society Press.
4. ORA. *Romulus Theories*. Technical Report TM-94-0016, Odyssey Research Associates, Ithaca, NY, March 1994.
5. ORA. *Romulus User's Manual*. Technical Report TM-94-0018, Odyssey Research Associates, Ithaca, NY, March 1994.
6. R. Needham and M. Schroeder. Using Encryption for Authentication in Large Networks of Computers. *CACM*, 21(12):993–999, December 1978.
7. R. Needham and M. Schroeder. Authentication Revisited. *CACM*, 21(1):7, January 1987.
8. J. Steiner and C. Neuman and J. Schiller. An Authentication Service for Open Network Systems. In *Proceedings of the USENIX Winter Conference*, pages 191–202, February, 1988.

# A Practical Method for Reasoning about Distributed Systems in a Theorem Prover

Holger Busch

SIEMENS AG
Corporate Research and Development
Otto–Hahn–Ring 6, 81739 München, Germany.
E–Mail: holger.busch@zfe.siemens.de

**Abstract.** The Temporal Language of Transitions is the basis of a formal methodology for the verification of distributed systems. To complement model-checking, which forms an integral part of the methodology, we contribute a proof calculus in a higher-order logic theorem prover. A method for embedding TLA[17] in the system LAMBDA is described. Motivated by the importance of practicality in an industrial setting, a simple representation of TLA is combined with extended reasoning functions. Translation of user programs, safety, liveness, and refinement proofs are depicted by way of a parametric mutual exclusion algorithm.

## 1 Introduction

The increasing importance of distributed and concurrent software necessitates suitable verification methodologies, which has led to a variety of promising formal approaches. In an industrial context, however, practicality is mandatory; a formal methodology should be applicable by engineers without additional qualification, and it must supply sufficient and efficient automation of all low-level reasoning.

This is the main purpose in designing a formal methodology[12, 13] which combines and extends several valuable characteristics of existing formalisms: *Temporal Language of Transitions*. The appealing simplicity of Unity[9] suggested it as a paradigm for the "programming" language TLT, which additionally provides an elaborate scoping of variables, an explicit notion of an environment, a distinction of local progress, strong and weak fairness, constructs for specifying concurrency and communication, and more. Transition predicates with primed program variables correspond to actions in TLA[17][1].

While finite systems are verified with the model checker SVE [6], we contribute an interactive proof environment to the TLT framework. By taking advantage of a theorem prover, we augment the scope of verifiability to parameterization,

---

[1] Temporal Logic of Actions

abstraction and recursive datatypes. Our reasoning is based on a typed variant of TLA. We actually modify the pure TLA-calculus to support further features of TLT. Unlike purely definitional approaches our embedding of TLA in higher-order logic comprises a limited amount of axioms and functions for evaluating expressions externally. Thus we obtain concise embedded terms and more efficient mechanical reasoning, but we have to take care of the soundness of proofs at the meta-level.

Fundamental research into formalisms for distributed systems has suggested the use of theorem provers. F. Anderson and his group have developed an advanced proof environment for Unity[4]; tactic-based automation, a graphical interface for structuring proofs of liveness properties, and a variety of case studies are results. Other Unity based research is described in [2]. Addressing the refinement of concurrent programs, T. Långbacka and J. v. Wright[18] have embedded TLA in HOL. C.-T. Chou has devised a theory of cause-event relations which provides an abstract yet simple view of distributed algorithms[11]. Choosing Zermelo-Fraenkel Set Theory, S. Kalvala has mechanized TLA in Isabelle[16]. The tool TLP[14] supplies a specific TLA interface on top of the Larch theorem prover.

Our approach is based on, but not restricted to the proof system LAMBDA[10]; its facilities for adding sophisticated tactics and rewrite strategies are utilized, along with the expressiveness of higher-order logic. The syntax of logical terms and user definitions of datatypes and functions in L2, the logic language of LAMBDA, is very similar to ML[10, 15]. LAMBDA supports rules and higher-order unification like Isabelle. As automation is crucial to a practical proof environment, we have spent much effort on strengthening the general automatic theorem proving capabilities of LAMBDA[8], which are still being extended.

This paper is organized as follows. First TLT and its translation into TLA are summarized. The embedding of TLA is detailed in Sect. 3. Then LAMBDA proofs of a safety and a liveness property are exemplified. Refinement is addressed in Sect. 5. A basic knowledge of TLA[17] and ML[15] is assumed.

## 2 From User Programs to Formulas

By way of a parametric mutual exclusion algorithm (Fig.1), the basic features of the user language of TLT[12] are introduced. Common predicate logic notation is used in this paper for quantifiers. A TLT specification consists of various sections:

- A program is first given a name, possibly with a parameter list defining rigid variables which, for instance, describe ranges of arrays.
- Type declarations introduce types either by providing primitive and recursive constructors in the style of ML, or by using a set notation.
- Variable declarations distinguish local (default), global, read, private, and communication variables according to permissions of programs and their environments to alter program variables.

- Program invariants are specified for which proof obligations are generated.
- A specification of state predicates which are initially satisfied follows.
- In contrast to Unity, program instructions are not assignments but relations between primed (next value) and unprimed (current value) variables according to TLA. Instructions may be labelled, multiplied and furnished with indices ($[instr_{i:index\_set}]$), guarded ($\ldots \longrightarrow$), and annotated with pre- and post-conditions ($\{\ldots\}$). Like in Unity a non-deterministic choice of an enabled instruction is assumed, yet finitely many stutter steps may occur.
- Explicit assumptions of weak or strong fairness are specified by the user.[2]
- Properties which are to be proven are specified in the last section.

```
PROGRAM Mutex
TYPES
     token_type := {i: natural | 0 ≤ i ∧ i ≤ n}
     n_type := {j: natural | 0 < j ∧ j ≤ n}
DECLARATIONS
     VAR: req,cs: n_type → Boolean,token: token_type
INVARIANT
     ∀ i:n_type csᵢ ⇒ token = i
INITIALLY
     ∀ i:n_type ¬ reqᵢ ∧ ¬ csᵢ ∧ token = 0
INSTRUCTIONS
     [requestᵢ:n_type] ¬ csᵢ ∧ ¬ reqᵢ ⟶ req'ᵢ
     [getᵢ:n_type] reqᵢ ∧ token = 0 ⟶ token' = i
     [enterᵢ:n_type]{reqᵢ} token = i ∧ ¬ csᵢ ⟶ cs'ᵢ
     [leaveᵢ:n_type] {token = i } csᵢ ⟶ ¬ cs'ᵢ ∧ ¬ req'ᵢ ∧ token' = 0
FAIRNESS
     [fair1] ∀ᵢ:n_type SF (getᵢ)
     [fair2] ∀ᵢ:n_type WF(enterᵢ) ∧ WF(leaveᵢ)
PROPERTIES
     [safe] always-true(¬ ∃ᵢ,ⱼ:n_type i≠j ∧ csᵢ ∧ csⱼ)
     [live1] ∀ i:n_type. reqᵢ ↦ csᵢ
END.
```

Fig. 1. An n-process Mutual Exclusion Example

Interfaces, connectors, modules, and link constraints are constructs for specifying structure, composition and the interaction of a program with its environment. Link constraints are assumptions on the behaviour of the environment; if there is no such specification, the environment is assumed to stutter non-deterministically for variables which it does not have the permission to alter.

TLT instructions are translated into TLA actions in two steps. First, pre- and post-conditions, guards, and kernel transition predicates are separated and kept almost literally because of the close similarity of TLT and TLA at that level.

---

[2] In addition to explicit fairness declarations, local progress is implicit, i.e., a program sometime has to execute some instruction (if at least one instruction is enabled).

In the second step, each transition predicate is transformed into disjunctive normal form which reflects separate alternatives if a guarded assignment contains conditional sub-cases. For each disjunctive subpredicate an explicit *Unchanged*-predicate is generated over all program variables which do not belong to its output variables. Technically speaking, program variables which do not occur primed are assumed to stutter.[3] For each TLT transition, a composite predicate is generated which is the conjunction of guard, pre-condition, augmented disjunctive transition predicate and primed post-condition. It is abbreviated as a function application $s(\ldots)$.

The instructions of the mutual exclusion program are thus translated into

$$s(\text{"request"}, i{:}\text{n\_type}) \triangleq \neg\, cs_i \wedge \neg\, req_i \wedge req'_i \wedge$$
$$Unchanged\ \langle req_{j:n\_type \wedge j \neq i}, token, cs\rangle$$

$$s(\text{"get"}, i{:}\text{n\_type}) \triangleq req_i \wedge token = 0 \wedge token' = i \wedge$$
$$Unchanged\ \langle req, cs\rangle$$

$$s(\text{"enter"}, i{:}\text{n\_type}) \triangleq req_i \wedge token = i \wedge \neg\, cs_i \wedge cs'_i \wedge$$
$$Unchanged\ \langle req, cs_{j:n\_type \wedge j \neq i}\rangle$$

$$s(\text{"leave"}, i{:}\text{n\_type}) \triangleq token = i \wedge cs_i \wedge \neg\, cs'_i \wedge token' = 0 \wedge$$
$$Unchanged\ \langle req_{j:n\_type \wedge j \neq i}, cs_{j:n\_type \wedge j \neq i}\rangle \quad .$$

In total, after initial consistency checks have been performed (Sect. 4.1) a TLT program is represented by the following formula, which corresponds to the standard form of program formulas in TLA[17]:

$$\Phi \triangleq Init_\Phi \wedge \Box[\bigvee_{l \in LB} s\, l]_f \wedge \Box Invariant_\Phi \wedge$$
$$\forall_{l \in LB_{WF}} \text{WF}_f(s\, l) \wedge \forall_{l \in LB_{SF}} \text{SF}_f(s\, l) \quad ,$$

where $LB$ are the identifiers of all program instructions, $LB_{SF}$, $LB_{WF}$ identify strongly and weakly fair transition predicates, and $f$ is the tuple of flexible (state-dependent) program variables. The subformula $\Box[\bigvee_{l \in LB} s\, l]_f$ states that in each step[4] either a transition predicate is true or a stutter step without any changes of program variables in $f$ occurs. WF and SF are fairness assumptions.

For Program Mutex, the parameters are

$$LB \triangleq \{l \mid \exists_{i:n\_type}\ l = (\text{"request"},i) \vee \exists_{i:n\_type}\ l = (\text{"get"},i) \vee$$
$$\exists_{i:n\_type}\ l = (\text{"enter"},i) \vee \exists_{i:n\_type}\ l = (\text{"leave"},i)\}$$

$$LB_{SF} \triangleq \{l \mid \exists_{i:n\_type}\ l = (\text{"get"},i)\}$$

$$LB_{WF} \triangleq \{l \mid \exists_{i:n\_type}\ l = (\text{"enter"},i) \vee \exists_{i:n\_type}\ l = (\text{"leave"},i)\}$$

$$Init_{Mutex} \triangleq Init(\forall_{i:n\_type} \neg\, req_i \wedge \neg\, cs_i \wedge token = 0)$$

$$Invariant_{Mutex} \triangleq \forall_{i:n\_type}\ cs_i \Rightarrow token = i$$

$$f \triangleq \langle req, token, cs\rangle \quad .$$

---

[3] This additional stutter predicate is called the *frame predicate*.

[4] In TLA, $\Box$ and $\Diamond$ denote that a predicate holds *always* and *eventually*, respectively.

# 3 Representing TLA in LAMBDA

Two basic methods of reasoning about programs in higher-order logic have been identified [7]: *deep* and *shallow embedding*. Deep embedding includes the modelling of semantic functions in logic, whereas shallow embedding uses an external interpretation of the constructs of a formal language by directly mapping them to logic terms. In both methods all reasoning is done in logic.

Aiming at productive verification of individual system specifications, we chose an approach weaker than shallow embedding: we not only perform parsing and semantic interpretation, but also a controlled amount of reasoning outside of the logic. Our decision is based on the perception that, although purely definitional approaches preclude logical inconsistencies - provided that the basic mechanisms for adding definitions are sound, they inherently add complexity as far as the embedding of imperative languages is concerned.

Accordingly, we specify a small number of operators which cannot be embedded semantically in our simple TLA representation as uninterpreted constants. They are evaluated through axioms, some of which are even generated dynamically in the course of proofs. This approach allows a less rigid term structure to be used than prescribed by the standard inference mechanism; in particular the quantification of program variables is considerably simplified.

## 3.1 The Semantic Levels of TLA

TLA is a coherent formalism for expressing and reasoning about local and global properties of transition systems. Accordingly, TLA expressions are classified into Boolean formulas and non-Boolean terms at four semantic levels:

|  | *temporal* | *action* | *state* | *constant* |
|---|---|---|---|---|
| *term*: | — | transition function | state function | constant |
| *formula*: | temporal formula | action | state predicate | boolean |

At the *constant* level there are no state-dependencies; the *state* level refers to one state, the *action* level to two adjacent states, and the *temporal* level to behaviours, i.e. complete traces of states.

A full embedding of TLA requires distinct instances of essentially the same logical operator at each temporal level and lifting functions[18]. In terms of ML or L2 (cf. Sect.1), TLA formulas could be typed as

```
state → bool          for state predicates,
state → state → bool  for actions,
behaviour → bool      for temporal formulas,   where
behaviour = natural → state, state = variable → value
```

The standard representation of the TLA calculus does not explicitly use lifting functions. In fact, the state dependency of the values of program variables is implicitly contained in the separately defined semantics of the TLA operators. We keep this implicitness by handling variables of TLA programs as uninterpreted L2 constants and controlling the distinction of the semantic levels outside of the formal embedding. Temporal formulas, state and transition predicates, are uniformly typed as **Booleans**. Dummy operators serve as recognizers if considered useful. On the other hand, while in TLA types are assigned to program variables by way of invariants, we exploit the polymorphic type-system of LAMBDA.

## 3.2  Primed Expressions

In TLA, a prime operator denotes reference to the successor of a current state. A primed TLA term is evaluated by priming all occurring flexible program variables:
$$p' \triangleq p(\forall \text{ '}v\text{' } : v'/v).$$

As our embedding does not distinguish program variables from constants, we supply axiomatic ML tactics; they apply the prime operator to embedded terms by identifying and substituting all occurring program variables with primed program variables defined as separate constants. For this purpose, basic equations are axiomatised when a TLT program is translated. They are combined with program-independent conversions which recursively propagate primes down to atomic subterms. The resulting compound conversions[5] are efficient implementations of evaluation functions for primed embedded TLA terms.

## 3.3  Basic TLA Operators

Presuming the standard operators of TLA[17], the embedded representations are given in Table 1; typed dummies simply introduce operators as uninterpreted constants which are specified separately in terms of permanent (X), program-specific (P), or dynamically generated axioms (ML). The sum type **mty** is beyond the polymorphic type system of LAMBDA; its meta-level implementation is not described. A list of **mty**-elements can be considered to be a tuple. The type variable 'a expresses polymorphism.

*Unchanged*-predicates are evaluated at the meta-level by replacing primed with unprimed terms in corresponding transition predicates. Non-scalar program variables which are partly unchanged are included.

Another meta-tactic applies the *enabled* operator to a transition predicate; as a result, all occurring primed program variables are replaced with existentially quantified bound variables to denote that program variables can take values in the next state which satisfy the transition predicate.

---

[5] In LAMBDA conversions are implementations of equation-based rewrite strategies.

| TLA | LAMBDA Representation | |
|---|---|---|
| $p'$ | $\vdash_{def} pr\,(p : {}'a) = DUMMY\!:\,{}'a$ | X,P |
| $Unchanged\ f$ | $\vdash_{def} Unchanged\,(\textbf{f:mty list}) \equiv DUMMY\!:\,\textbf{bool}$ | X,P |
| $\langle A \rangle_f$ | $\vdash_{def} DAct\,(\textbf{f:mty list})\,(\textbf{A:bool}) \equiv A \wedge NOT\,(Unchanged\,\textbf{f})$ | |
| $[A]_f$ | $\vdash_{def} BAct\,(\textbf{f:mty list})\,(\textbf{A:bool}) \equiv A \vee Unchanged\,\textbf{f}$ | |
| $\Box\ F$ | $\vdash_{def} Box\,(F : \textbf{bool}) \equiv DUMMY\!:\,\textbf{bool}$ | X |
| $\Diamond\ F$ | $\vdash_{def} Dmd\,(F : \textbf{bool}) \equiv NOT\,(Box\,(NOT\,F))$ | |
| $Enabled\ A$ | $\vdash_{def} Enabled\,\textbf{A} \equiv DUMMY\!:\,\textbf{bool}$ | ML,P |
| $F \rightsquigarrow G$ | $\vdash_{def} F\ leadsto\ G \equiv Box\,(F \Rightarrow Dmd\,G)$ | |
| $\mathrm{WF}_f(A)$ | $\vdash_{def} WF\,\textbf{f}\,\textbf{A} \equiv Box\,(Dmd\,(DAct\,\textbf{f}\,\textbf{A}))\ \vee$ $\qquad\qquad Box\,(Dmd\,(NOT\,(Enabled\,(DAct\,\textbf{f}\,\textbf{A}))))$ | |
| $\mathrm{SF}_f(A)$ | $\vdash_{def} SF\,\textbf{f}\,\textbf{A} \equiv Box\,(Dmd\,(DAct\,\textbf{f}\,\textbf{A}))\ \vee$ $\qquad\qquad Dmd\,(Box\,(NOT\,(Enabled\,(DAct\,\textbf{f}\,\textbf{A}))))$ | |
| $\bar{t}$ | $\vdash_{def} Bar\,(\textbf{m} : (\textbf{mty} * \textbf{mty})\textbf{list})\,(t : {}'a) = DUMMY\!:\,{}'a$ | X,P |
| $\exists\ x,y : F$ | $\vdash_{def} Bexists\,([\textbf{x,y}] : \textbf{mty list})\,(F : \textbf{bool}) \equiv DUMMY\!:\,\textbf{bool}$ | X |

**Table 1.** Standard Operators of TLA

A few operators are added to give TLA expressions more structure and ease the application of calculus rules, such as *Init* whose argument defines a predicate for the initial state. The refinement operators *Bar*, *Bexists* are discussed in Sect.5.1.

## 3.4  TLA Calculus

The proof calculus of TLA is based on rules whose format closely matches the representation of rules in LAMBDA. The LAMBDA facilities for applying rule schemes, unifying expressions, and controlling instantiations are therefore suited for efficient transformation of embedded TLA formulas.

Table 2 illustrates that the calculus rules of our LAMBDA embedding remain close to TLA[17].[6] *LATTICEnat$_\lambda$* is a specialised version of an induction rule for proving that the post-condition of a *Leads-to*-relation is either reached immediately or after a finite number of transitions. According to the semantics of strong fairness, Rule *SF1$_\lambda$* tests whether an enabled strongly fair transition ensures the post-condition and whether it is eventually enabled. Rule *INV$_\lambda$* breaks an invariant proof down to checking whether the invariant is initially satisfied and whether it is not affected by any transition. Rule *STL4$_\lambda$* reduces an implication on temporal formulas to an implication at a lower semantic level. Rule *AT$_\lambda$* is used for proving safety properties which are not immediate invariants.

Further rules for Unity-like reasoning are supplied for liveness proofs (Sect.4.3).

---

[6] There are hooks in LAMBDA to add specific pretty-printing tools that let the rules appear exactly as they would in TLA. The symbol $\Gamma$ may be ignored for the moment.

$LATTICEnat_\lambda$

| | |
|---|---|
| $1$: $F, \Gamma \vdash (H \wedge x = c)$ $leadsto$ $(H \wedge x = d \wedge d < c \vee G)$ | |

$$F, \Gamma \vdash H \; leadsto \; G$$

$INV_\lambda$

$2$: $\qquad\qquad$ I1, $\Gamma \vdash$ I

$1$: $\qquad$ I, $BAct$ $f$ $N$, $\Gamma \vdash_{pr}$ I

$$Init \; I1, Box\,(BAct \; f \; N), \; \Gamma \vdash Box \; I$$

$SF1_\lambda$

$3$: $Box\,(BAct \; f \; N), SF \; f \; A, Box \; F, \; \Gamma \vdash P \; leadsto\,(Enabled\,(DAct \; f \; A))$

$2$: $\qquad\qquad$ P, $NOT$ Q, $DAct$ $f$ $(A \wedge N)$, F, $\Gamma \vdash_{pr}$ Q

$1$: $\qquad$ P, $NOT$ Q, $BAct$ $f$ $(NOT$ A $\wedge$ N$)$, F $\Gamma \vdash_{pr}$ P $\vee_{pr}$ Q

$$Box\,(BAct \; f \; N), SF \; f \; A, Box \; F, \; \Gamma \vdash P \; leadsto \; Q$$

$STL4_\lambda$

$1$: $\qquad$ F, $\Gamma \vdash$ G

$$Box \; F, \; \Gamma \vdash Box \; G$$

$AT_\lambda$

$2$: $\qquad$ $\Gamma \vdash Box$ R

$1$: $\Gamma \vdash Box$ $(R \Rightarrow A)$

$$\Gamma \vdash Box \; A$$

Table 2. TLA Rules in LAMBDA

## 3.5 Safety of Reasoning

Having chosen a non-definitional approach, our embedding of TLA depends on the externally defined semantics and calculus of TLA, and on the implementation of the functions for the meta-level evaluation of TLA operators in LAMBDA. We discuss this second aspect.

Particular diligence is spent on avoiding predicate logic reasoning across different semantic levels. In the deduction calculus of LAMBDA, within one rule the symbol $\Gamma$ represents the same set of hypotheses and axioms. In TLA reasoning, this is not always correct, in particular, if sequents refer to different levels. This difficulty is handled by controlling the instantiation of $\Gamma$ when rules are applied. The following example illustrates a potential problem related to the priming operator. If the rewrite rule $x = y$, $\Gamma \vdash x = y$ were applied to the goal

$\ldots, \text{token} = i, \ldots, \Gamma \vdash_{pr} (\text{token} = i)$ ,

a wrong value would be used for the state variable token in the successor state. Obviously, this replacement is illegal as the argument of the prime function is treated as if it were a constant logic expression. The correctness is preserved by first evaluating the $pr$ expression, which leads to the goal

$\ldots, \text{token} = i, \ldots, \Gamma \vdash \text{token}' = i$ ,

where now the values of two adjacent states are distinct objects.

The proof environment for TLA in LAMBDA avoids such inconsistencies by sequencing transformations correctly at meta-level, protecting critical subterms from being transformed, and pruning goals from irrelevant predicates that pertain to other semantic levels than the key assertion. These precautions prevent general-purpose tactics for predicate-logic from corrupting proofs.

# 4 Proving Temporal Properties

Safety properties, which are expressed in terms of the *Box*-operator ($\Box$), must hold in all states during a program execution. The *leadsto*-operator specifies liveness properties. Proofs are discussed by way of the **Mutex** program.

## 4.1 Consistency Checks

Explicit invariants and annotations of transitions with pre- and post-conditions are optional hints of a programmer, which may significantly aid proof routines. It is to be proven that they are actually fulfilled before using them. Fortunately, such proofs are frequently accomplished fully automatically. Each pre-condition is proven to be fulfilled whenever the corresponding program instruction is enabled. Post-conditions are shown to get true with the execution of a transition. By means of Rule $INV_\lambda$, explicit invariants are proven to hold forever.

Proof obligations are generated for type invariants which are not handled automatically by the polymorphic type system of **LAMBDA**. This concerns subtype predicates, such as: ..., $\Gamma \vdash Box\,(\text{req } i \Rightarrow \text{n\_type } i)$,
where we have restricted ranges of natural numbers as indices of arrays.

Proven pre- and post-conditions are included in the corresponding transition predicate. Proven invariants are managed internally and added to the safety part of a program formula on request.

## 4.2 Safety

Safety properties are either invariants, which we prove with Rule $INV_\lambda$, or they are always true but cannot be deduced by the invariant approach which checks locally whether a predicate is fulfilled in a successor state; in that case such properties have to be derived from direct program invariants (Rule $AT_\lambda$).

In the n-process mutual exclusion program, the specified *always-true* property is such an example. The instruction **enter** allows a process i to enter the critical section, but it does not care whether another process is currently in it:

$$\exists_i\, s\,(\text{"enter"}, i), NOT\,(\exists_{i,j}\, i \neq j \land \text{n\_type } i \land \text{n\_type } j \land \text{cs } i \land \text{cs } j),$$
$$\Gamma \vdash NOT\,(\exists_{i,j}\, i \neq j \land \text{n\_type } i \land \text{n\_type } j \land \text{cs' } i \land \text{cs' } j)$$

Only within the reachable state space the guard of **enter** is guaranteed to be false as long as another process is still in its critical section. The explicit invariant $(\forall_i\, \text{cs } i \Rightarrow \text{token} = i)$ is needed; it implies the *always-true* property, as the program variable **token** cannot take two different values at a time.

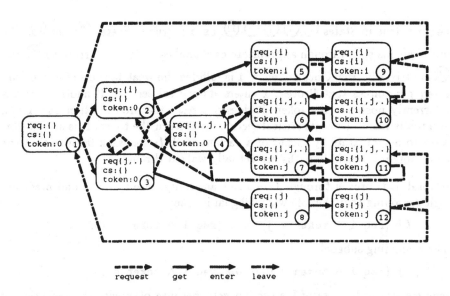

Fig. 2. State View of Mutex

## 4.3 Liveness

The proofs of liveness properties very much depend on an appropriate construction of *Leads-to* relations. In comparison to Unity, the *Leads-to*-operator has a semantical definition, apart from the additional notion of *strong* fairness of TLA, while analogous rules to insert intermediate states, split and replace antecedents and consequences of *Leads-to* relations, are valid.

| $LTTrans_\lambda$ | $LTDj1_\lambda$ | $LTSubstHyp_\lambda$ |
|---|---|---|
| 2: $\Gamma \vdash$ Q *leadsto* R | 2: $\Gamma \vdash$ Q *leadsto* R | 2: $\Gamma \vdash$ Q *leadsto* R |
| 1: $\Gamma \vdash$ P *leadsto* Q | 1: $\Gamma \vdash$ P *leadsto* R | 1: P, $\Gamma \vdash$ Q |
| $\Gamma \vdash$ P *leadsto* R | $\Gamma \vdash$ (P $\vee$ Q) *leadsto* R | $\Gamma \vdash$ P *leadsto* R |

The **Kripke** structure with the transition relations between the reachable states (Fig.2) of **Mutex** elucidates the proof of the specified liveness property, where we use the abbreviations introduced in Sect. 2:

$$_{Init} \mathbf{Init_{Mutex}}, {}^{Box}(BAct\, \mathbf{f}(\bigvee_{1 \in LB} s1)),$$
$$\bigwedge_{1 \in LB_{WF}} {}^{WF}\mathbf{f}(s1), \forall_i\, SF\,\mathbf{f}(s("\text{get}",i)), \quad \Gamma \vdash (\mathbf{req}\ i)\,_{leadsto}(\mathbf{cs}\ i) \ .$$

We take the point of view of some process i; thus the distinction of processes j with i $\neq$ j is irrelevant. State ① is the initial state. According to the picture,

**req** i is true in States ②,④-⑦,⑨-⑪; **cs** i is true in States ⑨ and ⑩. The initial goal is split through appropriate case analyses. Goals referring to ⑨ and ⑩ are trivially true. Once **token = i** is fulfilled the weak fairness of instruction **enter$_i$** guarantees that Process i enters its critical section. If another process j currently possesses the token, it does not reset the pre-condition **req** i and eventually has to leave its critical section and return the token. It is straightforward to demonstrate by means of a variant of $SF1_\lambda$ for weak fairness that the corresponding *Leads-to* paths are actually walked through.

In total, it is proven through *Leads-to*-transitivity, case splitting and predicate logic reasoning that the following subgoal is true:

$\dots, \Gamma \vdash$ (**req** i $\wedge$ **token = j**) *leadsto* (**req** i $\wedge$ **token = 0**).

The remaining subgoal

$\dots, \Gamma \vdash$ (**req** i $\wedge$ **token = 0**) *leadsto* (**req** i $\wedge$ **token = i**)

requires strong fairness of Instruction **get$_i$**, because otherwise the execution of **get$_i$** could always be preempted by another **get$_j$**. We prove the combined goal

$\dots, \Gamma \vdash$ (**req** i $\wedge$ *NOT* (**cs** i) $\wedge$ *NOT* (**token=i**)) *leadsto* (**req** i $\wedge$ **token=i**) .

Three subgoals are obtained through Rule $SF1_\lambda$. Subgoal *1*,

*1*:(**req** i $\wedge$ *NOT* (**cs** i) $\wedge$ *NOT* (**token = i**)), *NOT* (**req** i $\wedge$ **token = i**),

$_{BAct}$ **f** (*NOT* (**s**(**"get"**,i)) $\wedge$ $\bigvee_{l \in LB}$ **s**l), **I1**, $\Gamma \vdash$

(**req'** i $\wedge$ *NOT* (**cs'** i) $\wedge$ *NOT* (**token' = i**)) $\vee$ (**req'** i $\wedge$ **token' = i**) ,

which essentially states that the pre-condition of the *Leads-to*-relation is either maintained or the post-condition is satisfied, is proven by case analysis upon all instructions without **get$_i$**. The extra invariant **I1 := n_type token** $\vee$ **token = 0** follows directly from available subtype invariants.

Subgoal *2* states that instruction **get$_i$** ensures the post-condition:

*2*:**req** i $\wedge$ *NOT* (**cs** i) $\wedge$ *NOT* (**token = i**), *NOT* (**req** i $\wedge$ **token = i**),

$_{DAct}$ **f** (**s**(**"get"**,i)), **I1**, $\Gamma \vdash$ (**req'** i $\wedge$ **token' = i**)   .

The subgoal regarding eventual enabledness of **get$_i$** follows from previous results.

On the whole, once a goal has been broken down to predicate-logic conjectures, automatic general-purpose tactics [8] do the rest without further interaction. Meanwhile, customized TLA tactics pave the way for hiding most of the TLA level reasoning from the user.

# 5 Refinement Proofs

Two programs satisfy a refinement relation with reference to externally visible program variables, if the following implication is provable

$$(\exists\, x_1, \ldots, x_m : \Psi) \;\Rightarrow\; (\exists\, y_1, \ldots, y_n : \Phi)$$

where $\Psi$ and $\Phi$ are the TLA formulas of a concrete and an abstract program with initialization, safety and fairness assumptions, and stutter steps are allowed. Existentially-quantified program variables are hidden, i.e., the quantified formulas are true, if there exist sequences of values for these variables which satisfy the program formulas. An appropriate refinement mapping is needed to express hidden program variables of the abstract program in terms of concrete program variables, such that the implication between the instantiated program gets true.

A version of the mutual exclusion algorithm with priorities is specified as a concrete program to illustrate refinement.[7]

```
PROGRAM PMutex
INVARIANT ∀i:n_type pcsi ⇒ 0 < preqi
INITIALLY ∀ i:n_type preqi = 0 ∧ ¬ pcsi
INSTRUCTIONS
    [prequesti:n_type] ¬ pcsi ∧ preqi = 0  ⟶  preq'i = 1
    [penteri:n_type] 0 < preqi ∧ ¬ pcsi ∧ (∀j:n_type preqj ≤ preqi)  ⟶
                pcs'i ‖ ∀j:n_type IF 0 < preqj THEN preq'j = preqj + 1
    [pleavei:n_type] {0 < preqi} pcsi  ⟶  ¬ pcs'i ‖ preq'i = 0
FAIRNESS
    [pfair] ∀i:n_type WF(penteri) ∧ WF(pleavei)
END
```

Fig. 3. An n-process Mutual Exclusion Example with Priorities

While Mutex does not have any arbitration between several requesting processes, PMutex only permits a process with the maximum priority to enter the critical section. The priority of all other requesting processes is incremented.

## 5.1 Refinement Operators

In addition to the hiding operator *Bexists* ($\exists$), the operator *Bar* ($^-$) is used to denote the substitution of program variables according to some refinement mapping. Their LAMBDA implementation is summarized without expanding on their formal semantics (cf. [17]).

---

[7] For succinctness, we combine array variables that record requests and priorities. Although both programs thus do not have any variables in common, this is in fact inessential for our discussion and is no simplification of the illustrated procedure.

The hiding operator is eliminated with a tactic which uses axiomatic rules like

$$E1cons_\lambda$$
$$1: \Gamma \vdash {}_{Bar} [(\mathbf{v},\mathbf{r})] \, (_{Bexists} \mathbf{vs} \, \mathbf{F})$$
$$\overline{\Gamma \vdash {}_{Bexists} (\mathbf{v} :: \mathbf{vs}) \mathbf{F}}$$

$$E1nil_\lambda$$
$$1: \qquad \Gamma \vdash \mathbf{F}$$
$$\overline{\Gamma \vdash {}_{Bexists} \, [] \, \mathbf{F}}$$

$$E2_\lambda$$
$$1: \qquad \mathbf{F}, \Gamma \vdash \mathbf{Q}$$
$$\overline{{}_{Bexists} \mathbf{vs} \, \mathbf{F}, \Gamma \vdash \mathbf{Q}} \; .$$

The tactic disables $E2_\lambda$, if the hidden variable occurs on the right-hand side of the turnstile; users do not have direct access to $E2_\lambda$.

The conversion for the *Bar*-operator resembles the prime-conversion, as it propagates the *Bar*-operator down to program variables and applies substitutions according to a specified refinement mapping. It is ensured that the implicit existential quantification of the *Enabled* operator is not affected ([17]). In contrast to its TLA counterpart, the *Bar*-operator is furnished with an explicit parameter for pairs of program variables and their individual mappings.

## 5.2  Specifying a Refinement Mapping

The concrete version of the mutual exclusion algorithm allows a prioritized process immediately to enter the critical section, while the abstract program requires an intermediate step to pass the token. This is one of three possible cases, in which refinement mappings cannot be specified directly, even though a program semantically implements another one [3]:

1. The abstract program contains variables which record parts of abstract states, while there is no such notion of history in the refined program.
2. Several abstract transition predicates correspond to fewer concrete ones.
3. A non-deterministic choice is made in the abstract program at an earlier step than in the concrete program.

History, stutter and prophecy variables, if added to the concrete program in such a way that the external behaviour is unaffected except for possible stutter steps, may enable a refinement mapping in these cases. LAMBDA rules and tactics elegantly allow such auxiliary variables to be added to program formulas; automatic checks of the augmentation are included.

Considering PMutex we introduce a stutter variable (scs) which blocks all transitions for one step when a process has entered the critical section. An additional history variable (pid) records the identification of that process.

The augmented program PMutexA corresponds to the formula

$$Init \, (\mathbf{scs} = 0 \land \mathbf{pid} = 0 \land \forall_\mathbf{i} \, \mathbf{preq} \; \mathbf{i} = 0 \land NOT \, (\mathbf{pcs} \; \mathbf{i})) \land$$
$$Box \, (BAct \, [\mathbf{pid}, \mathbf{scs}, \mathbf{preq}, \mathbf{pcs}] \, (\exists_\mathbf{i}$$
$$\qquad\qquad s \, (\text{"prequest"}, \mathbf{i}) \land \mathbf{tp1} \lor s \, (\text{"penter"}, \mathbf{i}) \land \mathbf{tp2} \; \mathbf{i} \lor$$
$$\qquad\qquad \mathbf{tp3} \lor s \, (\text{"pleave"}, \mathbf{i}) \land \mathbf{tp1})) \land$$
$$WF \, [\mathbf{pid}, \mathbf{scs}, \mathbf{preq}, \mathbf{pcs}] \, \mathbf{tp3} \land \forall_\mathbf{i} \, WF \, [\mathbf{pid}, \mathbf{scs}, \mathbf{preq}, \mathbf{pcs}] \, (..\mathbf{i}..)$$

Transition **tp2** disables all other transitions until **tp3** has become true:

$$\begin{aligned}
\textbf{tp1} \quad &:= Unchanged\,[\textbf{scs},\textbf{pid}] \\
\textbf{tp2}\ \text{i} &:= \textbf{scs} = 0 \wedge \textbf{scs'} = 1 \wedge \textbf{pid'} = \text{i} \\
\textbf{tp3} \quad &:= \textbf{scs} = 1 \wedge \textbf{scs'} = 0 \wedge Unchanged\,[\textbf{pid},\textbf{preq},\textbf{pcs}]\ .
\end{aligned}$$

Based on the auxiliary variables, the following refinement mapping is used, where the variable **cs** is updated to the value of **pcs** after one stutter step:

$$\begin{aligned}
\textbf{token} &:= \textit{if}\ \exists_\text{i}\,\textbf{pcs}\ \text{i}\ \textit{then}\ \textbf{pid}\ \textit{else}\ 0 \\
\textbf{req}\ \text{i} &:= 0 < \textbf{preq}\ \text{i} \\
\textbf{cs}\ \text{i} &:= \textbf{scs} = 0 \wedge \textbf{pcs}\ \text{i}\ .
\end{aligned}$$

## 5.3 A Refinement Proof

The refinement proof is broken down to subproofs for initialization, transitions and fairness. In total, after refinement mappings have been determined and corresponding transitions of both programs have been identified, the **LAMBDA** proof is analogous to safety and liveness proofs which we already discussed.

**Initialization.** The initialization goal is proven automatically:

$\forall_\text{i}\,\textbf{preq}\ \text{i} = 0 \wedge NOT\,(\textbf{pcs}\ \text{i}),\textbf{scs} = 0,\textbf{pid} = 0,$
$\Gamma \vdash \forall_\text{i}\,NOT\,(0 < \textbf{preq}\ \text{i}) \wedge \forall_\text{i}\,NOT\,(\textbf{scs} = 0 \wedge \textbf{pcs}\ \text{i}) \wedge 0 = 0$

**Transitions.** As each transition of **Mutex** corresponds to one of **PMutexA**, four subgoals are to be proven:

$4: s\,(\texttt{"pleave"},\text{i}) \wedge \textbf{tp1},\ \Gamma \vdash\ Bar\,[\ldots]\,(s\,(\texttt{"leave"},\text{i}))$
$3: \textbf{tp3},\ \Gamma \vdash \exists_\text{i}\ Bar\,[\ldots]\,(s\,(\texttt{"enter"},\text{i}))$
$2: s\,(\texttt{"penter"},\text{i}) \wedge \textbf{tp2}\ \text{i},\ \Gamma \vdash\ Bar\,[\ldots]\,(s\,(\texttt{"get"},\text{i}))$
$1: s\,(\texttt{"prequest"},\text{i}) \wedge \textbf{tp1},\ \Gamma \vdash\ Bar\,[\ldots]\,(s\,(\texttt{"request"},\text{i}))$

The case that the refined program just stutters is trivial as the abstract program can do the same. The most interesting (expanded) goals are:

$3: \textbf{scs} = 1 \wedge \textbf{scs'} = 0,\text{I},\ \Gamma \vdash \exists_\text{i}\ \textbf{pid} = \text{i} \wedge 0 < \textbf{preq}\ \text{i} \wedge \textbf{scs'} = 0 \wedge \textbf{pcs'}\ \text{i}$
$2: \textbf{scs'} = 1,\textbf{pcs'}\ \text{i},\textbf{pid'} = \text{i},\ \Gamma \vdash\ (\textit{if}\ \exists_\text{i}\,\textbf{pcs'}\ \text{i}\ \textit{then}\ \textbf{pid'}\ \textit{else}\ 0) = \text{i}\ ,$

where I $:= \textbf{scs} = 1 \Rightarrow \exists_\text{i}\,\textbf{pcs}\ \text{i} \wedge \textbf{pid} = \text{i} \wedge 0 < \textbf{preq}\ \text{i}$ is an additional invariant of the extended program **PMutexA**, which is proven through the usual invariant approach. The remainder is straightforward.

**Fairness.** As the priorities allowed us to specify just weak fairness in the concrete program, we centre on the strong fairness proof for $s\,(\texttt{"get"},\text{i})$. The case that eventually there is no further request is trivial. The other case is essentially

$$\ldots,\ \Gamma \vdash\ Bar\,[\ldots]\,(Box\,(Dmd\,(\textbf{token} = 0 \wedge \textbf{req}\ \text{i} \wedge \textbf{token'} = \text{i})))\ .$$

The refinement mappings are applied first. Using the weak fairness of **penter**, the proof includes case analysis and induction for the case that several processes have highest priority.

# 6 Conclusion

We have discussed a method for reasoning about concurrent programs in the higher-order logic theorem prover LAMBDA on the basis of an embedding of TLA. By way of a mutual exclusion algorithm for an arbitrary number of processes, proofs of safety and liveness properties and refinement have been depicted.

Embedding a formal language in predicate logic has three aspects: parsing, semantic interpretation, and formal reasoning. The last two of these can be entirely formalized in higher-order logic. Depending on the degree of formality[5], there is a trade-off between the security of tool implementation and practicality, which has been rated differently in the communities of theorem proving and algorithmic verification or simulation. As practicality is a prerequisite for industrial acceptance, the simplicity of a verification methodology is crucial. Therefore we deliberately chose to perform not only semantic interpretation, but also a controlled and limited amount of reasoning outside of the logic.

The examples of this paper and other small case-studies done so far demonstrate that a simple and uniform representation of all semantic levels of TLA in predicate logic is obtained. Consequently the overhead due to the embedding is negligible. In comparison to TLA paper proofs, we not only provide automatic machine support for lemmas which are typically presupposed yet error-prone, but we also devise high-level tactics for automating standard sequences of TLA proofs and even some proof decisions. Underneath, automatic general-purpose tactics of a higher-order logic theorem prover are utilized. Notably, proof functions for predicate logic and arithmetical reasoning [8] turned out to be useful.

We further investigate composition, communication between distributed processes, projections of infinite to finite state spaces, construction of invariants, Leads-to-paths and refinement mappings, real-time aspects, and full integration with model-checking. Large case studies are to follow. In addition to prohibiting illegal interactions, maximum safety could be ensured without giving up the simplicity of the reasoning machinery for practical applications, if our embedding was formally verified once forever, based on a deep embedding.

TLA is a favourable internal formalism, but it is not adequate for engineers in industrial development. However, we believe that interaction with the envisaged proof environment can be disguised in application-oriented user menus and reduced to the level of user specifications of algorithms, such that no specific knowledge is needed for guiding highly automatic proof functions. Experimental results show that information just supplied in terms of state predicates, e.g., assertions attached to program instructions, invariants and Leads-to-relations, greatly aids mechanical verification. Moreover, the degree of automation increases with the proximity of problems to the domain of model-checking. An incremental strategy therefore appears to be appropriate for extending the scope of industrially applicable formal verification to high-level data types and parameterization.

**Acknowledgements** I would like to thank my colleagues C. Cuéllar, D. Barnard, G. Gouverneur, M. Huber, S. Prummer, I. Wildgruber, as well as S. Merz and J. Hu for fruitful discussions and helpful comments on drafts of this paper.

# References

1. 'Higher Order Logic Theorem Proving and Its Applications', 6th International Workshop, HUG'93, Vancouver, B.C., Canada, August 11-13 1993, Proceedings, J.J. Joyce and C.-J.H. Seger (Eds.), LNCS 780, Springer, 1994.
2. 'Higher Order Logic Theorem Proving and Its Applications', 7th International Workshop, HUG'94, Valetta, Malta, September 1994, Proceedings, T. Melham, and J. Camilleri (Eds.), LNCS 859, Springer, 1994.
3. M. Abadi and L. Lamport, 'The Existence of Refinement Mappings', *Theoretical Computer Science*, 81(2),pp. 253–284, May 1991.
4. F. Andersen, K.D. Petersen, and J.S. Pettersson, 'A Graphical Tool for Proving UNITY Progress', in [2], pp. 1–15.
5. C.M. Angelo, L. Claesen, H. De Man, 'Degrees of Formality in Shallow Embedding Hardware Description Languages in HOL', in [1], pp. 89–100.
6. D. Barnard and S. Crosby, 'The Specification and Verification of an ATM Signalling Protocol', PSTV'95, Proceedings, 1995.
7. R. Boulton, A. Gordon, M. Gordon, J. Harrison, J. Herbert, and J.v. Tassel, 'Experience with Embedding Hardware Description Languages in HOL', in *THEOREM PROVERS IN CIRCUIT DESIGN*, IFIP Transactions A-10, edited by V. Stavridou, T.F. Melham, and R.T. Boute, pp. 129–156, North-Holland, 1992.
8. H. Busch, 'First-Order Automation for Higher-Order-Logic Theorem Proving', in [2], pp. 97–112.
9. K.M. Chandy and J. Misra, 'Parallel Program Design - A Foundation', Addison-Wesley,1988.
10. N. Chapman, S. Finn. and M. Fourman, 'Datatypes in L2', in [2], pp. 128–143.
11. C.-T. Chou, 'A Formal Theory of Distributed Algorithms in Higher-Order Logic', in [2], pp. 144–157.
12. J. R. Cuéllar, I. Wildgruber, and D. Barnard, 'Combining the Design of Industrial Systems with Effective Verification Techniques', in *Proc. of FME'94, Barcelona, Spain*, pp. 639–658, Springer LNCS 873, M. Naftalin, T. Denvir, and M. Betran (Eds.), October 1994.
13. J. Cuéllar and M. Huber, 'The FZI Production Cell Case Study: A Distributed Solution using TLT', Springer LNCS 891, 1994.
14. U. Engberg, P. Grønning, and L. Lamport, 'Mechanical Verification of Concurrent Systems with TLA', in *Proc. of CAV'92*, pp. 44–55, July 1992.
15. R. Harper, R. Milner, and M. Tofte, 'The Definition of Standard ML, Version 3', University of Edinburgh, LFCS Report Series, ECS-LFCS-89-81, May 1989.
16. S. Kalvala, 'A formulation of TLA in Isabelle', in these proceedings, 1995.
17. L. Lamport, 'The Temporal Logic of Actions', *ACM Transactions on Programming Languages and Systems*, Vol.16, No.3, pp. 872–923, May 1994.
18. T. Långbacka, 'A HOL Formalization of the Temporal Logic of Actions', in [2], pp. 332–345.

# A Theory of Finite Maps

Graham Collins[1] and Donald Syme[2]

[1] Department of Computing Science,University of Glasgow, Glasgow,
Scotland, G12 8QQ. email: grmc@dcs.gla.ac.uk.
[2] The Computer Laboratory, New Museums Site, University of Cambridge,
Cambridge, UK, CB2 3QG. email: Donald.Syme@cl.cam.ac.uk

**Abstract.** Finite maps, functions defined on only a finite domain, occur
often, particularly when reasoning about programming languages. This
paper presents a theory of finite maps in HOL. We discuss the choice
of representation, present the theory we have defined, and discuss the
issue of defining recursive types containing finite maps. We also discuss
decision procedures and give an example of the use of finite maps in
developing the semantics of a small language.

## 1 Introduction

Functions defined on only a finite domain occur frequently in computing science.
One field in which these functions, commonly referred to as finite maps or finite
partial functions, are used is in reasoning about the semantics of programming
languages, where they can model semantic objects such as type contexts and
environments.

A commonly-used representation for finite maps is simply the theory of lists; a
finite map can be represented by a list of pairs, and functions to update and apply
maps can be defined easily and will behave correctly when used. Unfortunately
this simple use of lists is flawed because two lists that behave the same when
used as finite maps may not be logically equal, a property that is essential where
reasoning about the equality of finite maps is required. These issues are discussed
in more detail later.

This paper presents a theory of finite maps that will be the basis for a finite
maps library in the HOL theorem prover [1]. This work follows the HOL tradition
of taking a purely definitional approach. We characterize the theory in terms of
a small set of axioms that are sufficient to capture the intended meaning of finite
maps. The choice of these axioms is discussed in section 2. A model for these
properties is then constructed using types and constants that already exist in
HOL. Section 3 describes the representation used for this model and how the
characteristic theorems are proved. Section 4 describes how the theory can be
enriched with more theorems and concepts using those already defined. Section
5 addresses the issue of defining recursive types containing finite maps. This
potentially difficult problem has been the motivation for using lists to model
finite maps in the past, as this provides a means to define such types. Section 6
describes some decision procedures we have implemented and section 7 gives an

example that uses finite maps to represent contexts for the type system for a small language.

# 2 Finite maps

This section describes the choice of axioms for the theory of finite maps. These axioms are intended to fully characterise the type, and must of course be chosen so as to be consistent. We determine that this is the case by providing a model for the axioms in later sections. It is also necessary that the choice of axioms be complete; any property we wish to prove of finite maps should be provable from these axioms. A further result, interesting for theoretical reasons, is that the axioms should be independent of each other; if any axiom is removed or weakened then the set of axioms will fail to completely specify finite maps.

The type of finite maps will be introduced by a new binary type operator *fmap*. A finite map from type $\alpha$ to type $\beta$ has type $(\alpha, \beta)$*fmap*. An important concept is the domain of a finite map. This is the finite set of values over which the application of a finite map will be specified.

## 2.1 Axioms for the constants

Four constants are introduced, with informal definitions as follows:

- Empty : The finite map with no elements in its domain.
- Update $f$ $(x, y)$ : The basic operation to allow the extension of a finite map $f$ with a new mapping from $x$ to $y$. There should be no restriction on whether or not $x$ is already in the domain of $f$. If $x$ is in the domain then the value to which $x$ is mapped will be updated to be $y$. Some other formalisms of finite maps restrict Update to extension of a finite map only with elements not in the domain.
- Apply $f$ $x$ : If $x$ is in the domain of $f$ then Apply $f$ $x$ denotes the value to which $x$ is mapped.
- Domain $f$ $x$ : The function Domain tests whether an element $x$ is a member of the domain of $f$. The domain is formulated in terms of a boolean function rather than a set so that the resulting theory does not depend on a particular variety of set theory. It is a relatively trivial task to construct the domain set from the definitions given below in the user's choice of set theory.

The above informal definitions still leave some ambiguities to be resolved. In particular nothing has been said about the outcome of applying a finite map to an element not in its domain. (Apply $f$) : $\alpha \rightarrow \beta$ should be a partial function, only defined on the domain of $f$. But all functions in HOL are total and so are defined for all elements of the correct type. The traditional solution in HOL is to leave a partial function unspecified for values not in the correct domain. Thus applying a finite map $f$ of type $(\alpha, \beta)$*fmap* to a value that is not in the domain of $f$ will return a value of type $\beta$, but this value will be unspecified and it will not be possible to prove which member of the type $\beta$ has been returned.

An alternative approach, similar to that used in [3, 9], is to define Apply to return a result of type $\beta + one$ where $one$ is the type with only one element, namely the value denoted by one. Returning one indicates that the finite map is undefined for that element. This has the advantage of reducing the number of constants that need to be "axiomatised" but the disadvantage of complicating the type of the value produced by Apply. This modified apply function can, however, be defined in terms of the constants Apply and Domain introduced above.

It is claimed that the intended meaning of the constants Empty, Apply, Update and Domain can be formalised by the six basic axioms

$\vdash \forall f\ a\ b.\ \text{Apply}\ (\text{Update}\ f\ (a,b))\ a\ =\ b$

$\vdash \forall x\ a.\ (x \neq a)\ \supset\ \forall f\ b.(\text{Apply}\ (\text{Update}\ f\ (a,b))\ x\ =\ \text{Apply}\ f\ x)$

$\vdash \forall a\ c.\ (a \neq c)\ \supset$
$\qquad \forall f\ b\ d.$
$\qquad\qquad (\text{Update}\ (\text{Update}\ f\ (a,b))\ (c,d)\ =\ \text{Update}\ (\text{Update}\ f\ (c,d))\ (a,b))$

$\vdash \forall f\ a\ b\ c.\ \text{Update}\ (\text{Update}\ f\ (a,b))\ (a,c)\ =\ \text{Update}\ f\ (a,c)$

$\vdash \forall a.\ \neg(\text{Domain}\ \text{Empty}\ a)$

$\vdash \forall f\ a\ b\ x.\ \text{Domain}\ (\text{Update}\ f\ (a,b))\ x\ =\ (x\ =\ a)\ \vee\ \text{Domain}\ f\ x$

together with a further induction axiom which is explained in the next section.

## 2.2 Induction

The axioms in the previous section do not express the fact that the partial functions being considered are finite. In addition to these axioms, an induction principle is needed:

$\vdash \forall P.$
$\qquad P\ \text{Empty} \wedge (\forall f.\ P\ f \supset (\forall x\ y.\ P\ (\text{Update}\ f\ (x,y))))$
$\qquad \supset$
$\qquad \forall f.\ P\ f$

This gives us the property that any finite map can be formed by a finite number of updates of the empty map. It easily follows from this that the domain of a finite map can be enumerated by some initial fragment of the natural numbers.

This induction principle is not strong enough to derive the natural characterisation of equality for finite maps (see below). Indeed, it is possible to formulate a model in which the six basic axioms hold along with this induction axiom but the characterisation of equality shown below is false.

The following stronger induction principle is sufficient to derive the characterisation of equality for finite maps. In the step case induction, this stronger principle allows us to assume that the element being added is not in the domain of the finite map.

$\vdash \forall P.$
    $P$ Empty $\wedge$
    $(\forall f.\ P\ f \supset (\forall x.\ \neg(\text{Domain } f\ x) \supset \forall y.\ P\ (\text{Update } f\ (x, y))))$
    $\supset$
    $\forall f.\ P\ f$

An alternative to adding to this stronger induction axiom is to add the weaker induction theorem and the equality theorem below as axioms and derive the strong induction theorem. This would also replace other basic axioms for the constants discussed earlier. It was felt that it was better to use the basic axioms and derive the equality theorem as the basic axioms capture the intended meaning of the various constants more precisely.

## 2.3 Equality

We now consider a theorem characterising when two finite maps are equal. If the axioms above provide a complete characterisation of finite maps then it should be possible to derive such a theorem from them. We begin by considering how to formulate equality. The naive formulation

$$\vdash \forall f\ g.\ (\forall x.\ \text{Apply } f\ x\ =\ \text{Apply } g\ x)\ =\ (f = g)$$

does not hold because of a problem with the application of finite maps to elements not in their domain. Consider the two finite maps

$$f = \text{Empty}$$
$$g = \text{Update Empty } (x,\ \text{Apply Empty } x)$$

where $x$ is some arbitrary value. Then $f$ and $g$ are different finite maps, with different domains; but for any $y$

Apply $g$ $y$
$=$ Apply (Update Empty $(x$, Apply Empty $x$)) $y$
$=$ Apply Empty $y$           by a case split on $y = x$ and the axioms for Apply
$=$ Apply $f$ $y$

So we have two different finite maps that agree on all elements to which they are applied and hence are equal by the naive formulation of equality. To fix this we modify the characterisation of equality to say that two finite maps are equal only if, in addition to agreeing on all elements, their domains are equal. The following theorems can be proved

$\vdash \forall f\ g.\ ((\text{Domain } f = \text{Domain } g) \wedge (\text{Apply } f = \text{Apply } g))\ =\ (f = g)$

$\vdash \forall f\ g.$
    $((\text{Domain } f = \text{Domain } g) \wedge (\forall x.\ \text{Domain } f\ x \supset (\text{Apply } f = \text{Apply } g))$
      $=\ (f = g)$

# 3   The logical definition of finite maps

## 3.1   Possible representations

Having decided on the characteristic axioms we now must supply a model from which these can be derived. From the view of a programmer the obvious choice is a list of pairs. This has practical merits since lists are well supported in HOL. But we soon run into difficulties if we take this path. Consider the two lists $[(x, int), (x, int)]$ and $[(x, int)]$. These are clearly equal when considered as the finite maps mapping $x$ to $int$, but are not logically equal lists.

In the theory for HOL-ML [4, 11] this problem was overcome by defining the update function so that only ordered lists with every element appearing at most once in the domain can be constructed. Thus the lists representing two equal finite maps will also be equal. The disadvantage here is that an ordering over the domain of the map is needed, an ordering that should not be needed.

A variation is to define an equivalence relation relating all lists that are equal when considered as finite maps and then define the type of finite maps to be the quotient of lists with this relation. Each element in the defined type will be represented by an equivalence class of lists generated by the defined equivalence relation.

Another possible representation is sets of pairs. This solves some of the problems indicated above but forces us to ensure that no two elements of the set have the same first element. In set theory functions are represented as sets of pairs. This is a good representation of functions in terms of the fundamental object, namely sets. In HOL functions are the fundamental object and therefore sets offer us no advantage. The representation discussed in the next section therefore uses functions instead. This also avoids the need to restrict the theory to some particular set library.

## 3.2   The representation used

The representation used is a function from the type of the domain, $\alpha$, to the type $\beta + one$. This function maps an element to one if it is not in the domain of the map and to the image of the element if it is in the domain.

What remains is to define a notion of finiteness for functions of this type. A predicate is_fmap can be defined inductively by the following rules:

$$\frac{}{\text{is\_fmap } (\lambda a.\ (\text{InR one}))}$$

$$\frac{\text{is\_fmap } f}{\text{is\_fmap } (\lambda x.\ (x = a) \Rightarrow \text{InL } b \mid f\ x)}$$

This gives rise to an induction principle that expresses the finiteness of the functions for which is_fmap holds.

$$\vdash \forall P.$$
$$P \ (\lambda x. \ \mathsf{InR} \ \mathsf{one}) \ \wedge$$
$$(\forall f. \ P \ f \ \supset \ (\forall a \ b. \ P(\lambda x. \ (x = a) \Rightarrow (\mathsf{InL} \ b) \mid (f \ x))))$$
$$\supset$$
$$(\forall f. \ \mathsf{is\_fmap} \ f \ \supset \ P \ f)$$

The type $(\alpha, \beta)fmap$ can then be defined to be the set of functions of type $\alpha \rightarrow (\beta + one)$ for which is_fmap holds. The witness that this new type is non-empty is the function $\lambda x. \ \mathsf{InR} \ \mathsf{one}$, which represents the empty finite map. A bijection between the the type $(\alpha, \beta)fmap$ and the representation is defined by the functions fmap_ABS, of type $(\alpha \rightarrow (\beta + one)) \rightarrow (\alpha, \beta)fmap$ and fmap_REP, of type $(\alpha, \beta)fmap \rightarrow (\alpha \rightarrow (\beta + one))$. This process of defining a new type and the bijection is described in [1].

The constants Empty, Update, Apply, and Domain can be defined in terms of the representation as follows:

$$\vdash \mathsf{Empty} = \mathsf{fmap\_ABS} \ (\lambda a. \ \mathsf{InR} \ \mathsf{one})$$

$$\vdash \mathsf{Update} \ f \ (a, b) = \mathsf{fmap\_ABS} \ (\lambda x. \ (x = a) \Rightarrow \mathsf{InL} \ b \mid (\mathsf{fmap\_REP} \ f) \ x)$$

$$\vdash \mathsf{Apply} \ f \ x = \mathsf{OutL} \ ((\mathsf{fmap\_REP} \ f) \ x)$$

$$\vdash \mathsf{Domain} \ f \ x = \mathsf{IsL} \ ((\mathsf{fmap\_REP} \ f) \ x)$$

From these definitions we can derive all the axioms given in section 2. Having done so the representation is not used again; all other theorems in the theory can be proved from just these seven axioms.

The derivation of the characteristic axioms for each constant is straightforward. The axioms can be proved easily at the representation level and then "lifted" to the abstract level. The derivation for the stronger induction principle requires an induction over the size of the domain. This proof involves formalising the concept of the size of the domain. The HOL implementation of these proofs currently employs a set library but this dependency on sets will be removed for the final finite maps library.

## 3.3 Consistence, independence and completeness

The axioms listed in section 2 are consistent because they are derivable from the model just described. Completeness of the axioms with respect to the model can be shown by assuming that the axioms hold and showing that the type specified by the axioms is isomorphic to the model. We show that the function rep with the defining property

$$\vdash \forall f : (\alpha, \beta)fmap. \ \mathsf{rep} \ f \ = \ \lambda x. \ (\mathsf{Domain} \ f \ x \ \Rightarrow \ \mathsf{InL} \ (\mathsf{Apply} \ f \ x) \mid \mathsf{InR} \ \mathsf{one})$$

is a bijection from the type $(\alpha, \beta)fmap$ to the subset of the type $\alpha \rightarrow (\beta + one)$ satisfying the predicate is_fmap, using only the axioms and not the underlying model. This gives us a means by which to reconstruct the model from the axioms.

This function rep is onto and one to one:

$$\vdash \forall f : \alpha \to (\beta + one). \text{ is\_fmap } f \;\supset\; (\exists g. \text{ rep } g = f)$$
$$\vdash \forall (f : (\alpha, \beta)\text{fmap}) (g : (\alpha, \beta)\text{fmap}). (\text{rep } f = \text{rep } g) \;\supset\; (f = g)$$

and its image is contained in the subset of $\alpha \to (\beta + one)$ defined by is\_fmap:

$$\vdash \forall f : (\alpha, \beta)\text{fmap}. \text{ is\_fmap } (\text{rep } f)$$

This is in effect an redefinition of the function fmap\_REP using only the axioms and not referring to either fmap\_ABS or fmap\_REP.

We believe that the axioms are also independent but have not attempted a formal proof. That is, we have not shown that a model can be found for each possible set of axioms with one axiom replaced by its negation. While still important for theoretical reasons, this property is not as important in practice; it does not affect either what can be proved or the consistency of the system.

## 4    Enriching the theory

The seven theorems that characterise finite maps are sufficient to build a rich and useful theory. The most important theorems that can be proved are those characterising equality, as discussed above. Many more theorems can also be proved about the basic constants, but this section concentrates on how the theory can be extended with new concepts built up from the seven axioms.

The only method introduced so far for constructing finite maps is Update. In practice, functions are needed to update finite maps by extending them with other finite maps and to allow the domain over which a finite map is defined to be reduced.

The constant Extend is defined so that

$$\text{Apply (Extend } f\ g)\ x \;=\; \begin{cases} \text{Apply } f\ x \text{ if Domain } f\ x \\ \text{Apply } g\ x \text{ otherwise} \end{cases}$$

Formally, the defining property of Extend is:

$\vdash \forall f\ g.$
    $(\forall x. \text{ Domain (Extend } f\ g)\ x = \text{Domain } f\ x \lor \text{Domain } g\ x) \land$
    $(\forall x. \text{ Apply (Extend } f\ g)\ x = ((\text{Domain } f\ x) \Rightarrow (\text{Apply } f\ x) \mid (\text{Apply } g\ x)))$

This definition is made by first proving, by a straightforward induction over $f$, that a function with this property exists and then using the principle of constant specification to define Extend. More useful theorems about Extend can be derived from this definition. Some examples are

$$\vdash \forall g.\ \text{Extend Empty } g\ =\ g$$

$$\vdash \forall f.\ \text{Extend } f \text{ Empty}\ =\ f$$

$$\vdash \forall f\ g\ x\ y.\ \text{Extend (Update } f\ (x,y))\ g\ =\ \text{Update (Extend } f\ g)\ (x,y)$$

$$\vdash \forall f\ g\ x\ y.\ \text{Extend } f\ (\text{Update } g\ (x,y))\ =$$
$$((\text{Domain } f\ x)\ \Rightarrow\ (\text{Extend } f\ g)\ |\ (\text{Update (Extend } f\ g)\ (x,y)))$$

$$\vdash \forall f\ g\ x.\ \text{Domain (Extend } f\ g)\ x\ =\ \text{Domain } f\ x\ \vee\ \text{Domain } g\ x$$

These results all follow by simple proofs using the basic axioms and the definition of Extend.

All the constants discussed above either increase or preserve the domain of a finite map. A constant DRestrict can be defined which reduces the domain of a finite map to those elements satisfying some predicate. DRestrict is again defined by proving the existence of a function with the appropriate properties and then using constant specification. The function is characterised by the theorem

$$\vdash \forall f\ p.$$
$$(\forall x.\ \text{Domain (DRestrict } f\ p)\ x\ =\ \text{Domain } f\ x\ \wedge\ p\ x)\ \wedge$$
$$(\forall x.\ \text{Domain } f\ x\ \wedge\ p\ x\ \supset\ (\text{Apply (DRestrict } f\ p)\ x\ =\ \text{Apply } f\ x))$$

Some useful properties that can be proved of DRestrict are

$$\vdash \forall p.\ \text{Restrict Empty } p\ =\ \text{Empty}$$

$$\vdash \forall f\ p\ a\ b.$$
$$\text{DRestrict (Update } f\ (a,b))\ p\ =$$
$$((p\ a)\ \Rightarrow\ (\text{Update (DRestrict } f\ p)\ (a,b))\ |\ (\text{DRestrict } f\ p))$$

$$\vdash \forall f\ p\ q.\ \text{DRestrict } f\ (\lambda x.\ p\ x\ \vee\ q\ x)\ =\ \text{Extend (DRestrict } f\ p)\ (\text{DRestrict } f\ q)$$

The proofs of these theorems are again straightforward. The function Delete, to remove a single element from a finite map, can be defined in terms of DRestrict.

A related concept to the domain of a finite map is the range. An element is in the range if there is some element in the domain which is mapped to it. The function Range and RRestrict are defined with the same functionality as Domain and DRestrict but relating to the the range rather the domain.

Another important concept is composition, either of two finite maps or a finite map and a function. Three infix composition functions are defined:

$$f\_o\_f\ :\ (\beta,\gamma)\text{fmap} \rightarrow (\alpha,\beta)\text{fmap} \rightarrow (\alpha,\gamma)\text{fmap}$$
$$o\_f\ \ :\ (\beta \rightarrow \gamma) \rightarrow (\alpha,\beta)\text{fmap} \rightarrow (\alpha,\gamma)\text{fmap}$$
$$f\_o\ \ :\ (\beta,\gamma)\text{fmap} \rightarrow (\alpha \rightarrow \beta) \rightarrow (\alpha,\gamma)\text{fmap}$$

The notation is designed to show the link with composition of functions

$$o\ :\ (\beta \rightarrow \gamma) \rightarrow (\alpha \rightarrow \beta) \rightarrow (\alpha \rightarrow \gamma)$$

Two of the other functions defined are Submap, a mapping that is defined on a subset of the domain of another map but maps the elements for which it is defined to the same values and EveryMap, a function that takes a predicate and tests whether it holds of every pair of (domain,range) elements inserted into the finite map.

# 5   Finite maps and recursive types

In this section we consider the problem of defining recursive types that include finite maps. It is a well known difficulty with the HOL system that types such as

$$Val = \text{CONST} \mid \text{RECORD } num \rightarrow Val \qquad (1)$$

may not be introduced into the system easily by automated or manual techniques. This is because the presence of the function $num \rightarrow Val$ results in a type that is too large for the trees used in the system's automatic type definition package [6]. The user must therefore carry out a tiresome type construction manually. A possible solution is described in [3].

A related problem is that of defining types of the form

$$Val = \text{CONST} \mid \text{RECORD } (num, Val) fmap \qquad (2)$$

Types such as this are often used to represent expression values in programming language formalisations such as the definition of Standard ML [8]. This section describes how types of this form may be introduced into HOL manually, without needing a solution to the more general problem (1) above.

The method we will use is as follows. First, we shall define a mapping between the type $(\alpha, \beta) fmap$ and a subset of the type $(\alpha, \beta) list$. We then show that this mapping gives a unique list for each finite map, which we call its *canonical representation*. Next we manually introduce the concrete type

$$ListVal = \text{List\_CONST} \mid \text{List\_RECORD } (num \times ListVal) list \qquad (3)$$

into HOL. We use a subset of this type as the representation for type (2). Finally we develop the necessary theorems which characterise type (2) independently of its representation.

## 5.1   Canonical representations for finite maps

We introduce two operators FFst and FRest which decompose a finite map into a single element and a remainder. This does not need an ordering of elements in the finite map, since we appeal to the HOL choice operator.

$$\vdash \forall f.$$
$$\text{FFst } f = (\varepsilon p. \text{ Domain } f \text{ (FST } p) \wedge (\text{SND } p = \text{Apply } f \text{ (FST } p))))$$
$$\vdash \forall f. \text{ FRest } f = \text{ Delete (FST (FFst } f)) f$$

A relation Canon_Rel between finite maps and paired lists can now be defined by primitive recursion on lists:

$$\vdash (\forall f. \; \mathsf{Canon\_Rel} \; f \; [\,] \; = \; (f = \mathsf{Empty})) \; \wedge$$
$$(\forall f \; h \; t.$$
$$\quad \mathsf{Canon\_Rel} \; f \; (\mathsf{Cons} \; h \; t) \; =$$
$$\quad \neg(f \; = \; \mathsf{Empty}) \; \wedge \; (h \; = \; \mathsf{FFst} \; f) \; \wedge \; \mathsf{Canon\_Rel} \; (\mathsf{FRest} \; f) \; t)$$

Intuitively this definition ensures that $f$ is related to $l$ if and only if $l$ has the form [FFst $f$, FFst (FRest $f$), FFst (FRest (FRest $f$)), ...]. Thus precisely one list is defined for each finite map. It is a lengthy process, but fairly straightforward, to prove that Canon_Rel defines a unique list for every map.

Based on Canon_Rel, two functions Canon_of_Fmap and Fmap_of_Canon can then be defined, giving an isomorphism between finite maps and their canonical representations.

$$\vdash (\forall f. \; \mathsf{Fmap\_of\_Canon} \; (\mathsf{Canon\_of\_Fmap} \; f) \; = \; f) \; \wedge$$
$$(\forall l.$$
$$\quad \mathsf{Canon\_Rel} \; (\mathsf{Fmap\_of\_Canon} \; l) \; l \; \supset$$
$$\quad (\mathsf{Canon\_of\_Fmap} \; (\mathsf{Fmap\_of\_Canon} \; l) \; = \; l))$$

## 5.2 Introducing the type *ListVal*

The type

$$ListVal = \mathsf{List\_CONST} \; | \; \mathsf{List\_RECORD} \; (num \times ListVal)list \qquad (4)$$

can be introduced manually using a technique similar to that for the type

$$data = \mathsf{List\_CONST} \; | \; \mathsf{List\_RECORD} \; (data)list \qquad (5)$$

The manual method for doing this was described on the info-hol mailing list [5]. The only real complication is that we are defining a type where a recursive reference to the type occurs nested within a product type on the right-hand side. We derive the following characteristic theorem for the type, which states that a unique function exists for every primitive recursive specification over the type:

$$\vdash \forall e \; f. \; \exists! \; fn.$$
$$(fn \; \mathsf{List\_CONST} \; = \; e) \; \wedge$$
$$(\forall l. \; fn \; (\mathsf{List\_RECORD} \; l) \; = \; f \; (\mathsf{MAP} \; (fn \circ \mathsf{SND}) \; l) \; l)$$

## 5.3 Introducing the type *Val*

A subset of *ListVal*, isomorphic to our required type *Val*, is defined by a predicate Is_Val introduced by a primitive recursive definition:

$$\vdash \mathsf{Is\_Val} \; \mathsf{List\_CONST} \; \wedge$$
$$(\forall l. \; \mathsf{Is\_Val} \; (\mathsf{List\_RECORD} \; l) \; =$$
$$\quad (\exists f. \; \mathsf{Canon\_Rel} \; f \; l) \; \wedge \; \mathsf{ALL\_EL} \; (\mathsf{Is\_Val} \circ \mathsf{SND}) \; l))$$

where ALL_EL tests if a predicate is true of every element of a list. This definition ensures that each list within a *ListVal* value is a canonical representation of some finite map.

The new type *Val* is now introduced based on the subset defined by Is_Val. The constructors CONST and RECORD are defined in terms of List_CONST and List_RECORD . The characteristic theorem for the type is derived from the characteristic theorem for *ListVal*. The derivation is a lengthy forward proof, and relies on properties of FFst, FRest, EveryMap and the functions Canon_of_FMap and FMap_of_Canon. The resulting theorem is:

$$\vdash \forall e \; f. \; \exists! \; fn.$$
$$(fn \; \mathsf{CONST} = e) \land$$
$$(\forall fmap. \; fn \; (\mathsf{RECORD} \; fmap) = f \; (fn \; \mathsf{o\_f} \; fmap) \; fmap)$$

where o_f composes a function with a finite map. This gives a full characterisation of the type *Val*. As with other HOL recursive types, an induction principle for *Val* may be derived from this theorem, along with theorems proving that the constructors CONST and RECORD are one to one and distinct.

# 6 Decision procedures

HOL theories typically come with a set of tools for reasoning about the constructs defined in them. This section discusses some decision procedures for finite maps, and in particular a conversion for determining the result of applying a finite map to an element. The decision procedures for finite maps fall into two categories; those which simplify terms as far as possible and return a single theorem capturing the result of this simplification; and those which will perform case splits and make additional assumptions to return more information. We first discuss an example of the former.

## 6.1 Simplifying terms

Function evaluation conversions can be introduced for all the constants defined in the theory. For example, the conversion Extend_CONV : *conv* simplifies the addition of finite maps as far as possible. For the finite map,

$$\mathsf{Extend} \; (\mathsf{Update} \; \mathsf{Empty} \; (1, \mathsf{T})) \; (\mathsf{Update} \; \mathsf{Empty} \; (2, \mathsf{F}))$$

Extend_CONV will return the theorem

$$\vdash \mathsf{Extend} \; (\mathsf{Update} \; \mathsf{Empty} \; (1, \mathsf{T})) \; (\mathsf{Update} \; \mathsf{Empty} \; (2, \mathsf{F})) =$$
$$\mathsf{Update} \; (\mathsf{Update} \; \mathsf{Empty} \; (2, \mathsf{F})) \; (1, \mathsf{T})$$

Note that in this case the conversion has been able to remove the Extend operator completely. The conversion Restrict_CONV : *conv* $\rightarrow$ *conv* simplifies terms of the form Restrict $f \; p$ as far as possible. The first argument must be a conversion to evaluate applications of the restriction function $p$ to individual elements of the domain of $f$.

## 6.2 A reducer for Apply

The most common decision procedure needed for finite maps is one to determine the result of applying a finite map to an element. A similar conversion computes whether an element is in the domain of a map.

Finite maps are frequently used in formal descriptions of programming languages, and it is common to write *symbolic evaluators* for these languages once they have been embedded into HOL. Such a symbolic evaluator was constructed for the Standard ML Core language in [10]. When writing a symbolic evaluator it is useful to have an application reducer capable of handling applications of finite maps containing HOL variables, as in the case Apply (Update $E$ $(x, 200)$) $y$. This lookup may arise if the evaluator were reducing a program expression such as "let x = 200 in y" in an arbitrary variable environment $E$. It is useful if the symbolic evaluator can make the necessary assumption that the variable $y$ has some value in this environment, which can later be proved by type inference or some other method. We rarely want the symbolic evaluator to fail just because it cannot determine exactly the result of applying an arbitrary finite map to an element.

The conversion described in this section does not halt when it cannot determine the result of an application. Instead it makes additional assumptions in order to compute a result. These are accumulated in the assumption list of the returned theorem. Sometimes the most important use of these assumptions is to help the users of the conversion find mistakes in their input.

The apply reduction conversion, Apply_CONV, has type *conv* → *convl*. The type *convl* is a function from a term to a list of theorems corresponding to the results of evaluation under different assumptions. [3] The first argument should be a conversion that decides equality between members of the domain of the finite map. For example, applying Apply_CONV to num_EQ_CONV and the term

$$\text{Apply (Update Empty } (2, \text{F))} \ x$$

returns the theorems

$$(2 = x) \vdash \text{Apply (Update Empty } (2, \text{F))}x = \text{F}$$
$$(2 \neq x) \vdash \text{Apply (Update Empty } (2, \text{F))}x = \text{Apply Empty } x$$

In this example the equality conversion num_EQ_CONV has been unable to determine whether $2 = x$. Thus, Apply_CONV has returned two theorems with different assumptions.

Applying Apply_CONV to num_EQ_CONV and the term

$$\text{Apply (Update (Update } (E : (num, bool)fmap) \ (x, y)) \ (1, \text{T))} \ 3$$

returns the theorems

$$(x = 3) \vdash \text{Apply (Update (Update } E \ (x, y)) \ (1, \text{T))} \ 3 \ = y$$
$$(x \neq 3) \vdash \text{Apply (Update (Update } E \ (x, y)) \ (1, \text{T))} \ 3 \ = \text{Apply } E \ 3$$

---

[3] In the current implementation a lazy list or sequence is used. since the function could potentially return a large list of theorems.

Here, the arbitrary finite map $E$ extended with an arbitrary pair $(x, y)$ and the pair $(1, \mathsf{T})$ has been applied to 3. Apply_CONV has reduced the application as far as possible, making assumptions about whether or not $x = 3$. In practice Apply_CONV would be used in conjunction with the conversion Domain_CONV.

Another kind of term we would like to be able to reduce are those of the form Apply (Extend $f$ $g$) $x$. Adding two variable environments together is common in formal programming language descriptions, and hence the construct Extend $f$ $g$ will often arise. The conversion Apply_CONV is able to reduce applications of this form also, even if the Extend operators are nested arbitrarily.

# 7 An example

Much of the motivation for this work has come from research on embedding the semantics of programming languages in HOL. In this section we give an example of using finite maps to reason about a small language (a simply-typed $\lambda$-calculus). Finite maps can be used in two important places here. The first is in the type system, where a finite map can be used to store the context in which a typing judgement holds. Here the finite map used is a mapping from identifiers to types. Strictly speaking, there is no need for the mappings used here to be finite; for our purposes an infinite map would suffice. But there is also no need to allow infinite maps, and the restriction to finite maps provides an induction principle that is useful in proofs.

The second place where finite maps are useful is in the definition of the evaluation relation. Here substitution functions or environments mapping identifiers to expressions can be represented by finite maps.

For the purpose of this example we concentrate on the type system.

## 7.1 A small language

Our aim is to construct a small language that includes variable binding. We define a type, *ty*, to be either an atomic type or a function type:

$$ty \quad ::= \quad \mathsf{Atom}\ string$$
$$| \quad ty \rightarrow ty$$

and an expression, *exp*, to be either an identifier, function abstraction, or function application:

$$exp \quad ::= \quad \mathsf{Id}\ string$$
$$| \quad \mathsf{Lambda}\ string\ ty\ exp$$
$$| \quad \mathsf{App}\ exp\ exp$$

The typing rules are defined as a relation of the form Type $C$ $e$ $t$ where Type has type $(string, ty)fmap \rightarrow exp \rightarrow ty \rightarrow bool$. This denotes true if the expression $e$ has type $t$ in the context $C$. The rules for this relation are:

$$\overline{\mathsf{Type}\ (\mathsf{Update}\ C\ (v, t))\ (\mathsf{Id}\ v)\ t}$$

$$\frac{\text{Type} \quad (\text{Update } C \ (y, t_1)) \ e \ t_2}{\text{Type} \quad C \quad (\text{Lambda } y \ t_1 \ e) \ t_1 \rightarrow t_2}$$

$$\frac{\text{Type} \quad C \ e_1 \ (t_1 \rightarrow t_2) \qquad \text{Type} \quad C \ e_2 \ t_1}{\text{Type} \quad C \ (\text{App } e_1 \ e_2) \ t_2}$$

An important point to note about these rules is that the expression **Update** $C \ (v, t)$ is used to denote any finite map that maps $v$ to $t$. The use of this expression does not imply that $(v, t)$ must be the "last update" used to build the type context. This works in our theory because any finite map that maps $v$ to $t$ is equal to a finite map in which the last update was $(v, t)$. This gives another illustration of the usefulness of a theory in which equality does not depend on the order of the updates.

## 7.2   Context extension

This section presents some theorems about how the context of a valid typing judgement can be extended. In semantics this is referred to as weakening the context, as we are adding surplus information.

The first theorem shows that the context can be extended by any mapping from an element not already present in the domain.

$$\vdash \forall C \ e \ t.$$
$$\text{Type } C \ e \ t \ \supset$$
$$(\forall x. \ \neg(\text{Domain } C \ x) \ \supset \ (\forall y. \ \text{Type}(\text{Update } C \ (x, y)) \ e \ t))$$

This is proved by an induction over the rules for the **Type** relation. The proof makes use of several of the theorems about finite maps, including the theorem that asserts the equality of maps with two elements inserted in a different order provided they are updating different elements of the domain.

Further theorems, which can be proved by a simple induction over one of the contexts, show under what conditions extending the context with another context will preserve typing judgements. The simplest such theorem is

$$\vdash \forall C \ e \ t. \ \text{Type } C \ e \ t \ \supset \ \forall C'. \text{Type } (\text{Extend } C \ C') \ e \ t$$

This says the context can be weakened by extension with any context and the type judgement will still be preserved.

## 7.3   Restriction of the context

An important and practical theorem about this language is that type judgements are preserved by restricting the context to the free variables in the expression being typed. A function **Fv** can be defined to test if a variable is free in an expression

$$\vdash (\forall i \; x. \; \mathsf{Fv} \; (\mathsf{Id} \; i) \; x \; = \; i = x) \; \wedge$$
$$(\forall y \; t \; e \; x. \; \mathsf{Fv} \; (\mathsf{Lambda} \; y \; t \; e) \; x \; = \; (y \neq x) \wedge \mathsf{Fv} \; e \; x) \; \wedge$$
$$(\forall e_1 \; e_2 \; x. \; \mathsf{Fv} \; (\mathsf{App} \; e_1 \; e_2) \; x \; = \; \mathsf{Fv} \; e_1 \; x \; \vee \; \mathsf{Fv} \; e_2 \; x)$$

The theorem that can then be proved is

$$\vdash \forall C \; e \; t. \; \mathsf{Type} \; C \; e \; t = \mathsf{Type} \; (\mathsf{DRestrict} \; C \; (\mathsf{Fv} \; e)) \; e \; t$$

This theorem follows by using the two lemmas below:

$$\vdash \forall C \; e \; t. \; \mathsf{Type} \; (\mathsf{DRestrict} \; C \; (\mathsf{Fv} \; e)) \; e \; t \; \supset \; \mathsf{Type} \; C \; e \; t$$

$$\vdash \forall C \; e \; t. \; \mathsf{Type} \; C \; e \; t \; \supset \; \mathsf{Type} \; (\mathsf{DRestrict} \; C \; (\mathsf{Fv} \; e)) \; e \; t$$

The first of these follows by observing that the expression $\mathsf{Fv} \; e \; x \; \vee \; \neg(\mathsf{Fv} \; e \; x)$ is true for any $e$ and $x$. This is used to show that

$$\mathsf{Type} \; C \; e \; t$$
$$= \; \mathsf{Type} \; (\mathsf{DRestrict} \; C \; ((\lambda x. \; (\mathsf{Fv} \; e \; x) \; \vee \; ((\lambda y. \; \neg(\mathsf{Fv} \; e \; y)) \; x))) \; e \; t$$
$$= \; \mathsf{Type} \; (\mathsf{Extend} \; (\mathsf{DRestrict} \; C \; (\mathsf{Fv} \; e) \; (\mathsf{DRestrict} \; C \; (\lambda y. \; \neg(\mathsf{Fv} \; e \; y)) \; e \; t$$

The result then follows from the theorem

$$\vdash \forall C \; e \; t. \; \mathsf{Type} \; C \; e \; t \; \supset \; \forall C' \; \mathsf{Type} \; (\mathsf{Extend} \; C \; C') \; e \; t$$

The other implication requires an induction over the rules for the relation Type. This decomposes the goal into three subgoals, each of which can be solved by manipulation of the contexts similar to that used above.

Both the free variables in a term and the restriction of a context can be computed easily using simple conversions. The last theorem then provides a means to reduce the problem of type checking in a large context.

The importance of this simple example is that much of the task of formalising and proving these results has been removed by the use of the finite maps library and the concepts and theorems developed there.

## 8 Conclusions

Any theory based on a set of axioms must satisfy two essential properties. First, it must be consistent, a fact guaranteed here by the derivation of the axioms from a representation in terms of functions. The axioms should also be complete with respect to the model. This has been proved and so the equivalent of any property provable in the model will be provable from our axioms.

The principal motivation for this work was the need for a practical tool to aid in the development, within HOL, of semantics for programming languages. The theory makes the task of meta-reasoning, such as that in the example, significantly easier than it would be otherwise. Enough theorems have been proved to allow conversions to be written to reduce a variety of expressions involving finite maps to simpler forms. This has practical benefits when building systems like partial evaluators or type checkers using HOL.

One reason that others have used lists to represent finite maps was the ability to define recursive types such as those discussed in section 5. We have shown here that such types can also be defined with the finite maps presented here, although this is not automated. The derivation of this type justifies the axiomatisation of a similar type used in [10].

## Acknowledgements

Thanks are due to Tom Melham for advice on all aspects of this work, from the theory to the presentation style. Thanks must also go to the Engineering and Physical Sciences Research Council for financial support of the first author.

## References

1. M. J. C. Gordon and T. F. Melham, editors. *Introduction to HOL: A theorem proving environment for higher order logic.* Cambridge University Press, 1993.
2. M.J.C. Gordon. Merging HOL with Set Theory: preliminary experiments. Technical Report 353, University of Cambridge Computer Laboratory, 1994.
3. Elsa Gunter. A Broader Class of Trees for Recursive Type Definitions for HOL. In J. J. Joyce and C. J. H. Seger, editors, *Higher Order Logic Theorem Proving and its Applications*, volume 780 of *Lecture Notes in Computer Science*, pages 141–154. Springer-Verlag, 1993.
4. Savi Maharaj and Elsa Gunter. Studying the ML Module System in HOL. In Tom Melham and Juanito Camilleri, editors, *Higher Order Logic Theorem Proving and its Applications*, volume 859 of *Lecture Notes in Computer Science*. Springer-Verlag, September 1994.
5. Tom F. Melham. Recursive Data Types. Message on *info-hol* mailing list, 9th November 1991.
6. Tom F. Melham. Automating Recursive Type Definitions in HOL. In Graham Birtwistle and P. A. Subrahmanyam, editors, *Current Trends in Hardware Verification and Automated Theorem Proving*, pages 341–386. Springer-Verlag, 1989.
7. Tom F. Melham. A Package for Inductive Relation Definitions in HOL. In M. Archer, J J Joyce, K N Levitt, and P J Windley, editors, *Proceedings of the 1991 International Workshop on the HOL Theorem Proving System and its Applications, Davis, August 1992*, pages 350–357. IEEE Computer Society Press, 1992.
8. Robin Milner, Mads Tofte, and Robert Harper. *The Definition of Standard ML.* The MIT Press, 1990.
9. Donald Syme. Reasoning with the Formal Definition of Standard ML in HOL. In *Higher Order Logic Theorem Proving and Its Applications*, volume 780 of *Lecture Notes in Computer Science*, pages 43–60. Springer-Verlag.
10. Donald Syme. Supporting Formal Reasoning about Standard ML. Honours Thesis, Australian National University, 1992.
11. Myra VanInwegen and Elsa Gunter. HOL-ML. In J. J. Joyce and C. J. H. Seger, editors, *Higher Order Logic Theorem Proving and its Applications*, volume 780 of *Lecture Notes in Computer Science*, pages 61–74. Springer-Verlag, 1993.

# Virtual Theories

Paul Curzon

University of Cambridge, Computer Laboratory
New Museums Site, Pembroke Street,
Cambridge. CB2 3QG. United Kingdom
Email: pc@cl.cam.ac.uk

**Abstract.** Proof is a programming activity. Consequently programming
environments which support proof in the large are required. We describe
an environment which supports one area of proof-in-the-large: that of
theory management. We present the notion of virtual theories. They give
the illusion of multiple active theories allowing the user to switch between
different theories at will, proving theorems and making definitions in
each. The system ensures that proofs only use resources that are available
in the environment of the current virtual theory. The code has been
implemented on top of the HOL90 system. A side effect is that a version
of autoloading is obtained for HOL90. A more radical feature that is
obtained is the autoloading of tools. The system has been tested on part
of a real hardware verification proof.

> *Who controls the past controls the future,*
> *Who controls the present controls the past.*
> George Orwell,
> Nineteen Eighty-Four

## 1 Introduction

Interactive, machine-checked proof is essentially a programming activity. Proofs
are programs in the meta-language of the theorem prover. Commands correspond
to calls to proof tools. Thus many of the problems of managing large program
developments apply to the development of proofs. Indeed the problems can be
enhanced: with current state-of-the-art theorem proving technology, proofs are
much longer than the system they verify. Most theorem proving systems provide some way of structuring proofs in the large. For example the HOL system
provides a mechanism for grouping theorems about related definitions or constants into separate theories. This corresponds roughly to a module system of a
programming language. As with programming, language features for structuring
proofs are not the whole solution: programming environments must be provided
to support the development process. Such a proof environment must supply tools
to support many different activities. Typical systems give support for proof in
the small. For example, they manage the proof obligations generated preventing
them from being neglected and ensure that the proof programmer cannot cor-
rupt the soundness of the system. They may also provide a means for working

on several theorems at once: which is useful if a new lemma must be proved in the middle of some other proof. Environments that support proof in the large are also needed. In particular, similar tools for managing theories are needed. We describe a tool for managing one such aspect of theory development, that of switching between different theories in the middle of the proof effort. It is described in the context of the HOL system. This work was motivated by problems which arose on a real, large hardware verification project. The lack of proof support of this kind was severely hampering the proof effort.

## 2   Multiple Active Theories

When working on large proof efforts, it is often desirable to work on several theories at once, creating new theories and switching between old ones at will. For example, theorems about basic datatypes such as lists are often proved as lemmas to the main results of a theory. Ideally, such lemmas should be placed in a more appropriate theory. The current provisions of HOL (extend_theory, close_theory, etc) are at best clumsy to use. They essentially only allow theories to be developed in a linear fashion. For large proofs this is insufficient. As a consequence, either theorems and definitions are left in inappropriate theories, or a time consuming effort is required to tidy up the theory hierarchy after the proof is finished.

A common way to use HOL is to type the proof commands into an emacs file to maintain a record of the proof, then cut and paste these commands into a HOL session. One way to overcome the above problems would therefore be to record the proofs of lemmas directly into the source file of the theory they will ultimately reside in. The whole theory hierarchy can then be rebuilt in batch mode overnight so that it is in a consistent state for the next days session. The problem with this approach is that it is very difficult to keep track of the context (essentially the ML environment) in which each theorem was developed. This is necessary to ensure the appropriate context is recreated in the different source files. It is easy to use resources inadvertently that were believed only to be used by other theories. This leads to a great deal of time being wasted fixing all the problems that arise when the source files are replayed. If the theory hierarchy requires several hours to rebuild (which is not uncommon) this is not practical.

In this paper we present a virtual theory mechanism which was developed to overcome these problems. The aim was to provide simple automated help that would enhance our current way of working, without changing the core HOL system. In particular, we wanted to keep track of the dependencies of each separate theory so that the above problems would not be encountered. We wanted to provide a simple practical way of combining new proofs with old theories. The system was intended to ensure that no necessary context would be missed from the source files.

The system and corresponding methodology was intended to obey three main design criteria.

1. It should allow multiple theories to be developed in one session.

2. It should be possible to later rebuild each theory independently from its source file, with no additional user intervention.

3. Extending a theory in a new session should be no different from extending it in the original session in which it was created.

The last criteria is not met by the current HOL system because normally to extend a theory at least part of the ML state used originally must be rebuilt: loading tools, theorems, etc.

# 3  Virtual Theories

Virtual theories are just views of normal theories that can coexist and be extended within a single session. A virtual theory is created for each separate theory that is to be modified in a session (whether it previously exists or not). It is represented by a datastructure recording the resources available to the theory. Only a single theory is active at any time, but the user can view and switch between different ones at will. Resources are accessed through this datastructure using *censor* functions, rather than being stored in ML variables.

The use of the virtual theory system does not significantly alter the way HOL is used. As theorems and definitions are added, the user creates and switches between (virtual) theories as appropriate. The source commands that allow the theories to be recreated are stored in files specific to those theories.

The system does not modify the real theory files (ie, the permanent record of the theories) during the session. This must be done later by reloading the source files in batch mode using an appropriate makefile. This ensures that the source files are consistent before a real theory is created. Because of the general nature of the ML programming language (the meta-language of HOL) the consistency of the system cannot otherwise be guaranteed: theorems could be leaked between theories [2]. The virtual theory system forces the user to repeat commands if they must appear in more than one source file. It is assumed that the user is disciplined in ensuring that the source files do contain an accurate record of the session. This does not impose any new burden. Furthermore, user interfaces are now being developed that could maintain the source files automatically, so this is not a problem in the long term. To prevent the user from getting confused over which is the current virtual theory (and thus storing resources to the wrong place) most commands take the current virtual theory name as an argument, and fail if it is incorrect.

## 3.1  An Example Session

As an illustration, we give below a session to develop a theory "wombat" about constant WOMBAT. In the middle of the session an extension to an existing theory "mylist" is made. The use of virtual theories ensures that a clear separation of the two theories is made.

On starting a session, we first create the initial virtual theory we will work in. A dummy real theory will also be created automatically. We cut and paste these commands from a source file for theory "wombat".

```
new_vtheory "wombat";
```

We wish the new theory "wombat" to depend on a previous real theory "mylist", so we make the latter theory a virtual parent. This makes the basic list operators and theorems in that theory available.

```
new_vparent "wombat" "mylist";
```

We then make a definition in virtual theory "wombat". It will ultimately be stored in the real theory, "wombat".

```
new_vdefinition("wombat","WOMBAT", (--'WOMBAT l1 l2 = SUM l1 + SUM l2'--));
```

Having made the definition we can prove theorems about it, accessing theorems from ancestor theories using censor THM.

```
vstore_thm("wombat","WOMBAT_CONS",
      (--'WOMBAT (CONS x l1) l2 = x + SUM l1 + SUM l2'--),
      REWRITE_TAC[THM "WOMBAT",THM "SUM"]);
```

For the main theorem we realize we need a new definition and lemma about lists, which we decide ought really be placed in theory "mylist". We therefore open a virtual theory for the existing real theory "mylist" to allow us to extend it.

```
extend_vtheory "mylist";
```

At this point the definition, new constant and theorem in theory "wombat" are no longer visible. They cannot inadvertently be used in definitions or proofs of theorems within the theory "mylist".

We make the new definition and prove the theorem. Commands are now cut and pasted to and from the source file of theory "mylist".

```
new_vdefinition("mylist","SAPP",
      (--'SAPP l1 l2 = SUM (APPEND l1 l2)'--));
```

```
vstore_thm("mylist","SAPP_SUM",
      (--'SAPP l1 l2 = SUM l1 + SUM l2'--),
      REWRITE_TAC[THM "SUM_APPEND", THM "SAPP"]);
```

We then switch back to our main virtual theory, switching source files once more.

```
extend_vtheory "wombat";
```

The resources of theory "wombat" are visible again. So are the theorems of theory "mylist" because it was made a virtual parent. Thus we can prove our main theorem.

```
vstore_thm("wombat","WOMBAT_SAPP",
      (--'WOMBAT l1 l2 = SAPP l1 l2'--),
      REWRITE_TAC[THM "SAPP_SUM", THM "WOMBAT"]);
```

Finally we close the theories, and end the session.

The source file of theory "wombat" will consist of the following commands which were cut and pasted into the HOL session.

```
new_vtheory "wombat";

new_vparent "wombat" "mylist";

new_vdefinition("wombat","WOMBAT",
(--'WOMBAT l1 l2 = SUM l1 + SUM l2'--));

vstore_thm"wombat"("WOMBAT_CONS",
     (--'WOMBAT (CONS x l1) l2 = x + SUM l1 + SUM l2'--),
     REWRITE_TAC[THM "WOMBAT",THM "SUM"]);

vstore_thm("wombat","WOMBAT_SAPP",
     (--'SAPP l1 l2 = WOMBAT l1 l2'--),
     REWRITE_TAC[THM "SAPP_SUM", THM "WOMBAT"]);

close_vtheory();
```

The following commands will have been added to the source file for "mylist", prior to its call to close_vtheory.

```
new_vdefinition("mylist","SAPP",
     (--'SAPP l1 l2 = SUM (APPEND l1 l2)'--));

vstore_thm("mylist","SAPP_SUM",
     (--'SAPP l1 l2 = SUM l1 + SUM l2'--),
     REWRITE_TAC[THM "SUM_APPEND", THM "SAPP"]);
```

After updating the makefile for these entries, we remake the theories from the source files. It is only at this point that the theory files for "wombat" and "mylist" are created and updated, respectively.

In practice the need for the addition of a theorem to an existing theory would often be discovered in mid proof, assuming a top down proof methodology is followed. This would not pose any problems. The original proof would just be suspended and then resumed once the other theory had been extended and the correct theory returned to.

## 3.2  Censors

All non-global resources are accessed using *censors*. Each kind of resource (theorem, tool, etc) has an associated censor which is used to access individual resources of that kind. For example THM, used in the example session, censors theorems. Censors are functions which ensure that only resources that have been declared to the current virtual theory can be used.

Whenever a new resource is created, such as when a new theorem is proved, it must be declared to the censor. This is done using commands such as `vstore_thm`. The resource is recorded in the current virtual theory's datastructure. In the current implementation, the virtual theory datastructure stores the following resources of a theory,

- its parents (whether real or virtual),
- its theorems and definitions,
- its constants (term and type),
- its tools (tactics, inference rules, etc), and
- a cache of the resources accessed from other theories in this session.

For a complete system, other kinds of resources, such as libraries, need to be added. We have so far included only those needed for the particular case study we were concerned with.

When the censor is called to release a resource it attempts to recover it from the current virtual theory datastructure or that of its ancestors as determined from the datastructure. If the ancestors are real theories this may involve reading the theory file. If the resource cannot be found by the censor an exception is raised instead. If a resource is needed but not visible, either it must be redeclared, or the appropriate theory must be made an ancestor of the current theory. Thus if all declarations are recorded in the source file of the theory which is current at the time, the source file will be able to access all the necessary resources when it is replayed.

The virtual theory system requires a certain degree of discipline by the user. Resources must not be bound to top level ML identifiers, as otherwise the virtual theory manager will not be able to keep track of their use. Furthermore, the user must switch to an appropriate virtual theory before, for example, saving a theorem, or adding a new virtual parent so that a theorem can be accessed. If this is not done the same anarchy that results from working only with real theories can result.

## 4    Navigating Around Virtual Theories

New theories are created using `new_vtheory`. It adds a new empty virtual theory data object to a table of such objects maintained by the system. It also changes the current theory to be this new virtual theory. `new_vtheory` can be called as many times as necessary in a session as new theories are needed.

If an old theory, whether real or virtual, is required to be extended, a call to `extend_vtheory` is made. If the theory already exists as a virtual theory it is made current. Otherwise a new virtual theory is created. Real resources are only moved from the real theory (ie disk) into the virtual theory datastructure as they are needed since this is time-consuming. Read-only system theories can also be extended in this way. The real theory will not then be modifiable, of course, unless the source file can be written to by the user and the Makefile reflects the fact that the system will need to be rebuilt.

**new_vtheory s**

Start a new virtual theory called **s**. If no working real theory has been created, then create one with the given name.

**extend_vtheory s**

Switch to virtual theory **s** if it exists. If it is not an existing virtual theory but is a real theory, open a virtual theory corresponding to it. If no working real theory has been created, then create one with the name "working".

**new_vparent c s**

Make the named theory (real or virtual) a virtual parent of **c** which should be the current virtual theory. If the named theory is not an ancestor of the current real theory, it will be made a real parent of that theory.

**current_vtheory()**

Return the name of the current virtual theory.

**print_vtheory s**

Give details of the named virtual theory.

**close_vtheory()**

Save the theorems proved in the current virtual theory to the working real theory.

**Fig. 1.** Navigating Around Virtual Theories

When the resources of some other theory that is not an ancestor of the current virtual theory are required, **new_vparent** is called. The ancestry of that parent is then made available to the censors when checking resources.

The user works in a dummy real theory which is created automatically by the first call to **new_vtheory** or **extend_vtheory**. Its purpose is to provide a working area within which the virtual theories live. All the definitions are saved to the dummy theory during the working session (since they must be saved somewhere). Similarly, all the real theories accessed in the session are made real ancestors of this theory. Its theory file is discarded after the working session, since everything of value is transferred to the source files of other theories. It is only later when the source files are replayed that the definitions and theorems are put into the real theory files.

## 5   Theorems

Theorems are saved, using **vsave_thm** or **vstore_thm**, in the virtual theory corresponding to the real theory in which they will ultimately reside. They are stored in a table mapping theorem names to stored theorems. The commands to prove them must be stored in its source file. Theorems are not bound to ML identifiers. They can instead be accessed using a call to the censor LOAD_THM, giving the theorem's name and the ancestor virtual theory where it resides. LOAD_THM raises an exception if the theorem is not visible from the current virtual theory. Theorems should not be bound to ML identifiers. The only means used to access

**LOAD_THM thy s**

Return the theorem named s from virtual theory thy. Raise an exception if the theorem is not there or the theory is not a virtual ancestor. LOAD_DEF is similar.

**THM s**

Return the theorem or definition named s. It may be in the current virtual theory or in one of its virtual parents. In the latter case it will be autoloaded into the current virtual theory. DEF is similar.

**vstore_thm(c,s,tm,tac)**

Prove using tactic tac the goal given by tm. Store the resulting theorem in the current virtual theory with name s. Check c is the current virtual theory.

**vsave_thm(c,s,th)**

Store the theorem th in the current virtual theory with name s. Check c is the current virtual theory.

**Fig. 2.** Commands for manipulating theorems

them should be via the censor. It will then only be possible to use those that are accessible in the current virtual theory.

## 5.1 Autoloading Theorems

HOL90 cannot do true HOL88 autoloading: ie, loading a theorem from an ancestor theory into the system when the the ML variable with its name is first used. This is because it would require the SML interpreter itself to recognise when an unknown SML variable had been referred to and take appropriate action to look for the theorem. In HOL90 the theorems must be manually loaded, though all the theorems of a theory can be loaded at once. This is time consuming and results in more theorems being loaded than are required. It also means a separate declaration must be made for each ancestor theory whose theorems are to be accessed in a session, even though from a logical perspective they are visible. Also, when a theory is extended in a new session, these declarations must be remade. The theory development cannot just be continued as though it had never been suspended. Autoloading is very useful for the large theory hierarchies encountered on large proofs, where remembering the names of theories where a theorem resides can be difficult, and theories can contain large numbers of theorems. Autoloading also helps with theorem mobility. If a theorem has been moved from one theory to another, then it can be automatically found. No change to other theories is needed provided it has been moved to one of their ancestor theories. Virtual theories do not give a complete solution to moving theorems, which would require a more complete tracking of dependency information, however.

With virtual theories, all theorems are obtained by a function call to a censor. Therefore if the named theorem has not previously been loaded, the function can automatically search for it throughout the theory's ancestry, and load it once

**new_vdefinition(c, s, tm)**
    Make the given definition tm in the current real theory. Also store it in the current virtual theory with name s. A check is made to ensure that only constants defined in the current virtual theory or its ancestors are used. c must be the current virtual theory.

**Fig. 3.** Making virtual definitions

found. This is implemented by a separate censor, THM. The autoloaded theorem can be loaded from both virtual and real theories. All that is required is that the theory is a (possibly virtual) ancestor of the current one. When the source files are rebuilt to create real theories, on first access, the theorem will be autoloaded from the real theory instead.

## 6  Constants and Definitions

Constants (term and type) are only visible to descendant theories of the theory they are created within. Thus with multiple active theories, censors must be used to ensure that constants are only visible to appropriate theories. We have implemented this by redefining the term and type parsers to check that the constants within the term or type are visible. This is sufficient provided the parsers are the only way used to create terms and types. Clearly, this could be circumvented using for example the term constructors directly, so in a complete system they should also be censored. However, censoring the parsers was sufficient for our immediate purposes.

Since the censors are likely to have to access many real theories when a term is first checked, the first use can be very slow. However, the names of all the constants of each theory visited are cached so that future accesses are quicker.

New constants are created using definitions. Thus the definition declaration functions must register the new constant they create as well as the corresponding theorem. Only simple term constant definitions have so far been implemented in the virtual theory system (see Figure 3) as this was all we immediately needed. However, the other forms of definition should not be problematic. The constants used in the definition must be checked by the constant censor, but as they are parsed by the term parser, this is done automatically.

Unlike the theorem censor, the constant censor does not need to return the resource (ie constant) on a successful check, thus the virtual theory datastructure records only the names of visible constants.

A minor advantage of not making ML bindings is that when giving definitions, the user no longer has to give the definition name twice as at present – once as the ML variable name and once as the name it will be stored as in the theory. This removes the possibility of incompatibilities. It is also less confusing for learners of the system.

**new_vtool c s tool**

Declare function `tool` to be available to the current virtual theory with name **s**. Check **c** is the current virtual theory.

**LOAD_*TOOL* thy s**

Return the function named **s** after checking that it is in the given theory and that that theory is an ancestor of the current virtual theory. Fail if it is not. *TOOL* is a name describing the type of the function (for example `THM_TACTIC` to find tools of type `:thm -> tactic`)

*TOOL* s

As LOAD_*TOOL* except that the tool is sought through the ancestors of the theory.

**declare_tool_type s t**

Create a censor for tools of type given by string **t**, using name **s**.

**Fig. 4.** Virtual Tools

# 7 Tools

A theory's environment not only consists of its theorems and constants, but also the tools (tactics, inference rules etc) that it has access to. Thus tools need to be managed by the virtual theory system. Tools which exist in the core HOL system are not a problem since they can be accessed by all theories. Similarly, user defined tools that only refer to global resources can be made global and placed in a file that is loaded by all theories.

There are two issues which must be dealt with concerning other tools. Firstly, the code of the tool must be loaded before it can be used. Secondly, any resources it is dependent on (such as other tools and theorems) must be accessible. These issues must be accounted for in both the working session and the source files.

For tools dependent on theorems, this means that the appropriate theory must be an ancestor of the current theory. This is dealt with by ensuring that the tools only access theorems using censors. If the tool is used in a virtual theory where the theorem is not available, an exception will result.

Managing the loading of tools requires that the tools themselves be censored. If they are bound to ML variables when loaded, then all virtual theories in a session will be able to use them unchecked. They will not, consequently, necessarily be loaded when the source files are replayed.

As with theorems, when a tool is loaded it is declared to the censor (by a call to `new_vtool`) rather than being bound to an ML variable. Tools are accessed via function call, to a function `TOOL` say, rather than just by name. Unfortunately, the censor cannot store the tools themselves directly in a single datastructure. The problem is that it would not type check if say both a tactic and derived rule were stored. One way round this would be to have a separate datastructure and related function for each tool type. This is obviously undesirable given the wide range of types that tools have and the flexibility HOL gives for tools with new types to be defined. Adding tools of types not envisaged by the virtual theory implementor would require reprogramming the virtual theory system.

The problem can be overcome using a method described by Larry Paulson [3]. It allows all tools of whatever type to be stored in a single datastructure. It uses the fact that exceptions can take arguments of different types, but the type of the exception does not reflect that argument's type. Its type is just :exn. A separate exception is declared for each tool type to be stored. Tools are then wrapped in their appropriate exceptions before being stored in the database. A separate function must be provided for each tool type to recover the tools (a single function will not type check). They take an exception argument which they raise and immediately handle, returning the embedded tool.

Thus to add a tool of a new type, the user must declare the exception, write a handler (the censor) and then declare the tool itself. This only involves adding functions, not changing existing code. Furthermore, the code required is of a very standard form. We have provided a function which creates the appropriate ML functions given strings representing the name of the censor to be created and the type it censors. The function does this by writing the code out into a file and then reloading it into the ML system. A concrete example is given below for theorem tactics.

To declare a new tool type for theorem tactics, declare_tool_type is called:

```
declare_tool_type "THM_TACTIC" "thm -> tactic";
```

This call loads in the ML code below, declaring an exception and two censors: one for loading from a given theory and one for autoloading (see Section 7.1). The functions declare_new_tool_type and declare_to_load_new_tool_type contain the code which does the censoring. Their argument is a raise-handle function for a specific exception.

```
exception DECLARE_THM_TACTIC of (thm -> tactic);

val THM_TACTIC = declare_new_tool_type
    (fn e => (raise e) handle (DECLARE_THM_TACTIC ttac) => ttac);

val LOAD_THM_TACTIC = declare_to_load_new_tool_type
    (fn e => (raise e) handle (DECLARE_THM_TACTIC ttac) => ttac);
```

New theorem tactics can now be declared to the censor.

```
new_vtool "wombat" "IMP_RES_REWRITE_TAC" (DECLARE_THM_TACTIC
    (fn th => IMP_RES_THEN (fn th1 => REWRITE_TAC[th1]) th));
```

The censor can then be used to recover and use the theorem tactic.

```
e(THM_TACTIC "IMP_RES_REWRITE_TAC" (THM "CONS"));
```

Censors for common tool types are pre-declared, so it is only unusual types that would require user intervention. The calls to recover tools are verbose, but this could be hidden by a user-interface.

The declaration of new tool types should also strictly be treated as a resource, since it involves ML commands which may need to be placed in more than one source file. However, for our purposes it was sufficient to treat them as global resources loaded in all sessions.

An advantage of using censors is that only a single copy of the code for a tool is needed. Frequently tools are developed that are both based on the theorems of a theory and used to prove theorems within that theory. This often means two copies of the tool's code are stored. One copy is placed in the source file for the theory. A second copy is placed in the file that is loaded when the tools are to be used in the development of other theories. Since censors are lazy, only one copy is now needed in the separate file. This file can be loaded before the theorems it uses have been proved. Only when a tool is actually used must its resources exist.

## 7.1 Autoloading Tools

Once accessed by censors, tools can be treated like theorems with respect to autoloading. Within the virtual world, any tool that has been declared in an ancestor virtual theory can be made accessible in the current theory. This poses a problem when migrating to real theories, however, as tools are not recorded in the theory files like the other resources. The source file of the descendent theory will have no record of the tools, and no way to access them from the real ancestor theory. A way is thus needed to associate tools with real theories as well as virtual ones. The most obvious place to make this association is in the theory files. However, that would involve changing core HOL. Instead, the virtual declarations of tools associated with a theory are placed in a file with the same name as the theory, but with a .tsml extension. Each such file also has an associated signature file (with .tsig extension). It just contains the names of the tools in the .tsml file and is used when searching for a tool. Only when the tool is found is the corresponding file of declarations loaded. A call to an unknown tool initiates a search through the ancestor theories of the current virtual theories to find it. Once found, all the code for that theory is loaded.

The granularity of loading is not very fine in this scheme. However, this could easily be overcome using multiple source files, with the signature file giving the file name for each tool. Tools could then be individually loaded. This has not been implemented yet, however.

If tools are to be used by theories other than their own, their declaration must not explicitly refer to a specific theory, but instead should refer to the current theory.

By associating tools with theories in this way, theories essentially become a form of lightweight mini-library. They have the advantage that tools are available, though not loaded, whenever the theorems and definitions they reason about are available. The location of the tools does not need to be known, they are found by the system when used. If this system were adopted for the HOL system as a whole, the tools related to each datatype: booleans, lists, numbers, pairs etc, could be associated with the theorems for it, giving a more structured core system.

Autoloading is not to everyone's taste. Explicit loading is quicker, avoids name clash problems and leaves explicit dependency information in the source files. As with theorems, tools do not need to be autoloaded, however. Censors which take the name of the theory associated with the tool are provided. Furthermore, if the non-autoloading censor is used on the first call of a session, subsequent calls can then use the autoloaded form to quickly obtain the same tool because it will be cached in the virtual theory.

## 8 Use on the Fairisle Switching Fabric Proof

We were motivated to develop virtual theories because of difficulties encountered on a real proof of the Fairisle 16 by 16 switching fabric [1]. We had resorted to working in a single theory which we then tidied into separate theories at intervals. As a consequence the resulting theories were not tidied as well as they might, leaving a maintenance headache that was slowing the proof effort. The theory had grown very large and was desperately in need of tidying. The longer this had been put off, the worse the situation became. The result was that significant effort was required to tidy the proof.

The virtual theory mechanism was developed to prevent the same situation from recurring in subsequent proofs. It was also hoped that it would provide a way to tidy the existing theory. Initial results suggest that the latter aim has been achieved. We have tested the system by using it to tidy part of the theory. The script was about 6000 lines long and contained a wide range of theorems that should have been placed in many different theories. So far approximately 15% of the file has been tidied using the virtual theory system. The remainder is in progress. The effort involved was less than would have been required without virtual theories. The latter would have involved working with multiple HOL sessions and would probably have involved rebuilding parts of the theory hierarchy several times. From past experience of tidying scripts in that way, the virtual theory approach is easier, faster and more accurate (in the sense that only necessary dependencies are added).

## 9 Further Work

### 9.1 Other Resources

The current implementation of virtual theories is a prototype, and as such it only tracks the main resources such as theorems and tools. Various other resources would need to be dealt with to obtain a complete system. Essentially, anything that is part of the environment of a theory needs to be censored. Whenever a new resource is created it must be declared to a censor, and every time it is used it must be accessed via that censor. The resources we considered were those relevant to our case study.

One important resource we did not explicitly consider is the libraries that a theory has access to. If a library is loaded for one virtual theory it will be available to (and so might unwittingly be used by) other virtual theories. It is not sufficient just to keep a list of libraries loaded by each theory, since some way is needed of detecting when the resources of the library are used. Also a way of declaring the resources when the library was loaded would be needed.

## 9.2 Interface Issues

We have used the virtual theory system in a cut and paste style, writing directly to a source file and copying the commands into the HOL session. This is typically the way HOL is currently used. It would be better if the source files were created automatically by the system. This can be done for normal theories using TkHolWorkbench [5]. It keeps a log of the session, yielding a file with the same contents that would be created using the cut-and-paste approach. It would be relatively simple to modify the interface for virtual theories and to direct the parts of sessions corresponding to different virtual theories to different log files.

We have experimented with a prototype interface for virtual theories based on TkHolWorkbench[1]. It graphically displays a tree of the virtual theory hierarchy, tracking any changes made. It also allows the user to switch to a new virtual theory within the HOL session by clicking on its icon in the tree. It could obviously be expanded to help visualise all the resources of a virtual theory. Such an interface could remove the problem of the verbosity of the virtual commands for retrieving resources. Once the interface knows of the source files it could provide means for hyper-text documentation of specifications, theorems and proofs to be displayed from the session.

The use of virtual theories would allow a dynamic theory browser such as TkHolWorkbench, to visualise all theories, rather than just the ancestors of the current theory. Other theories could be loaded into the session to be viewed without making them unwanted parents of an actual theory. This would be useful to allow theories to be searched prior to a decision to make them a parent.

For the purposes of exposition we have used distinct function names from those of real theories (for example, new_vtheory instead of new_theory). However, once working in the virtual world, there is no reason for the user to call the real versions (indeed this would be harmful). Thus the original functions could just be redefined when the virtual system is loaded.

## 9.3 ML Bindings

A key part of the virtual theory methodology as described is that resources such as theorems are not bound to ML identifiers and are only accessed via the censors. However, it may be possible to reintroduce variable bindings in a limited form. Instead of an actual resource being bound, a censor specialised for that resource is bound. For theorems, this means a function taking a unit argument

---

[1] With lots of help from Donald Syme

is bound (so they must be treated as tools). When the function is called the censor is checked before the resource is actually returned. Thus even though the variable bindings are visible to all virtual theories, their use in a theory where the resource is not visible will raise an exception

Autoloading is no longer possible as the ML binding must be made before it can be used. Also when a theory is extended, the ML bindings of the theory must first be set up, thus breaking our third design criteria. However, since this corresponds to the current situation in HOL90, a satisfactory methodology could perhaps be built around it, allowing multiple active theories, without the need to explicitly call censors.

### 9.4  Multiple Active Real Theories

In the long term it would be better if the sources did not need to be replayed to update the theory files. However, this would require a method of preventing leaks between theories due to ML bindings, intentional or otherwise. Such leaks are possible with virtual theories if the user does not follow the prescribed methodology of never making bindings of raw resources to ML objects.

One way real multiple active theories might be achieved is by using a meta-language which supports multiple processes so that each theory could have its own private meta-language environment. Alternatively, for theory development, rather than system development, a restricted meta-language interface could be provided which does not provide features that could lead to information leakage. This might be adequate in a commercial environment where the theorem prover was just being used to prove theorems rather than build new verification systems. A further possibility might be to build the censors into the primitives, tagging all resources with their environment. More research is needed in this area. In the meantime virtual theories provide a workable solution to the problem.

## 10  Related Work

Some of the functionality of virtual theories was implemented by Konrad Slind as part of his parametrised proof manager [4]. The main difference is one of focus. Slind was mainly concerned with tracking dependencies between theorems proved in a particular theory. We have been concerned with tracking all the resources used by a theory, and in particular the external resources, to allow a current proof session to be integrated with previous ones. Because in Slind's approach the proof manager keeps track of the dependencies, it is restricted to being used only with specially designed proof managers. In our approach, any proof manager, whether forwards or backwards can be used with no modification. The details of the implementation also differ. In particular, we keep track of dependencies by requiring that the resources are accessed by function calls, whereas Slind uses the proof manager to determine the resources used from the commands passed to it. This requires more work on the part of the proof manager. It is clear that the functionality of our interface could be provided within the parametrised proof manager framework.

The abstract theory mechanism of Elsa Gunter [2] shares similar problems to those of virtual theories: that of making resources visible to only some theories. With abstract theories the resources are the axioms of the abstract theory. Because of the similarity, it may be possible to provide a unified system supporting both. However, we have done no work in this area.

## 11 Conclusions

We have presented a mechanism for managing multiple theories in HOL. It is loaded on top of the existing system so it does not involve incompatible changes to HOL itself. It can be used with any proof manager and proof style and enhances existing ways of working. The virtual theory mechanism was developed to overcome a particular problem encountered on a real hardware verification proof: that of tidying the script. It appears to have solved that problem very successfully. In future, by using the virtual theory mechanism, the problem should not arise in the first place.

## Acknowledgements

I am grateful to the members of the Cambridge Automated Reasoning Group for their support. Donald Syme has been an invaluable source of encouragement. Mark Staples told me about the exception programming trick. Brian Graham proof read a draft of this paper. The work was funded by EPSRC grants GR/G23654 and GR/K10294.

## References

1. Paul Curzon. Tracking design changes with formal machine-checked proof. *The Computer Journal*, 38(2), 1995.
2. Elsa L. Gunter. The implementation and use of abstract theories in HOL. In *Proceedings of the Third HOL Users Meeting*, 1990.
3. Larry Paulson. exn as dynamic type. Message sent to the comp.lang.ml newsgroup, January 1995.
4. Konrad Slind. A parameterized proof manager. In Thomas F. Melham and Juanito Camilleri, editors, *Higher Order Logic Theorem Proving and Its Applications: 7th International Workshop*, volume 859 of *Lecture Notes in Computer Science*, pages 407–423. Springer-Verlag, September 1994.
5. Donald Syme. A new interface for HOL - ideas, issues and implementation. In *Proceedings of the 8th International Workshop on Higher Order Logic Theorem Proving and Its Applications*, Lecture Notes in Computer Science, 1995.

# An Automata Theory Dedicated towards Formal Circuit Synthesis

Dirk Eisenbiegler and Ramayya Kumar

Forschungszentrum Informatik
(Prof. Dr.-Ing. D. Schmid)
Haid–und–Neu–Straße 10-14, 76131 Karlsruhe, Germany
e–mail: {eisen,kumar}@fzi.de

**Abstract** We present a theory for automata in HOL, which is dedicated towards formal hardware synthesis. The theory contains definitions for formally representing and transforming automata. In this approach hardware is represented by automata descriptions and formal synthesis is performed by applying formally proven theorems. The approach presented is constructive — i.e. starting from specifications at higher levels of abstractions, synthesis can be performed by repeated applications of these transformations. Specialized refinements and optimizations at the RT and gate levels are discussed.

## 1 Introduction

This paper is dedicated towards formal correctness in hardware design at the RT (register transfer) and gate level. During RT and gate level synthesis the circuit description is altered step by step using specific well known transformations such as: state encoding, state minimization, boolean optimization, etc. Although these basic synthesis steps conform to simple logical derivation steps, post-synthesis-verification is exacting. Post-synthesis-verification techniques only have access to a specification and an implementation, i.e. the input and the output of the synthesis process. Usually, there is a big gap between specification and implementation: the state representation and the originally given partitioning may have changed completely. As a major drawback, the information on *how* the implementation was derived from the specification is lost. Much of this information is essential for verification: How were the control states encoded? Where is which data stored? Is a redundant data representation used (one-hot-encoding, signed-digit-encoding etc.)? Which control states were eliminated because of unreachability, or have some (unreachable) control states been added in order to get a more efficient/testable implementation? Which parts of the gate level implementation belong to the control path/data path of the RT-level description? etc.

This paper is part of our ongoing work for developing techniques to perform formally correct synthesis of synchronous circuit descriptions. The automata theory is intended to be used for simple synchronous circuit descriptions at the gate and RT level [EiSK93]. The theory provides theorems describing the above mentioned elementary RT- and gate level transformations (data encoding, state minimization etc.) in a logical manner. The automata theory builds a basis for formal synthesis programs where the entire process is described by a sequence of refinement steps within logic. As a result of the formal synthesis process, there is not only the implementation of a given specification but also the proof of its correctness. In contrast to other approaches towards formal synthesis, this approach is very close to conventional synthesis techniques. We do not intend to invent new synthesis algorithms but implement conventional ones in a formal manner.

The current state of the art about embedding automata in HOL is as follows: In [ScKK93] a specific set of formulae named hardware formulae is used for describing specifications and implementations of automata and appropriate proof procedures are defined. Although such descriptions are very useful for post-synthesis verification, they do not allow a constructive approach for performing formal synthesis. Similar to the approach taken in this paper, [Loew92, Day92] describe automata explicitly by means of expressions. This allows definitions and derivations of general theorems about automata. However, they allow more complex specifications such as non-deterministic automata and do not give constructive transformations which could lead to circuit implementations. In our work we consider only deterministic automata whose formalization is purely functional in nature and give transformations which can be used to perform refinements and optimizations, especially at the RT and gate levels. The overall theory can be regarded as a simple toolbox for formal synthesis algorithms at the RT and gate levels.

The outline of this paper is as follows: starting from the functional input/output definitions of the automata, we go on to describe the property of reachability. In section 5, we define the transformations which correspond to simple synthesis steps: state encoding, removal of unreachable states and the elimination of redundant memory parts. In section 6, we provide some encoding theorems for a small set of data types which is followed by an example in section 7.

## 2 Automata Representation

Usually an automaton is represented by a 6-tuple consisting of input alphabet, output alphabet, set of states, output function, transition function and initial state. In our approach, we use the concept of typed functions, available in HOL, for representing automata. Given that $\iota$, $\omega$ and $\sigma$ are the types corresponding to the inputs, outputs and states, respectively, the output and the transition function have been combined to a single function $f$. It is to be noted here, that the types $\iota$, $o$ and $\sigma$ can be compound — such as, tuples of basic data types.

The entire automata is represented by a pair $(f, q)$, where $f$ has the type $\iota \times \sigma \to \omega \times \sigma$ and $q$ represents the initial state and has the type $\sigma$. The various manipulations that can be performed using such a representation is the chief concern of this paper.

$f$ and $q$ unambiguously determine, how the automaton maps a time dependent input signal $i_{num\to\iota}$ to a time dependent output signal $o_{num\to\omega}$. The constant automaton maps a pair $(f,q)$ to a function mapping $i$ to $o$. the constant automaton has the following type

$$((\iota\times\sigma\to\omega\times\sigma)\times\sigma)\to(num\to\iota)\to(num\to\omega)$$

Figure (1) sketches, how some automaton$(f,q)$ could be "implemented" using a combinatorial component realizing $f$ and a memory unit $\mathcal{D}^q$, which stores data of type $\sigma$ and its initial value is $q$.

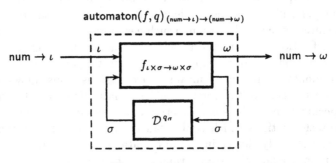

**Figure1.** Automaton

The constant automaton will formally be defined by means of another constant named automaton'. automaton' is similar to automaton except that the set of states are also visible (see figure 2). Hence the constant automaton' has the type:

$$((\iota\times\sigma\to\omega\times\sigma)\times\sigma)\to(num\to\iota)\to(num\to(\omega\times\sigma))$$

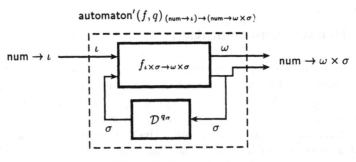

**Figure2.** Automaton'

automaton' is defined by means of primitive recursion over natural numbers, which represent time. For a given $i_{num\to\iota}$, the expression (automaton' $(f,q)$ $i$)

denotes the output and the present state and (automaton' $(f, q)$ $i$ $t$) denotes the output and the present state at some time $t$. The definition to follow is performed by using primitive recursion over $t$.

The output and the next state for some time $t$ can be obtained by applying $f$ to the pair of current input $i(t)$ and current state $s$. In the beginning $t$ is 0 and the automata is in the initial state $s = q$. For all other times $t = (\text{SUC } t')$, the next state of the output is defined using the current input $i(\text{SUC } t')$ and the current state $s$. Since (automaton' $(f, q)$ $i$ $t'$) produces a pair corresponding to the output and the state, the function SND is applied in order to extract the state from this result.

$$\vdash \ (\text{automaton'} \ (f, q) \ i \ 0 \ = \ f(i(0), q)) \ \wedge \tag{1}$$

$$(\text{automaton'} \ (f, q) \ i \ (\text{SUC } t') =$$
$$\text{let}$$
$$s = \text{SND}(\text{automaton'} \ (f, q) \ i \ t')$$
$$\text{in}$$
$$f(i(\text{SUC } t'), s) \ )$$

Now automaton can be defined as

$$\vdash \ \text{automaton} \ (f, q) \ i \ t \ = \ \text{FST}(\text{automaton'} \ (f, q) \ i \ t) \tag{2}$$

## Example

A simple traffic light controller is to be described based on the constant automaton. The controller has two boolean inputs *reset* and *up*. So $\iota$ becomes bool $\times$ bool. There are three outputs named *ron*, *yon* and *gon*. Each corresponds to one single light and determines whether this light is on or off. All outputs are of type bool and so $\omega$ becomes bool $\times$ bool $\times$ bool.

To represent the state $s$ of the traffic light controller, a simple enumeration type named ryg with values red, yellow and green is used. The type of the output and transition function $f$ is as follows:

$$((\text{bool} \times \text{bool}) \times \quad \text{ryg} \quad ) \ \to \ ((\text{bool} \times \text{bool} \times \text{bool}) \times \quad \text{ryg} \quad )$$

| $\underbrace{reset \quad up}$ | $\underbrace{old \ s}$ | $\underbrace{ron \quad yon \quad gon}$ | $\underbrace{new \ s}$ |
|---|---|---|---|
| input | old state | output | new state |

The definitions of f and q are as follows:

$$q \ = \ \text{green}$$

$$
\begin{aligned}
f((F, F), \text{red}) \quad &= ((T, F, F), \text{red}) \ \wedge \\
f((F, F), \text{yellow}) &= ((F, T, F), \text{yellow}) \ \wedge \\
f((F, F), \text{green}) &= ((F, F, T), \text{green}) \ \wedge \\
f((F, T), \text{red}) \quad &= ((T, F, F), \text{yellow}) \ \wedge \\
f((F, T), \text{yellow}) &= ((F, T, F), \text{green}) \ \wedge \\
f((F, T), \text{green}) &= ((F, F, T), \text{red}) \ \wedge \\
f((T, F), x) \quad &= ((T, F, F), \text{red}) \ \wedge \\
f((T, T), x) \quad &= ((T, F, F), \text{red})
\end{aligned}
$$

The expression automaton(f, q) has the following type:

$$(\text{num} \rightarrow (\text{bool} \times \text{bool})) \rightarrow (\text{num} \rightarrow (\text{bool} \times \text{bool} \times \text{bool}))$$

$$\underbrace{\text{time}}\ \ \underbrace{\text{reset}\ \ \text{up}}\qquad \underbrace{\text{time}}\ \ \underbrace{\text{ron}\ \ \text{yon}\ \ \text{gon}}$$

$$\underbrace{\qquad\text{input}\qquad}\qquad\qquad\underbrace{\qquad\text{output}\qquad}$$

## 3 Special Cases of Automata

As already mentioned, we intend to use automata to describe both combinatorial and sequential circuits. We will now define two constants named combinatorial_block (for purely combinatorial circuits) and memory_block (for memory parts) and we will explain, how they are related to the previously defined automaton.

A combinatorial cuircuit can unambiguously be defined by a function $e_{\iota \rightarrow \omega}$ mapping the current input to the current output. The constant combinatorial_block maps $e$ to a function mapping some time dependent input $i_{\text{num} \rightarrow \iota}$ to some time dependent output $o_{\text{num} \rightarrow \omega}$ with $o(t) = e(i(t))$ (see figure 3). Definition:

$$\vdash \text{combinatorial\_block}\ e\ i\ t\ =\ e(i(t)) \tag{3}$$

combinatorial_block$(e)$ $_{(\text{num} \rightarrow \iota) \rightarrow (\text{num} \rightarrow \omega)}$

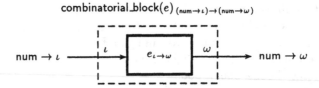

**Figure3.** Combinatorial Circuits

Memory parts delay the input by one clock cycle. The initial state is given as a parameter to the memory_block constant.

$$\vdash \text{memory\_block}\ init\ i\ 0 \quad = init\ \wedge \tag{4}$$
$$\text{memory\_block}\ init\ i\ (\text{SUC}t) = i(t)$$

One can represent a combinatorial circuit by an ordinary automaton, where the type of the state is one. one is a HOL standard data type with only one element. The constant one$_{\text{one}}$ represents its unique element.

$$\vdash \text{combinatorial\_block}\ e\ =\ \text{automaton}\ ((\lambda(x, y_{\text{one}}).\ (e(x), \text{one})),\ \text{one}) \tag{5}$$

Memory parts can be represented by automata, where the input is directly connected with the input of the internal memory and the output of internal memory is connected with the output of the automaton.

$$\vdash \text{memory\_block}\ init\ =\ \text{automaton}\ ((\lambda(x, y).\ (y, x)),\ init) \tag{6}$$

# 4 Reachability of States

Using the definition of an automaton given in section 2, we can define the concept of reachability. The constant reachable maps an automaton given by $(f, q)$ onto a predicate which indicates if some state $s$ may be reached or not. reachable has the following type:

$$((\iota \times \sigma \to \omega \times \sigma) \times \sigma) \to \sigma \to \text{bool}$$

reachable is defined by means of a constant definition using automaton'. The definition states that a state $s$ is reachable iff there is some input sequence $i_{\text{num} \to \iota}$ and some time $t$ such that the current state, i.e. SND(automaton'$(f, q) i t)$, becomes $s$.

$$\vdash \text{reachable } (f, q) s \;=\; (\exists i, t. \text{SND(automaton' } (f, q) i t) = s) \tag{7}$$

Theorem (8) states, that the initial state $q$ is reachable. Theorem (9) states, that if some $s$ is reachable then so is any successor state SND$(f(a, s))$ for arbitrary input $x$.

$$\vdash \text{reachable } (f, q) q \tag{8}$$

$$\vdash (\text{reachable } (f, q) s) \;\Rightarrow\; (\forall x. \text{reachable } (f, q) (\text{SND}(f(x, s)))) \tag{9}$$

When encoding states of automata later on in this paper, we will have to find subsets of states, that cover all reachable states. Given a predicate $P_{\sigma \to \text{bool}}$ indicating the chosen subset, we can prove the theorem, that $P$ covers all reachable states in an inductive manner using theorem (10).

$$\vdash \forall P. \tag{10}$$
$$(
$$
$$P(q) \wedge$$
$$(\forall s. P(s) \Rightarrow (\forall x. P(\text{SND}(f(x, s)))))$$
$$)$$
$$\Rightarrow (\forall s.(\text{reachable } (f, q) s) \Rightarrow P(s))$$

Theorem (10) states, that $P$ covers all reachable states if

1. the initial state $q$ is in the subset described by $P$, and
2. for all states $s$ within this subset any succeeding state SND$(f(x, s))$ (for arbitrary input $x$) is also in this subset.

# 5 Transformations on Automata

Equivalence of automata means that for a given input, they produce the same output. In other words, two automata $(f, q)$ and $(\tilde{f}, \tilde{q})$ are called equivalent iff automaton$(f, q) = $ automaton$(\tilde{f}, \tilde{q})$.

An automaton $(f, q)$ can be trivially turned into an equivalent automaton by substituting $f$ and $q$ by equivalent terms $\tilde{f} = f$ and $\tilde{q} = q$. All automata achievable by such transformations have one thing in common: the states are represented in the same way. In this section we will present automata transformations which go beyond this — namely those, where the states are represented in a different manner, the number of states differs, etc. .

In this section, we will first introduce a more general state encoding theorem, then derive two corollaries to this theorem and finally we introduce a theorem for removing redundant memory parts.

## 5.1   The State Encoding Theorem

The general state encoding theorem has two technical applications: encoding the data types of the state and elimination of unreachable states.

$$\vdash \ (\forall s. \ (\text{reachable} \ (f,q) \ s) \ \Rightarrow \ h(g(s)) = s) \tag{11}$$

$$\Rightarrow$$
$$($$
$$\text{automaton} \ (f,q) \ =$$
$$\text{let}$$
$$\tilde{f} = (\lambda(v,x).(\lambda(y,z).\,(y,g(z)))(f(v,h(x)))) \ \text{and}$$
$$\tilde{q} = g(q)$$
$$\text{in}$$
$$\text{automaton}(\tilde{f},\tilde{q})$$
$$)$$

The left hand side of the implication states in theorem (11), that there functions $g$ and $h$ fulfilling $h(g(s)) = s$ for all reachable states. $g$ maps a value of type $\sigma$ to a value of some type $\sigma'$ and $h$ maps this value back to the former one (see figure 4).

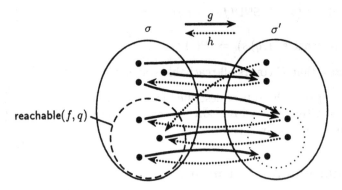

**Figure4.** Encoding from $\sigma$ to $\sigma'$

The right hand side of theorem (11) states, that the automata $\text{automaton}(f,q)$ and $\text{automaton}(\tilde{f},\tilde{q})$ are equivalent. $\tilde{f}$ and $\tilde{q}$ have been derived from $f$, $q$, $g$ and

$h$. The new initial state $\tilde{q}$ has been obtained by encoding $q$. The new output and transition function $\tilde{f}$ has been derived from $f$ by encoding every state input and decoding every state output.

Figure 5 illustrates, how the new automaton looks like. Theorem (11) states that provided the above mentioned assumption, the automata in figure 1 and 5 are equivalent.*

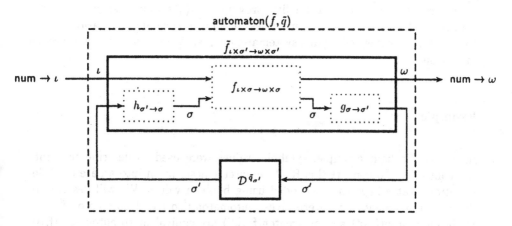

**Figure5.** State Encoding

**Corollary A** Determining reachability can only be performed for small sized automata and theorem (11) is applied to pure encoding problems. The following corollary is convenient for this purpose:

$$\vdash \quad (\forall s. \, h(g(s)) = s) \tag{12}$$

$$\Rightarrow$$

$$($$

automaton $(f, q)$ =

let

$\tilde{f} = (\lambda(v, x).(\lambda(y, z). \, (y, g(z)))(f(v, h(x))))$ and

$\tilde{q} = g(q)$

in

automaton$(\tilde{f}, \tilde{q})$

$$)$$

---

* This automata encoding transformation with its pair of encoding/decoding functions $(g, h)$ resembles the type definition mechanism of HOL [Melh88]. However, in state encoding of automata, the new type may have some extra elements. Furthermore, the subset of states to be encoded cannot be an arbitrary nonempty set as in type definitions but must cover at least all reachable states of the automaton.

In contrast to theorem 11, theorem 12 performs the state encoding for the entire set of states — reachability need not be considered.

Before this corollary can be applied, an appropriate encoding in terms of $h$ and $g$ has to be found and it has to be proven, that the encoding is correct, i.e. $\forall s.\, h(g(s))$ holds. The quality of the synthesis result (size of combinatorial logic, size of memory, etc.) very much depends on the encoding chosen. Usually there are lots of different encodings, and there already exist different techniques for determining good encodings according to different optimization criteria.

For types with a huge cardinality, proving $\forall s.\, h(g(s))$ may become exacting. Besides explicitly proving the correctness of a given encoding, it is also possible to derive a correct encoding in a systematic manner. We will present an approach in section 7.

## Example

In our traffic light example, symbolic values were used to describe the state the controller. To convert this RT-level circuit description into a gate level description, states have to be encoded using boolean values. We will describe to different implementation alternatives: automaton($f'$, $q'$) and automaton($f''$, $q''$). Both automaton($f'$, $q'$) and automaton($f''$, $q''$) are equivalent to automaton($f$, $q$). They are derived by means of state encoding using the encodings ($g'$, $h'$) and ($g''$, $h''$), respectively.

($g'$, $h'$) is a minimal bit encoding, where only two bits are used:

$$
\begin{aligned}
g'(\text{red}) &= (\mathsf{F}, \mathsf{F})\ \wedge \\
g'(\text{yellow}) &= (\mathsf{F}, \mathsf{T})\ \wedge \\
g'(\text{green}) &= (\mathsf{T}, \mathsf{F})
\end{aligned}
$$

$$
\begin{aligned}
h'(\mathsf{F}, \mathsf{F}) &= \text{red}\ \wedge \\
h'(\mathsf{F}, \mathsf{T}) &= \text{yellow}\ \wedge \\
h'(\mathsf{T}, \mathsf{F}) &= \text{green}\ \wedge \\
h'(\mathsf{T}, \mathsf{T}) &= \text{red}
\end{aligned}
$$

Obviously, the state $(\mathsf{T}, \mathsf{T})$ remains unused and $h'(g'\,s)$ is fulfilled no matter how the result of $h'$ is defined for $(\mathsf{T}, \mathsf{T})$. Besides red, every other value could have been chosen, and it would also be possible to leave this decision open at this moment and instantiate the value later on during boolean optimizations.

Applying the ($g'$, $h'$) state encoding leads to

$$\vdash \text{automaton}(f, q) = \text{automaton}(f', q')$$

with:

$q' = (T, F)$

$f'((F,F),(F,F)) = ((T,F,F),(F,F)) \wedge$
$f'((F,F),(F,T)) = ((F,T,F),(F,T)) \wedge$
$f'((F,F),(T,F)) = ((F,F,T),(T,F)) \wedge$
$f'((F,T),(F,F)) = ((T,F,F),(F,T)) \wedge$
$f'((F,T),(F,T)) = ((F,T,F),(T,F)) \wedge$
$f'((F,T),(T,F)) = ((F,F,T),(F,F)) \wedge$
$f'((T,F),(x,y)) = ((T,F,F),(F,F)) \wedge$
$f'((T,T),(x,y)) = ((T,F,F),(F,F))$

$(g'', h'')$ is a one hot encoding. For the one hot encoding three bits are required but only the states $(F,F,T)$, $(F,T,F)$ and $(T,F,F)$ are used. Since the outputs also correspond to the control states, this approach helps minimizing the combinatorial logic required for the implementation.

$g''(\text{red}) \quad= (F,F,T) \wedge$
$g''(\text{yellow}) = (F,T,F) \wedge$
$g''(\text{green}) = (T,F,F)$

$h''(F,F,T) = \text{red} \wedge$
$h''(F,T,F) = \text{yellow} \wedge$
$h''(T,F,F) = \text{green} \wedge$
$h''(F,F,F) = \text{red} \wedge$
$h''(F,T,T) = \text{red} \wedge$
$h''(T,F,T) = \text{red} \wedge$
$h''(T,T,F) = \text{red} \wedge$
$h''(T,T,T) = \text{red}$

Applying the $(g'', h'')$ state encoding leads to

$\vdash \text{automaton}(f,q) = \text{automaton}(f'',q'')$

with:

$q'' = (T, F)$

$f''((F,F),(F,F,T)) = ((T,F,F),(F,F,T)) \wedge$
$f''((F,T),(F,T,F)) = ((F,T,F),(F,T,F)) \wedge$
$f''((F,T),(T,F,F)) = ((F,F,T),(T,F,F)) \wedge$
$f''((F,T),(F,F,T)) = ((T,F,F),(F,T,F)) \wedge$
$f''((F,T),(F,T,F)) = ((F,T,F),(T,F,F)) \wedge$
$f''((F,T),(T,F,F)) = ((F,F,T),(F,F,T)) \wedge$
$f''((T,F),(x,y,z)) = ((T,F,F),(F,F,T)) \wedge$
$f''((T,T),(x,y,z)) = ((T,F,F),(F,F,T))$

**Corollary B** Corollary B to theorem (11) is dedicated to pure state reduction problems. It is assumed, that one has divided $\sigma$ into $\sigma^1 + \sigma^2$, where all the reachable states are in $\sigma^1$. In this situation, the state representation can be cut down to $\sigma^1$ using the following pair of encoding/decoding functions $g^B$ and $h^B$. $g^B$ is introduced by means of a constant specification. The variable $z$ may be instantiated in an arbitrary manner to derive some "concrete" $g^B$.

$$\vdash \exists z. (g^B(\mathsf{INL}\ x) = x) \wedge (g^B(\mathsf{INR}\ y) = z(y))$$
$$\vdash h^B = \mathsf{INL}$$

Remark: It is not demanded, that $\sigma^1$ represents exactly the set of all reachable states. It must cover all reachable states, but there may also be some unreachable states.

$$\vdash (\ (\forall s.\ (\text{reachable}\ (f,q)\ s))\ \Rightarrow\ \mathsf{ISL}(s)\ ) \tag{13}$$
$$\Rightarrow$$
$$($$
$$\quad \text{automaton}\ (f,q)\ =$$
$$\quad \text{let}$$
$$\quad\quad \tilde{f} = (\lambda(v,x).(\lambda(y,z).\ (y, g^B(z)))(f(v, h^B(x))))\ \text{and}$$
$$\quad\quad \tilde{q} = g^B(q)$$
$$\quad \text{in}$$
$$\quad\quad \text{automaton}(\tilde{f}, \tilde{q})$$
$$)$$

Usually $\sigma$ does not have the form $\sigma^1 + \sigma^2$ with all reachable states being on the left hand side. Conversions based on corollary A can be used to reach such a representation.

## 5.2 Elimination of Redundant Memory Parts

The last theorem to be introduced describes, how parts of the memory can be omitted if these parts are of no importance for the output and transition function $f$. This theorem can be used for removing flipflops with unconnected outputs from a synchronous circuit description.

Let us assume, that the type of the states $\sigma$ is a scalar product of two types $\sigma^1 \times \sigma^2$ and that $f$ is $(\lambda(x, (s^1, s^2)).\ f'(x, s^1))$ for some $f'$. In other words, $f$ depends on the input and on the left hand side of the pair $(s^1, s^2)_{\sigma^1 \times \sigma^2}$ representing the state but not on the right hand side. Theorem (14) states, that this automata $(f, q)$ is equivalent to the automaton $(f', q^1)$.

$$\vdash \quad \text{let} \tag{14}$$
$$\quad\quad f = (\lambda(x, (s^1, s^2)).\ f'(x, s^1))\ \text{and}$$
$$\quad\quad q = (q^1, q^2)$$
$$\quad \text{in}$$
$$\quad\quad \text{automaton}(f, q)$$
$$\quad =$$
$$\quad \text{automaton}(f', q^1)$$

# 6 Systematic Derivation of State Encodings

The automata theory provides several pairs of encoding/decoding functions for the following set of data types useful for RT and gate level circuit descriptions. These theorems are intended for pure encodings according to corollary B.

$$
\begin{array}{lcl}
\text{one} & = & \text{one} \\
\text{bool} & = & \text{T} \mid \text{F} \\
\text{num} & = & 0 \mid \text{SUC of num} \\
(\alpha)\text{option} & = & \text{none} \mid \text{any of } \alpha \\
\alpha \times \beta & = & \text{, of } \alpha \Rightarrow \beta \\
\alpha + \beta & = & \text{INL of } \alpha \mid \text{INR of } \beta
\end{array}
$$

On the gate level, booleans shall also be used for representing signal values and the scalar product shall be used for constructing compound signals. On the RT level, more complex data types such as enumeration types, natural numbers, records and variants can be used. Additionally, one, num , $(\alpha)$option and $\alpha + \beta$ shall also be used for representing data types at the RT level.

The automata theory provides some theorems with pairs of correct encoding/decoding functions for the data types mentioned above. They support conversions from RT level data type descriptions down to gate level data types. We will explain, which are the types these conversions come from and go to, rather then, explain them in detail.

We will use $\alpha \rightharpoonup \beta$ to indicate, that there is some encoding from type $\alpha$ to type $\beta$ and we will use $\alpha \rightleftharpoons \beta$ to indicate, that there are bijective encodings, i.e. encodings from $\alpha$ to $\beta$ and viceversa. Table 1 lists some useful encoding theorems and describes which types they are related to.

The theorems NUM_BOOL and NUM_PROD can be used to convert natural numbers with a limited range to tuples of booleans. NUM_PROD is used to split a boolean from a natural number and to halve the size of the number, and NUM_BOOL is used for encoding natural numbers less than 2.

Theorem OPTION_SUM states, that $(\alpha)$option can be encoded by means of + and one. Theorem BOOL_NEG states, that there is an encoding from booleans to booleans (turning T to F and viceversa).

option, + and $\times$ are all type operators. The theorems OPTION_TRANS, SUM_TRANS and PROD_TRANS derive encodings for these type operators, i.e. under the assumptions that there are encodings for their parameters — let us say some $\alpha \rightleftharpoons \alpha'$ and $\beta \rightleftharpoons \beta'$ — the encoding for the entire type expressions $(\alpha)$option, $\alpha + \beta$ and $\alpha \times \beta$, respectively, can be derived.

The binary type operators + and $\times$ are commutative and associative in the sense that there are bijective encodings between such type expressions (see theorems SUM_ASSOC, SUM_COM, PROD_ASSOC and PROD_COM).

All the encodings described until now, are bijective encodings. The encodings in the theorems OPTION_EXTEND, SUM_EXTEND and PROD_EXTEND are

---

[**] $\rightharpoonup$ only for natural numbers < 2

[***] under the assumption that $\alpha \rightleftharpoons \alpha'$ and $\beta \rightleftharpoons \beta'$

| Theorem Names | Encoding/Decoding |
|---|---|
| NUM_BOOL** | num $\rightleftharpoons$ bool |
| NUM_PROD | num $\rightleftharpoons$ num $\times$ bool |
| OPTION_SUM | $(\alpha)$option $\rightleftharpoons$ one $+\,\alpha$ |
| OPTION_TRANS*** | $(\alpha)$option $\rightleftharpoons$ $(\alpha')$option |
| OPTION_EXTEND | $\alpha$ $\rightharpoonup$ $(\alpha)$option |
| SUM_ASSOC | $(\alpha + \beta) + \gamma \rightleftharpoons \alpha + (\beta + \gamma)$ |
| SUM_COM | $\alpha + \beta$ $\rightleftharpoons$ $\beta + \alpha$ |
| SUM_TRANS*** | $\alpha + \beta$ $\rightleftharpoons$ $\alpha' + \beta'$ |
| SUM_EXTEND | $\alpha$ $\rightharpoonup$ $\alpha + \beta$ |
| SUM_PROD | $\alpha + \alpha$ $\rightleftharpoons$ bool $\times \alpha$ |
| PROD_ASSOC | $(\alpha \times \beta) \times \gamma \rightleftharpoons \alpha \times (\beta \times \gamma)$ |
| PROD_COM | $\alpha \times \beta$ $\rightleftharpoons$ $\beta \times \alpha$ |
| PROD_NEUTRAL | $\alpha \times$ one $\rightleftharpoons$ $\alpha$ |
| PROD_TRANS*** | $\alpha \times \beta$ $\rightleftharpoons$ $\alpha' \times \beta'$ |
| PROD_EXTEND | $\alpha$ $\rightharpoonup$ $\alpha \times \beta$ |
| BOOL_NEG | bool $\rightleftharpoons$ bool |

**Table1.** Encodings For Simple Data Types

applicable only in one direction. They all lead to "bigger" types in the sense that the new type contains some extra elements.

# 7 Algorithms for Deriving Correct Encodings

## 7.1 The Task

We have applied the automata theory to formally describe behavioural circuit descriptions of a synchronous VHDL subset. For a given behavioural description, we extracted the automata description in terms of its initial state $q$ and the output and transition function $f$. In these automata derived from synchronous VHDL, the state $\sigma = \sigma^c \times \sigma^d$ consists of two parts: control state $\sigma^c$ and data state $\sigma^d$. This section addresses the encoding of the control state part using the encodings given in the previous section.

The set of controller states is finite. To represent them, we used type expressions built with one, option and $+$. To derive a representation on the gate level, these types have to be mapped by tuples of booleans, i.e. data types bool and $\times$. There usually is a broad range of correct encodings. Let us assume, that only the number of bits is to be minimized and that every possible representation with a minimum number of bits is an appropriate encoding.

Each control state represents either the starting point or one of the `wait`-statement positions in the VHDL program. We will not go into the detail of how these type expressions have resulted. Here is just a brief hint on their meaning:

- one is used to represent single wait statement positions,
- $\alpha + \beta$ is used to represent the control states of a compound statement (sequence, if-then-else) consisting of two parts where $\alpha$ represents the set of wait-statement positions in the first part and $\beta$ is used to represent the wait-statement positions of the second part.
- $(\alpha)$option is used for expressing positions before or after (compound) statements. While any($s$) is used to represent wait-statement positions within a statement, none is used to indicate either the position before the statement or (in another context) the position immediately after the statement.

## 7.2   Derivation of a Minimal Bit Encoding

We will illustrate the minimal bit encoding algorithm by an example. Let us assume, that $\sigma^c$ is as follows:

$$(\text{one} + (\text{one})\text{option})\text{option} + (\text{one} + \text{one}) \tag{I}$$

**Substitution of option**   In the first step all occurances of $(\alpha)$option are replaced by one $+ \alpha$. Theorem OPTION_SUM is used to perform this encoding step. The type reached after the encoding:

$$(\text{one} + (\text{one} + (\text{one} + \text{one}))) + (\text{one} + \text{one}) \tag{II}$$

**Balancing**   Now the type expression consists of the type constant one and the binary type operator $+$ only. The cardinality of a set represented by such a type expression equals the number of one occurences. Such type expressions can be seen as binary trees, whose depth corresponds to the number of bits needed for encoding.

In this step, the depth of the tree is reduced by applying SUM_ASSOC. The algorithm balances the tree in a bottom up fashion. Let $\alpha + \beta$ be some node where the cardinalities of $\alpha$ and $\beta$ are $|\alpha|$ and $|\beta|$, respectively. If $|\alpha| > 2 * |\beta|$ holds, then SUM_ASSOC is applied and if $|\beta| > 2 * |\alpha|$ holds, then SUM_ASSOC is applied in the inverse direction.

In our example, there is only one position, where the tree has to be balanced: the subexpression (one $+$ (one $+$ (one $+$ one))). Here the cardinality of the left hand side is 1 and the cardinality of the right hand side is 3. So SUM_ASSOC is applied in the inverse direction. We obtain:

$$((\text{one} + \text{one}) + (\text{one} + \text{one})) + (\text{one} + \text{one}) \tag{III}$$

**Extension** Until now, the cardinality of the entire type has been left unchanged. In order to reach a symmetric tree and to be able to encode the type by scalar products of booleans, we will now add some redundant states. Theorem SUM_EXTEND is applied to encode one by one + one whenever one is a leaf with a depth less than the maximum depth of the tree.

In our example, there were 6 states. After the extension, there are 8. In the automaton the two extra states which have been added during the extension are unreachable.

$$((\text{one} + \text{one}) + (\text{one} + \text{one})) + ((\text{one} + \text{one}) + (\text{one} + \text{one})) \qquad \text{(IV)}$$

**Substitution of + and one** Now the type expression tree is symmetric, i.e. in every node the left hand side equals the right hand side. Theorem SUM_PROD is now applied repeatedly applied in a top down fashion.

$$\text{bool} \times (\text{bool} \times (\text{bool} \times \text{one})) \qquad \text{(V)}$$

Finally SUM_NEUTRAL is applied to encode bool × one by bool.

$$\text{bool} \times (\text{bool} \times \text{bool}) \qquad \text{(VI)}$$

## 7.3   Derivation of a One Hot Encoding

We use the same example $\sigma^c$ as in the minimal bit encoding example:

$$(\text{one} + (\text{one})\text{option})\text{option} + (\text{one} + \text{one}) \qquad \text{(I)}$$

**Substitution of option** As in the previous example, the option type operator is eliminated using OPTION_SUM:

$$(\text{one} + (\text{one} + (\text{one} + \text{one}))) + (\text{one} + \text{one}) \qquad \text{(II)}$$

**Flattening** Applying SUM_ASSOC repeatedly leads to:

$$\text{one} + (\text{one} + (\text{one} + (\text{one} + (\text{one} + \text{one})))) \qquad \text{(III)}$$

**Substitution of + and one** Combining the encodings SUM_TRANS (in forward direction), ONE_EXTEND and SUM_PROD leads to the following compound encoding:

$$\alpha + \text{one} \rightharpoonup \text{bool} \times \alpha$$

Applying this compound encoding encodes each repeatedly produces:

$$\text{bool} \times (\text{bool} \times (\text{bool} \times (\text{bool} \times (\text{bool} \times \text{bool})))) \qquad \text{(IV)}$$

The previous type expression consisted of 6 states, where each of them corresponds to one one-subexpression.

# 8 Conclusion and Future Work

We have introduced a theory for automata representation and transformation. The transformations defined are constructive and hence lead to refinements and optimizations on the automata through different levels of abstraction. An illustration of how state encodings can be derived in a formal synthesis fashion was also given. The state encoding algorithm presented is similar to conventional synthesis algorithms except that correctness is guaranteed implicitly, since the algorithm is based on HOL.

Such formal synthesis algorithms offer an alternative to the conventional synthesis/verification approach. We believe, that in general formal synthesis can be much more efficient than synthesis combined with an extra verification step. The result of a non-formal synthesis is just the implementation, the information on *how* the implementation is derived gets lost and cannot be used during the post-synthesis verification step.

We believe, that formal synthesis algorithms can also be exploited in other areas of hardware synthesis such as boolean optimization, scheduling, system level synthesis. The automata theory will be a basis for circuit descriptions on the algorithmic and system level.

# References

[Day92] Nancy Day. A comparison between statecharts and state transition assertions. In [hug92], pages 247–262.

[EiSK93] D. Eisenbiegler, K. Schneider, and R. Kumar. A functional approach for formalizing regular hardware structures. In [hug93], pages 101–114.

[hug92] Luc Claesen and Michael Gordon, editors. *Higher Order Logic Theorem Proving and Its Applications*, Leuven, Belgium, November 1992. North-Holland.

[hug93] Jeffrey J. Joyce and Carl-Johan H. Seger, editors. *Higher Order Logic Theorem Proving and Its Applications*, Vancouver, B.C., Canada, August 1993. Springer.

[Loew92] Paul Loewenstein. A formal theory of simulations between infinite automata. In [hug92], pages 227–246.

[Melh88] F. Melham. Automating recursive type definitions in higher order logic. Technical Report 140, University of Cambridge Computer Laboratory, 1988.

[ScKK93] R. Kumar K. Schneider and Thomas Kropf. Alternative proof procedures for finite-state machines in higher-order logic. In [hug93], pages 213–226.

# Interfacing HOL90 with a Functional Database Query Language

Elsa L. Gunter[1] and Leonid Libkin[2]

[1] AT&T Bell Laboratories, Rm.#2A-432
600 Mountain Ave., Murray Hill, N.J. 07974, USA
phone: 1 908 582 5613  email: elsa@research.att.com
[2] AT&T Bell Laboratories, Rm.#2A-422
phone: 1 908 582 7647  email: libkin@research.att.com

**Abstract.** We describe a functional database language OR-SML for handling disjunctive information in database queries, its implementation in Standard ML [10], and its interface to HOL90. The core language has the power of the nested relational algebra, and it is augmented with or-sets which are used to deal with disjunctive information. Sets, or-sets and tuples can be freely combined to create objects, which gives the language a greater flexibility. We give an example of queries over the "database" of HOL90 theories which require disjunctive information and show how to use the language to answer these queries. Since the system is running on top of Standard ML and all database objects are values in the latter, the system benefits from combining a sophisticated query language with the full power of a programming language. The language has been implemented as an HOL90-loadable library of modules in Standard ML.

## 1  Introduction

In this paper we describe a functional language, which we call OR-SML, for querying databases with incomplete and disjunctive information, and its application to querying HOL90 theories. Our language is based on the functional paradigm. Design of functional database query languages has been studied extensively in the past few years and proved very useful. (See, for example [1, 2, 11, 14, 9].) Functional query languages have clear syntax, they can be typechecked, their semantics is generally easy to define and they allow a limited form of polymorphism. We believe that such a powerful and general query language will greatly facilitate the user in interactively finding useful theorems in HOL90 theories, but will also allow for the development of tactics and other tools of proof development which make full use of stored ancestor theories.

The language we describe in this paper contains the nested relational algebra as a sublanguage. The nested relational algebra is a standard query language for database objects that freely combine values of base types, records and sets. Its standard presentations [4, 12, 13] have cumbersome syntax, so we have decided to follow the approach of [2] which gives a clean and simple language that has precisely the same expressive power.

In order to represent disjunctive information in our query language, we added a new type constructor for *or-sets* to the nested relational algebra. One of the problems addressed in the language is the difference between *structural queries* and *conceptual queries*. At the structural level, an *or-set* is a collection of objects, just as a set is. However, at the conceptual level, an or-set represents one element from the or-set, while the set continues to represent the whole collection. It was shown in [8, 7] that conceptually equivalent objects can be reduced in a canonical manner to the same object, called its *normal form*. The normal form is a disjunct of all usual objects (*i.e.* not involving disjunctive information) represented by the given object prior to normalization. Therefore, one can take the conceptual meaning of any object to be its normal form. Consequently, a conceptual query language can be built by extending a structural language with a single operator normal which takes the input object to its normal form. A query at the conceptual level is then simply a query performed on normal forms. In section 5 we give an example where it is desirable to be able to make queries at both the structural level and at the conceptual level. We make use of queries at the structural level to distinguish between recursive datatypes and other kinds of types while we make use of the conceptual level to determine what are the possibilities for induction principles for any recursive datatype.

The system OR-SML includes a subsystem which is equivalent to the nested relational algebra, but the whole system contains much more. First, normalization is present as a primitive. OR-SML also allows programming with structural recursion on sets and or-sets. It provides a mechanism for converting any user-defined functions on base types (integers, strings, HOL90 types, terms, and theorems) into functions that fit into the type system of OR-SML. It also gives a "way out" of database objects into SML values. This is useful, for example, if you wish to incorporate a query into a tactic or a derived rule of inference.

## 2 The core language

The theoretical language upon which OR-SML is based was developed by Libkin and Wong in [8]. In this section we describe this core language, called or-$\mathcal{NRA}$, and show how it is built on top of Standard ML. We have changed the names of all constructs of or-$\mathcal{NRA}$ to the names that are used in OR-SML.

Types of database objects (also called *complex objects* in the database literature) are given by the following grammar:

$$t ::= b \mid unit \mid bool \mid t \times t \mid \{t\} \mid \langle t \rangle$$

Here $b$ ranges over a collection of base types (which in OR-SML as interfaced to HOL90 consists of int, string and a datatype composed of hol_type, term, and thm); *unit* is a special type whose domain has a unique element denoted in OR-SML by (); *bool* is the type of booleans; $t \times s$ is the product type whose objects are pairs of objects of types $t$ and $s$. The set type $\{t\}$ denotes finite sets of elements of $t$ and the or-set type $\langle t \rangle$ denotes finite or-sets of elements of $t$. The core language of or-$\mathcal{NRA}$ consists of a family of operators, implemented as SML

functions which provides the basic interface to OR-SML. Their specific types as
or-$\mathcal{NRA}$ operators are given by the rules in Fig. 1. All occurrences of $s$, $t$ and $u$
in the table are object types.

---

**General operators**

$$\frac{g:u\rightarrow s \quad f:s\rightarrow t}{\text{comp}(f,g):u\rightarrow t} \qquad \frac{c:bool \quad f:s\rightarrow t \quad g:s\rightarrow t}{\text{cond}(c,f,g):s\rightarrow t} \qquad \frac{f:u\rightarrow s \quad g:u\rightarrow t}{\text{pair}(f,g):u\rightarrow s\times t}$$

$$\frac{}{\text{p1}:s\times t\rightarrow s} \qquad \frac{}{\text{p2}:s\times t\rightarrow t} \qquad \frac{}{\text{bang}:t\rightarrow unit}$$

$$\frac{}{\text{eq}:t\times t\rightarrow bool} \qquad \frac{}{\text{id}:t\rightarrow t}$$

---

**Operators on sets**

$$\frac{}{\text{emptyset}:unit\rightarrow \{t\}} \qquad \frac{}{\text{sng}:t\rightarrow \{t\}} \qquad \frac{}{\text{union}:\{t\}\times \{t\}\rightarrow \{t\}}$$

$$\frac{f:s\rightarrow t}{\text{smap } f:\{s\}\rightarrow \{t\}} \qquad \frac{}{\text{pairwith}:s\times \{t\}\rightarrow \{s\times t\}} \qquad \frac{}{\text{flat}:\{\{t\}\}\rightarrow \{t\}}$$

---

**Operators on or-sets**

$$\frac{}{\text{emptyorset}:unit\rightarrow \langle t\rangle} \qquad \frac{}{\text{orsng}:t\rightarrow \langle t\rangle} \qquad \frac{}{\text{orunion}:\langle t\rangle\times \langle t\rangle\rightarrow \langle t\rangle}$$

$$\frac{f:s\rightarrow t}{\text{orsmap } f:\langle s\rangle\rightarrow \langle t\rangle} \qquad \frac{}{\text{orpairwith}:s\times \langle t\rangle\rightarrow \langle s\times t\rangle} \qquad \frac{}{\text{orflat}:\langle\langle t\rangle\rangle\rightarrow \langle t\rangle}$$

---

**Interaction of sets and or-sets**

$$\frac{}{\text{alpha}:\{\langle t\rangle\}\rightarrow \langle\{t\}\rangle}$$

---

**Fig. 1.** or-$\mathcal{NRA}$ Type Inference of OR-SML Terms

Let us briefly recall the semantics of these operators. $\text{comp}(f,g)$ is composition of functions $f$ and $g$. $\text{pair}(f,g)$ is pair formation: $\text{pair}(f,g)(x) =$

$(f(x), g(x))$. First and second projections are called p1 and p2. id is the identity function. bang always returns the unique element of type *unit*, which has the name unit_co. $\text{cond}(c, f, g)(x)$ evaluates to $f(x)$ if condition $c$ is satisfied and to $g(x)$ otherwise.

The semantics of the set constructs is the following. emptyset() is the empty set. This value also has the name empty. Similarly, the constant emptyorset() is available under the name orempty. $\text{sng}(x)$ returns the singleton set $\{x\}$. $\text{union}(x, y)$ is union of two sets $x$ and $y$. $\text{smap}(f)$ maps $f$ over all elements of a set; that is, $\text{smap}(f)\{x_1, \ldots, x_n\} = \{f(x_1), \ldots, f(x_n)\}$. pairwith pairs the first component of its argument with every item in the second component: $\text{pairwith}(y, \{x_1, \ldots, x_n\}) = \{(y, x_1), \ldots, (y, x_n)\}$. Finally, flat is flattening: $\text{flat}\{X_1, \ldots, X_n\} = X_1 \cup \ldots \cup X_n$. The semantics of the or-set constructs is similar.

The operator alpha provides interaction between sets and or-sets. Given a set $\mathcal{A} = \{A_1, \ldots, A_n\}$, where each $A_i$ is an or-set $A_i = \langle a_1^i, \ldots, a_{n_i}^i \rangle$, let $\mathcal{F}$ denote the set of all functions $f : \{1, \ldots, n\} \to \mathbb{N}$ such that $1 \leq f(i) \leq n_i$ for all $i$. Then $\text{alpha}(\mathcal{A}) = \langle \{a_{f(i)}^i \mid i = 1, \ldots, n\} \mid f \in \mathcal{F} \rangle$.

These constructs are represented in SML as follows. Every complex object has SML type co. We shall refer to the type of an object or a function in *or-NRA* as its *true type*. True types of complex objects can be inferred using the function typeof. (This is much the same as the situation with the types and terms of HOL90.) They are SML values having type co_type. When OR-SML prints a complex object together with its type, it uses :: for the true type, as : co is used to show that the SML type of the object is co.

Besides integers, strings, and booleans, we may create complex objects from one other type, namely:

```
datatype hol_theory_data =
  Type of hol_type
| Term of term
| Thm of thm
| Parent of {thy_name : string, parent : string}
| TypeOp of {thy_name : string, tyop :{Name : string, Arity : int}}
| Constant_tm of {thy_name : string, constant : term}
| Infix_tm of {thy_name : string, constant : term}
| Binder_tm of {thy_name : string, constant : term}
| Axiom of {thy_name : string, theorem : (string * thm)}
| Definition of {thy_name : string, theorem : (string * thm)}
| StoredThm of {thy_name : string, theorem :(string * thm)}
```

The constructors Type, Term, and Thm are for injecting arbitrary HOL90 types, terms and theorems into the base type. The remaining constructors are intended to allow us to represent the components of HOL90 theories. The labels parent, constant, and theorem in the arguments to StoredThm, *etc.*, should not be confused with SML functions of the same name. To create complex objects from these types, the following basic functions are available:

```
val mkintco : int -> co
val mkboolco : bool -> co
```

```
val mkstringco : string -> co
val mkbaseco : hol_theory_data -> co
```

There are also some derived functions such as

```
val mksetint : int list -> co
val mkorsint : int list -> co
val mkprodco : co * co -> co
val mksetco : co list -> co
val mkorsco : co list -> co
val mk_theory_db : string -> co
val mk_all_theories_db : unit -> co
```

The string argument to `mk_theory_db` is the name of the HOL90 theory. It returns the set of entries from the theory. The function `mk_all_theories_db` creates a complex object that is the set of all the entires in all the theories available in the running HOL90 system. The reason for this being a function is that this is an extensible collection.

An example of the creation of some complex objects is as follows:

```
- val a = mksetint [1,3,5];
val a = {1, 3, 5} :: {int} : co
- val b = mkbaseco (StoredThm{thy_name = "prim_rec",
                              theorem = ("LESS_0",
                                         theorem "prim_rec" "LESS_0")});
val b =
  (StoredThm{theorem = ("LESS_0", |- !n. 0 < SUC n),
    thy_name = "prim_rec"}) :: hol_theory_data : co
- val c = itlist
            (fn x => (fn y => orunion(orsng(mkbaseco(Thm x)),y)))
            [EQ_SYM_EQ, TRUTH] orempty;
val c = <(Thm (|- T)), (Thm (|- !x y. (x = y) = y = x))
        > :: <hol_theory_data> : co
- val d = mkorsco (map (mkbaseco o Thm) [TRUTH, EQ_SYM_EQ]);
val d = <(Thm (|- !x y. (x = y) = y = x)), (Thm (|- T))
        > :: <hol_theory_data> : co
- eq(c,d);
val it = T :: bool : co
```

Output in the example above, as in the other examples in this paper was produced with a pretty-printer for the type co using the pretty-printer installation facility of the SML-NJ compiler. The pretty-printer for the type co is constructed from a pretty-printer for the type `hol_theory_data`, which has also been installed in SML-NJ.

In the second part of the example above, we give a sample derivation of `mkorsco` from primitives. Notice that eq returns T for c and d even though they don't print out in the same order. This is, of course, because we are dealing with or-sets, and order is disregarded.

The language we presented can express many functions commonly found in query languages. Among them are boolean *and*, *or* and negation, membership test, subset test, difference, selection, cartesian product and their counterparts for or-sets, see [2, 8]. These functions are included in OR-SML in the form of a structure called Set.

```
- val x1 = mksetint [1,2];
val x1 = {1, 2} :: {int} : co
- smap (pair(id,id)) x1;
val it = {(1, 1), (2, 2)} :: {(int * int)} : co
- val x2 = mksetint [3,4];
val x2 = {3, 4} :: {int} : co
- union(x1,x2);
val it = {1, 2, 3, 4} :: {int} : co
- Set.cartprod(x1,x2);
val it = {(1, 3), (1, 4), (2, 3), (2, 4)} :: {(int * int)} : co
```

# 3  Normalization

As we discussed before, while an object $\langle 1, 2, 3 \rangle$ is structurally just a set, conceptually it is a single integer which is either 1 or 2 or 3. Assume we are given an object $x : t$ where type $t$ contains some or-set brackets. What is this object conceptually? Since we want to list all possibilities explicitly, it must be an object $x' : \langle t' \rangle$ where $t'$ is derived from $t$ by erasing all or-set brackets from $t$. Intuitively, for any given object $x$ we can find the corresponding $x'$, but the question is whether there exists a coherent way of obtaining all objects which the given object can conceptually represent. Such a way was found in [8] and later refined in [7].

**Definition:** The *normal form* of a complex object is the or-set of all *conceptual representations* of the complex object, where

- for any $x$ of base type, $y$ is a conceptual representation of $x$ if and only if $y = x$;
- $(x', y')$ conceptually represents $(x, y)$ if and only if $x'$ conceptually represents $x$ and $y'$ conceptually represents $y$;
- $x$ conceptually represents $\langle x_1, \ldots, x_n \rangle$ if and only if $x$ conceptually represents one of $x_1, \ldots, x_n$;
- $\{x'_1, \ldots, x'_k\}$ conceptually represents $\{x_1, \ldots, x_n\}$ if and only if each $x' \in \{x'_1, \ldots, x'_k\}$ conceptually represents some $x \in \{x_1, \ldots, x_n\}$ and each $x \in \{x_1, \ldots, x_n\}$ is conceptually represented by some $x' \in \{x'_1, \ldots, x'_k\}$.

Normalization is supported in OR-SML by the addition of two new functions: normalize of SML type co_type -> co_type and normal of SML type co -> co. These two functions are sufficient to give OR-SML adequate power to work with conceptual representations of objects. The function normalize, when applied to a type $t$, returns the type of the or-set of conceptual representatives of objects of type $t$. As a primitive operator of OR-SML, normal has true type

$$normal : t \rightarrow normalize(t)$$

The semantics of normal x is the or-set of conceptual representatives of x. For example, if we construct a pair x of an HOL90 term coupled with selections of atomic subtypes of the term's type, this is conceptually the same as a selection of pairs of the term coupled with one of its atomic subtypes.

```
- val x = mkprodco(mkbaseco (Term(--'f:ind -> 'a'--)),
                   mkorsco (map (mkbaseco o Type)
                            [(=='':ind'==),(=='':'a'==)])));
val x =
  ((Term ((--'f'--))),
   <(Type ((=='':'a'==))), (Type ((=='':ind'==)))
    >) :: (hol_theory_data * <hol_theory_data>) : co
- normalize (typeof x);
val it = <(hol_theory_data * hol_theory_data)> : co_type
- normal x;
val it =
  <((Term ((--'f'--))), (Type ((=='':ind'==)))),
   ((Term ((--'f'--))), (Type ((=='':'a'==))))
   > :: <(hol_theory_data * hol_theory_data)> : co
```

Many conceptual queries do not require that all objects in the normal form be constructed at once, before the query can be asked. Instead, processing elements of the normal forms one-by-one will often suffice. For example, in order to select normal form entries that satisfy a given criterion, one need not construct the entire normal form first. This suggests a different evaluation scheme for conceptual queries that processes entries in the normal form one-by-one while accumulating the result. OR-SML provides a number of functions that implement this query evaluation mechanism. We do not discuss this feature of OR-SML here. The interested reader may consult [7]. Notice that such a scheme greatly reduces the space usage, which is important for large size databases.

# 4 Additional features of the system

## 4.1 Primitives involving hol_theory_data

So far we have seen ways to create objects of type co out of objects of type hol_theory_data, and how to perform various generic constructs, like pairing and union on the resulting objects. However, the system must also provide a way of making functions on hol_theory_data into functions that fit into the type system of OR-SML. For example, we want to be able to write a function thm_of_co : co -> co that extracts the theorem component (as a complex object) from complex object consisting of the information about a stored theorem. Furthermore, there is a need for a mechanism to translate predicates on hol_theory_data into predicates on complex objects which can be used with operators like cond and Set.select.

The solution to this problem is given by the function apply which takes a function f : hol_theory_data list -> hol_theory_data and returns a function from co to co representing the action of f on complex objects. For example, if val f_co = apply f, then f_co applied to a complex object $(r_1, (r_2, r_3))$

yields $f$ $[r_1, r_2, r_3]$ in the form of a complex object. If f_co is applied to a complex object which is not a tuple of hol_theory_data, then it raises the exception Cannotapply. Bundling the arguments to a function in a list allows us to apply functions of arbitrarily many arguments over hol_theory_data to the corresponding complex objects.

In practice, most of the functions we will wish to perform on hol_theory_data are unary or binary. Therefore, OR-SML has a special feature that allows you to apply unary and binary functions on hol_theory_data by using functions apply_unary and apply_binary. For predicates, apply_test takes a function of type (hol_theory_data -> bool) and returns it in the form of a function on complex objects. For example, we may define thm_of by

```
- fun thm_of (Axiom {theorem = (_,thm),...}) = Thm thm
  | thm_of (Definition {theorem = (_,thm),...}) = Thm thm
  | thm_of (StoredThm {theorem = (_,thm),...}) = Thm thm
  | thm_of (x as Thm _) = x
  | thm_of _ = raise HOL_ERR{origin_structure = "top",
                             origin_function = "thm_of",
                             message = "Has no theorem component"};
val thm_of = fn : hol_theory_data -> hol_theory_data
```

The function can when applied to a function f and then to an argument x returns true if (f x) does not raise an exception, and false if it does. Using can with thm_of gives us a test for whether an hol_theory_data object has a theorem component. With this we are able to write a query to extract all the theorems from the complex object representing an HOL90 theory as follows:

```
- val thm_of_co = apply_unary thm_of;
val thm_of_co = fn : co -> co
- val has_thm_component_co = apply_test (can thm_of);
val has_thm_component_co = fn : co -> co
- val one_db = mk_theory_db "one";
val one_db =
  {(StoredThm{theorem = ("one_axiom", |- !f g. f = g),
     thy_name = "one"}),
   (StoredThm{theorem = ("one", |- !v. v = one), thy_name = "one"}),
   (StoredThm{theorem = ("one_Axiom", |- !e. ?!fn. fn one = e),
     thy_name = "one"}),
   (Definition{theorem = ("one_TY_DEF",
    |- ?rep. TYPE_DEFINITION (\b. b) rep),
     thy_name = "one"}),
   (Definition{theorem = ("one_DEF", |- one = (@x. T)),
     thy_name = "one"}),
   (Constant_tm{constant = (--'one--)', thy_name = "one"}),
   (TypeOp{tyop = {Name = "one", Arity = 0}, thy_name = "one"}),
   (Parent {parent = "bool", thy_name = "one"})
  } :: {hol_theory_data} : co
- val one_stored_thms =
      smap thm_of_co (Set.select has_thm_component_co one_db);
val one_stored_thms =
```

```
{(Thm (|- one = (@x. T))),
 (Thm (|- ?rep. TYPE_DEFINITION (\b. b) rep)),
 (Thm (|- !e. ?!fn. fn one = e)), (Thm (|- !v. v = one)),
 (Thm (|- !f g. f = g))} :: {hol_theory_data} : co
```

## 4.2 Structural recursion

Structural recursion on sets [1] is a very powerful programming tool for query languages. Unfortunately, it is too powerful because it is often unsafe as an operation on sets. A function defined by structural recursion is not guaranteed to be well-defined as a function on sets or or-sets (*i.e.* two different presentations of the same set or or-set may yield unequal results), and well-definedness can not generally be checked by a compiler [3]. It is, however, often helpful in writing programs, so we have decided to include structural recursion in OR-SML. Structural recursion on sets and or-sets is available to the user by means of two constructs sr and orsr that take an object $e$ of type $t$ and a function $f$ of type $s \times t \to t$ and return a function $sr(e, f)$ of type $\{s\} \to t$ or a function $orsr(e, f)$ of type $\langle s \rangle \to t$ respectively. Their semantics is as follows: $sr(e, f)\{x_1, \ldots, x_n\} = f(x_1, f(x_2, f(x_3, \ldots f(x_n, e) \ldots)))$ and similarly for orsr. That is, sr and orsr behave on sets and or-sets in much the same way as fold or itlist behaves on lists. The two functions implementing structural recursion are contained in the SML structure SR. As an example of the use of structural recursion, functions for converting sets to or-sets and vice versa can be defined as follows:

```
- val set_to_or = SR.sr(orempty, (fn (x,y) => orunion(orsng x, y)));
val set_to_or = fn : co -> co
- val or_to_set = SR.orsr(empty, (fn (x,y) => union(sng x, y)));
val or_to_set = fn : co -> co
- val a = mksetint [1,2,3,4,5];
val a = {1, 2, 3, 4, 5} :: {int} : co
- val b = set_to_or a;
val b = <1, 2, 3, 4, 5> :: <int> : co
- eq (a, or_to_set b);
val it = T :: bool : co
```

## 4.3 Deconstruction of complex objects

While we can use the system as described so far to query the HOL90 theories interactively to find theorems that might be useful in solving goals, we really want to be able to incorporate the results of such queries into further computations, such as tactics and conversions. Since all operations of OR-SML described so far produce elements of type co, there is a need to have a way out of complex objects to the usual SML types. The structure DEST contains the following functions to deconstruct complex objects and obtain ordinary SML values.

```
exception Cannotdestroy
val co_to_base : co -> hol_theory_data
val co_to_bool : co -> bool
val co_to_int : co -> int
```

```
val co_to_list : co -> co list
val co_to_pair : co -> co * co
val co_to_string : co -> string
```

It should be noted that because DEST.co_to_list takes an object whose elements are supposed to be treated as unordered and orders them, deconstruction of complex objects is inherently as unsafe (in the sense of allowing ill-defined functions over sets and or-sets) as structural recursion is.

## 4.4 Derived functions for HOL90 queries

To support some of the kinds of queries that users are most likely to perform when browsing the HOL90 theories, we have provided a structure Hol_queries containing the following functions:

```
val mk_theory_db : string -> co
val mk_all_theories_db : unit -> co
val type_test : (hol_type -> bool) -> hol_theory_data -> bool
val term_test : (term -> bool) -> hol_theory_data -> bool
val thm_test : (thm -> bool) -> hol_theory_data -> bool
val data_to_type : hol_theory_data -> hol_type
val data_to_term : hol_theory_data -> term
val data_to_thm : hol_theory_data -> thm
val db_find_thms : {test:thm -> bool, theory:string} -> co
val db_find_all_thms : (thm -> bool) -> co
val seek : {pattern:term, theory:string} -> co
val seek_all : term -> co
```

The functions mk_theory_db and mk_all_theories_db were mentioned earlier. Care is taken with these two functions to memoize their results on any proper ancestor theories to avoid their subsequent recomputation. The functions type_test, term_test and thm_test convert predicates on HOL90 types, terms and theorems respectively into predicates over hol_theory_data. They can be composed with apply_test to create predicates over the corresponding complex objects. The functions data_to_type, data_to_term and data_to_thm extract the type, term or theorem from the given hol_theory_data. The function db_find_thms returns the set of all theorems satisfying the given test in the named theory, while db_find_all_thms looks in all currently available theories. The function seek returns the set of all theorems in the named theory that contain a subterm that match the given pattern. seek_all is the corresponding function for looking in all the currently available theories. The functions db_find_thms, db_find_all_thms, seek, and seek_all all return complex objects so that further queries may be performed on the result. These functions are not sufficient for complex queries, but will handle the simple lookups, and they can be the starting point of more complex queries.

# 5  Using OR-SML and HOL90 together — an example

The main use to which the OR-SML extension of HOL90 has been put so far is browsing the theories for theorems which might be relevant to the theorem proving task at hand. The power of the combination of OR-SML and HOL90 can be seen, however, with an example involving proof planning.

A very important class of user-defined types in HOL90 are those of recursive datatypes, including nested mutually recursive datatypes. Structural induction over these datatypes is often an important step in solving goals. Part of the process of defining a recursive datatype involves proving an "initiality theorem" (or pair of theorems) which states that a function over the datatype may be uniquely defined by cases over the constructors for the datatype. If a recursive datatype was defined by one of the automatic procedures for creating recursive datatypes, then such theorems have been stored in the theory database. Given a recursive datatype, there may or may not be a principle of structural induction for that type already stored in the theory database. However, one may test if a theorem is the principle of induction for a type that corresponds to a given initiality theorem. Moreover, if the principle of structural induction is not present, it may be automatically derived from the initiality theorems.

Given a goal to be proved, one often wants to proceed by structural induction over any recursive datatypes over which the goal is universally quantified. Thus, we would like to know all principles of induction stored in the HOL90 theory database that are relevant to a given goal. However, a given type may or may not be a recursive datatype. If the type is a recursive datatype, initiality may be stated as one or two theorems, one for existence of the function and the other for uniqueness. Moreover, a polymorphic datatype may have instances which are components of several mutually recursive datatypes. To see this, consider the two datatype specifications:

$$\sigma \text{ list } = \text{ Nil } | \text{ Cons of } \sigma \ (\sigma \text{ list})$$

and

$$\sigma \text{ Tree } = \text{ Lf of } \sigma \ | \ \text{ Nd of } (\sigma \text{ Tree}) \text{ list}$$

The type ($\sigma$ Tree) list is an instance of the type $\sigma$ list and is a component of the recursion defining the type $\sigma$ Tree. The types defined by these specifications provide us with the following two principles of induction:

$$\forall P. \ (P \text{ Nil } \wedge \ \forall t. \ P \ t \Longrightarrow \ \forall h. \ P \ (\text{Cons } h \ t)) \Longrightarrow \forall l. P \ l$$

and

$$\forall R \ P. \ (\ (\forall n. \ R \ (\text{Lf } n)) \wedge (\forall l. \ P \ l \ \Longrightarrow \ R \ (\text{Nd } l)) \wedge$$
$$P \text{ Nil } \wedge \ (\forall t \ l. \ (R \ t \wedge \ P \ l) \ \Longrightarrow \ P \ (\text{Cons } t \ l))\ ) \ \Longrightarrow$$
$$(\forall t. \ R \ t) \wedge (\forall l. \ P \ l)$$

The first principle says that to prove that any property $P$ holds for all lists, it suffices to show that it holds for the Nil list, and that, if it holds for the tail of a list, then it still holds when the head is put on. The second principle provides a similar reduction, but for proving properties over trees and tree lists jointly.

If we are trying to prove that a fact holds for all objects of type $\sigma$ Tree list, we could proceed by structural induction over lists, *or* we could proceed by mutual structural induction over both tree lists and trees. Our query for finding such information needs to be sensitive to the possibility of multiple choices, and thus to disjunctive information.

Assume we have the following:

all_theories_db : co is the OR-SML version of the theories database for HOL90, which is the set of entries from each ancestor theory, including the current theory;

universal_types : term -> hol_type list returns the list of the types of the leading universally quantified variables of a given term;

is_initial_theorem_for : hol_type -> hol_theory_data -> bool    tests whether a given theorem is an initiality theorem for a given type;

is_existential_for : hol_type -> hol_theory_data -> bool tests whether a given theorem is the existential half of a statement of initiality for the given type;

is_uniqueness_for : hol_theory_data -> hol_theory_data -> bool when applied to the existential half of a statement of initiality, tests whether the second argument is the the corresponding uniqueness theorem.

is_induction_for : hol_theory_data -> hol_theory_data -> bool takes an initiality theorem, or the existential half, as the first argument and tests whether the second argument is the corresponding induction theorem.

The testing functions given above may be converted into ones which work with OR-SML by composing them with apply_test. When we wish to apply a function over hol_theory_data to a complex object (*i.e.* an OR-SML object) that we know is the equivalent of an object of hol_theory_data type, we may accomplish this by composing the original function with DEST.co_to_base.

```
fun is_initial_co_for ty = apply_test (is_initial_theorem_for ty)
fun is_existential_co_for ty = apply_test (is_existential_for ty)
fun is_uniqueness_co_for thm_co =
    apply_test (is_uniqueness_for (DEST.co_to_base thm_co))
fun is_induction_co_for thm_co =
    apply_test (is_induction_for (DEST.co_to_base thm_co))
```

Using these functions together with some of the functions from OR-SML described previously, we may incrementally define the query for finding all possible sequences of relevant induction information as follows. We will gather each statement of initiality as a pair where either the first component is a theorem of initiality and the second component is the empty set, or the first component is the existential half of initiality and the second component is the set containing the uniqueness half. (Remember that OR-SML is a typed language, so we need to use the same type of representation in each case.)

```
fun mk_initial_co_for ty =
    comp (smap (fn thm_co => mkprodco(thm_co, empty)),
```

```
                  Set.select (is_initial_co_for ty))

fun mk_exist_uniq_co_for ty theory_db =
    smap (fn existential_co =>
            mkprodco(existential_co,
                     Set.select (is_uniqueness_co_for existential_co)
                                theory_db))
         (Set.select (is_existential_co_for ty) theory_db)

fun mk_initiality_options ty =
    set_to_or (union (mk_initial_co_for ty all_theories_db,
                      mk_exist_uniq_co_for ty all_theories_db))
```

For each initiality statement we find for a given type, we want to find an induction theorem, if it exists. Again, we will use sets to allow for the possibility that none exists.

```
fun get_induct_thm_co init_thms_co =
    let val init_co = p1 init_thms_co  (* Either the initiality theorem,*)
                                       (* or the existential half.       *)
        val induct_co =
    Set.select (is_induction_co_for init_co) all_theories_db
    in mkprodco (init_thms_co, induct_co)
    end

fun mk_induction_options ty =
    orsmap get_induct_thm_co (mk_initiality_options ty)
```

For any one given type ty, the query mk_induction_options ty returns the or-set of pairs of initiality theorems and possible induction theorems. We need to accumulate this information over the list of types over which the goal is universally quantified. We wish to preserve the order in which the information is gathered, to match it with the order in which the types are universally quantified. Therefore, the result we return will be a tuple. We would like each entry in the tuple to be the or-set of possibilities for viewing the type as a recursive datatype. However, some types simply are not recursive datatypes. Therefore, we take advantage of the structural level of OR-SML to replace the empty or-set by the empty tuple, that is, unit_co, the unique element of unit type, to represent that the type of the universally quantified variable does not admit induction. This allows us to switch to the conceptual level using normalization to acquire the collection of all possible sequences consisting of induction information when appropriate and a place holder of the empty tuple when induction is not appropriate.

```
fun fold_induction_options [] = unit_co
  | fold_induction_options (hd_ty :: tl_tys) =
    let val new_options = mk_induction_options hd_ty
    in cond(eq(new_options, orempty),
            (fn rem_co => mkprodco(unit_co,rem_co)),
            (fn rem_co => mkprodco(new_options,rem_co)))
```

```
                (fold_induction_options tl_tys)
    end

fun goal_induction_options goal =
    normal (fold_induction_options (universal_types goal))
```

Using a package for making nested recursive datatype definitions, we added to HOL90 the definition of the type Tree as given above. An example of finding the possible induction information for a goal over Tree list in this setting is as follows: (The output has been abbreviated for the sake of space.)

```
- val poss_ind = goal_induction_options
    (--'!(n:num) l. ((Nd (CONS (Lf n) l)) = Nd (APPEND l [(Lf n)])) =
            (EVERY (\x. x = Lf n) l)'--);
val poss_ind =
  <((((StoredThm{theorem = ("num_Axiom", (* ... the theorem ... *)),
        thy_name = "prim_rec"}),
      {}),
    {(StoredThm{theorem = ("INDUCTION", (* ... the theorem ... *)),
        thy_name = "num"})}),
   ((((StoredThm{theorem = ("list_Axiom", (* ... the theorem ... *)),
        {}),
     {(StoredThm{theorem = ("list_INDUCT", (* ... the theorem ... *),
     ())),
   (  . . . (* same first tuple as above *) . . . ,
   ((((StoredThm{theorem = ("Tree_existence",
            |- !Lf_case Nd_case Tree_NIL_Tree_case Tree_CONS_Tree_case.
               ?y y'.
                   (!x1. y (Lf x1) = Lf_case x1) /\
                   (!x1. y (Nd x1) = Nd_case (y' x1) x1) /\
                   (y' [] = Tree_NIL_Tree_case) /\
                   (!x1 x2.
                       y' (CONS x1 x2) =
                       Tree_CONS_Tree_case (y x1) (y' x2) x1 x2)),
         thy_name = "Tree"}),
      {(StoredThm{theorem = ("Tree_unique",
         . . . (* the uniqueness theorem corresponding *) . . .
         . . . (* to the above existence theorem       *) . . . ),
         thy_name = "Tree"})}),
     {(StoredThm{theorem = ("Tree_induct",
        |- !Tree_Prop Tree_list_Tree_Prop.
             (!y. Tree_Prop (Lf y)) /\
             (!y. Tree_list_Tree_Prop y ==> Tree_Prop (Nd y)) /\
             Tree_list_Tree_Prop [] /\
             (!y y'.
               Tree_Prop y /\ Tree_list_Tree_Prop y' ==>
               Tree_list_Tree_Prop (CONS y y')) ==>
             (!x1. Tree_Prop x1) /\ (!x2. Tree_list_Tree_Prop x2)),
         thy_name = "Tree"})}),
     ()))
   > :: <(((hol_theory_data * {'a}) * {hol_theory_data}) *
```

```
(((hol_theory_data * {hol_theory_data}) * {hol_theory_data}) *
    unit))> : co
- val number_of_options = length (DEST.co_to_list poss_ind);
val number_of_options = 2 : int
```

Thus there are two possible ways to proceed by induction. In each case, there is only one way to proceed by induction over the natural numbers. However, we have two different options for how to proceed by induction over lists of trees given to us by the second component of each tuple in poss_ind. We may either proceed by induction over lists, or by mutual induction over trees and lists of trees. Once having gathered this information in normal form, we may continue with further queries, such as which of the possibilities are missing the induction theorem and need to have it derived, or how many different ways are there to proceed.

In the above example, no further heap increases were required after the code from the library nested_rec was loaded. The heap at that time was 23 megabytes in HOL90 built using version 93 of SML-NJ. All database operations (including the calculation of all_theories_db) took place after the last major garbage collection. The time to run the query for poss_ind took 12 seconds of wall-clock time on a Sun Sparcstation 2. While more effort is needed to make queries more efficient in general, we feel that this is already acceptable performance for OR-SML to be a usable tool with HOL90 on moderately complicated queries.

In the above we have described a particular example of creating a query to find all possible principles of structural induction and related information relevant to a particular goal to be proved. Other examples exist which involve finding all possible sequences of equations and conditional equations for rewriting a goal towards a particular form. Our experience with using OR-SML in HOL90 is still limited. However, it is our belief that the ability to make queries involving conjunctive and disjunctive information using OR-SML within the theorem prover HOL90 will enhance the end-user's ability to gather information appropriate for planning the proof of goals.

# 6 Conclusion

We describe a functional database language built in Standard ML and interfaced to HOL90. The set part of the core language (*i.e.* the primitives not involving or-sets) is precisely the nested relational algebra. It is then extended with or-sets which are used to deal with disjunctive information. Normalization of objects, when added as a primitive, allows querying databases at the structural level and at the conceptual level. Moreover, representing objects as a single SML type allows the user to write queries using higher-order functions which are typically not present in query languages. In this paper we have described OR-SML as it connects to HOL90. OR-SML is also capable of being built as a stand-alone system, and as such, has certain features (for example file I/O, and the ability to handle multisets) that were not relevant to this setting and were not described here. A more complete description of the full system can be found in [6].

By interfacing HOL90 to a powerful, general purpose, database query language with good theoretical properties, such as a clear semantics, we believe we have provided a good tool for theory browsing as well as a solid platform for future work in constructing theorem-proving methodologies that make full use of previously developed theories.

# References

1. V. Breazu-Tannen, P. Buneman, and S. Naqvi. Structural recursion as a query language. In *Proc. of 3rd Int. Workshop on Database Programming Languages*, pages 9–19, Naphlion, Greece, August 1991.
2. V. Breazu-Tannen, P. Buneman, and L. Wong. Naturally embedded query languages. In *LNCS 646: Proc. ICDT, Berlin, Germany, October, 1992*, pages 140–154. Springer-Verlag, October 92.
3. V. Breazu-Tannen and R. Subrahmanyam. Logical and computational aspects of programming with sets/bags/lists. In *LNCS 510: Proc. of ICALP-1991*, Springer Verlag, 1991, pages 60–75.
4. L. Colby, A recursive algebra for nested relations, *Inform. Systems* 15 (1990), 567–582.
5. S. Grumbach, T. Milo, Towards tractable algebras for bags, *Proc. 12th Symposium on Principles of Database Systems*, Washington DC, 1993, pages 49–58.
6. E. Gunter and L. Libkin, OR-SML: A functional database programming language for disjunctive information and its applications, In: D. Karagiannis ed., *Proc. 5th Internat. Conf. on Database and Expert Systems Applications (DEXA'94)*, Athens, Greece, September 1994. Springer-Verlag LNCS vol. 856, 1994, pp. 641–650.
7. L. Libkin. Normalizing incomplete databases. In *Proceedings of the 14th Symp. on Principles of Database Systems*, San Jose CA, 1995, pages 219–230.
8. L. Libkin and L. Wong, Semantic representations and query languages for or-sets, *Proceedings of the 12th Symp. on Principles of Database Systems*, Washington DC, 1993, pages 37–48.
9. L. Libkin and L. Wong, Some properties of query languages for bags, In *Proceedings of the 4th International Workshop on Database Programming Languages, September 1993*, Springer Verlag, 1994, pages 97–114.
10. R. Milner, M. Tofte, R. Harper, *"The Definition of Standard ML"*, The MIT Press, Cambridge, Mass, 1990.
11. A. Ohori, V. Breazu-Tannen and P. Buneman, Database programming in Machiavelli: a polymorphic language with static type inference, In *SIGMOD 89*, pages 46–57.
12. H.-J. Schek and M. Scholl, The relational model with relation-valued attributes, *Inform. Systems* 11 (1986), 137–147.
13. S.J. Thomas and P. Fischer, Nested relational structures, in P. Kanellakis ed., *Advances in Computing Research: The Theory of Databases*, pages 269–307, JAI Press, 1986.
14. P.W. Trinder, Comprehension: A query notation for DBPLs, In *Proc. 3rd Int. Workshop on Database Progr. Languages*, 1991, pages 49–62, Morgan Kaufmann.

# Floating Point Verification in HOL

John Harrison

University of Cambridge Computer Laboratory
New Museums Site
Pembroke Street
Cambridge
CB2 3QG
England
jrh@cl.cam.ac.uk

**Abstract.** Floating-point verification is a very interesting application area for theorem provers. HOL is a general-purpose prover which is equipped with an extensive and rigorous theory of real analysis. We explain how it can be used in floating point verification, illustrating our remarks with complete verifications of simple square-root and (natural) logarithm algorithms.

## 1  Introduction

The correctness of floating point arithmetic operations is a topic of some current concern. A flaw in Intel's flagship Pentium processor's floating-point division instruction was discovered by a user and became public on 30th October 1994. After considerable vacillation, Intel eventually (on 21st December 1994) agreed to a policy of no-questions-asked replacement, and wrote off $306 million to cover the costs. (For a technical analysis of the bug, see [10].) This is a good illustration of how hardware correctness, even when not a matter of life and death, can be of tremendous financial, and public relations, significance.

Theorem proving seems a promising approach to verifying floating point arithmetic, at least when the theorem prover is equipped with a good theory of real numbers such as that developed by the present author [4]. We have two reasons in mind.

**Comprehensible specifications**

Floating-point numbers correspond to certain real numbers. This is the whole raison d'être of floating point arithmetic, and it is on this basis, rather than in terms of bitstrings, that one would like to specify the intended behaviour of floating-point operations. For example, suppose we have a set $F$ of floating point numbers. There is an obvious valuation function $v : F \to \mathbb{R}$ (for the sake of simplicity, we ignore special values like infinities and NaNs), and using this we may specify the intended behaviour of floating-point operations. For example, suppose $MUL : F \times F \to F$ is a floating-point multiplication function. We might demand that the result is the closest representable floating-point number to the true product, i.e.

$$\forall f_1, f_2 \in F. \neg \exists f' \in F. |v(f') - v(f_1)v(f_2)| < |v(MUL(f_1, f_2)) - v(f_1)v(f_2)|$$

This is too lax to specify the result of a floating point operation completely, since it may happen that there are two equally close representable values. Humans are usually taught to round $0.5000\ldots$ up to 1, but the IEEE standard for binary floating point arithmetic [5] mandates a (default) rounding mode of 'round to even', i.e. 'if the two nearest representable values are equally near, the one with its least significant bit zero shall be delivered'. The IEEE standard demands that the results of the operations of negation, addition, subtraction, multiplication, division, and square root are performed as if done with full accuracy then rounded using this scheme. This has the obvious merit that these operations will behave identically on all implementations, even if they use quite different (correct!) algorithms. It also ensures that the operations are symmetric, i.e. $MUL(f_1, f_2) = MUL(f_2, f_1)$ etc., and monotone, i.e. if $0 \leq v(f_1) \leq v(f_1')$ and $0 \leq v(f_2) \leq v(f_2')$ then $v(MUL(f_1, f_2)) \leq v(MUL(f_1', f_2'))$.

However the above correctness criterion is sometimes too *stringent* for other functions such as the transcendentals $sin, cos, exp, ln$ etc. The difficulty is known as the 'table maker's dilemma', and arises because being able to approximate a real number arbitrarily closely does *not* in general mean that one can decide the correctly rounded digits in a positional expansion. For example, a value $x$ one is approximating may be exactly the rational number 3.15. This being the case, knowing for any given $n \in N$ that $|x - 3.15| < 10^{-n}$ does not help to decide whether 3.1 or 3.2 is the correctly rounded result to one decimal place. For similar reasons, the original definition of 'computable real number' by Turing [12], based on a computable decimal expansion, turned out to be inadequate, and was subsequently modified, because the sum of two computable numbers may be uncomputable in this sense.

Now, if one knows that $x$ is irrational, then for any troublesomely close rational $q$, one must eventually be able to find $n$ such that $|x - q| > 10^{-n}$ and so decide the decimal expansion. The only trouble is, one does not know *a priori* that this is the case[1]. For example, it is still unknown whether Euler's constant:

$$\gamma = lim_{n \to \infty}(1 + 1/2 + \cdots + 1/n - ln(n))$$

is rational. In the case of the common transcendental functions like $sin$, there are results of number theory assuring us that for a rational argument (all floating-

---

[1] It's easy to dress up the Halting Problem as a question 'is this computable number rational?'. Define the $n$'th approximant to be zero if the machine is still going after $n$ steps, otherwise an approximant to some known irrational like $\pi$. Of course if $x$ is rational it has a trivially computable decimal expansion. This shows one must distinguish carefully between a function's being computable and having a computable graph.

point values *are* rational of course), the result is irrational[2]. However the appropriate bounds on the evaluation accuracy required to ensure correct rounding in all cases may be hard to find analytically (exhaustive search is hardly an option for double precision!), and even if they are found, they may be very much larger than the required accuracy in the result. Goldberg [3] gives the illustrative example of:

$$e^{1.626} = 5.083499996273\ldots$$

To round the result correctly to 4 decimal places, one needs to approximate it to within $10^{-9}$. In summary then, it is usually appropriate to settle for a weaker criterion. For example if $SIN : F \times F \to F$ is a floating point $SIN$ function, we might ask, for some suitable small $\epsilon$ that:

$$|v(SIN(f)) - sin(v(f))| < \epsilon$$

Since floating point values fluctuate wildly in order of magnitude, it is more plausible that one will want to limit the *relative* rather than the *absolute* error, i.e.

$$\left| \frac{v(SIN(f))}{sin(v(f))} - 1 \right| < \delta$$

## Mathematical infrastructure

The correctness of more complex floating-point algorithms often depends on some quite sophisticated mathematics. One of the attractions of using a theorem prover is that this can be integrated with the verification itself. In particular:

– One can avoid *mathematical* errors in the assumptions underlying the algorithm's correctness or efficiency.
– One can reliably generate any necessary precomputed constants by justifying them inside the theorem prover.

For example, the author is currently engaged in proving in HOL some theorems on polynomial and rational approximation. Define the Chebyshev polynomials as follows:

$$T_0(x) = 1$$
$$T_1(x) = x$$
$$T_2(x) = 2x^2 - 1$$
$$T_3(x) = 4x^3 - 3x$$
$$T_4(x) = 8x^4 - 8x^2 + 1$$

$$\cdots$$

---

[2] In fact for nonzero *algebraic* $x \in \mathbb{C}$ it follows from Lindemann's work [7] that all the following are *transcendental*: $e^x$, $sin(x)$, $cos(x)$, $tan(x)$, $sinh(x)$, $cosh(x)$, $sin^{-1}(x)$, and for $x \neq 1$ too, $cos^{-1}(x)$ and $ln(x)$. See [1] for a proof.

These have the property $T_n(x) = cos(n\ cos^{-1}(x))$, and can be generated by the recurrence $T_{n+1}(x) = 2xT_n(x) - T_{n-1}(x)$. They have numerous interesting properties, e.g. $T_m(T_n(x)) = T_{mn}(x)$, orthogonality:

$$\int_{-1}^{1} \frac{T_m(x)T_n(x)}{\sqrt{1-x^2}} = \begin{cases} 0 & \text{if } m \neq n \\ \pi/2 & \text{if } m = n \neq 0 \\ \pi & \text{if } m = n = 0 \end{cases}$$

and the fact that over the interval $[-1, 1]$ the scaled Chebyshev polynomials $\hat{T}_n(x) = T_n(x)/2^n$ have the smallest maximal absolute value of *any* polynomial of the same degree with leading coefficient 1, i.e. for any other monic polynomial $p(x)$ of degree $n$:

$$max_{x\in[-1,1]}|\hat{T}_n(x)| \leq max_{x\in[-1,1]}|p(x)|$$

with equality holding iff $p(x) = T_n(x)$. We can prove this in HOL, since the mathematical infrastructure easily encompasses such reasoning. Since $\hat{T}_n(x)$ has $n+1$ alternating maxima of $2^{1-n}$ and minima of $-2^{1-n}$, if $p(x)$ is any polynomial with strictly smaller extremal values, the difference $p(x) - \hat{T}_n(x)$ is alternately positive and negative at $\hat{T}_n(x)$'s extrema. By the Intermediate Value Theorem (proved in the HOL reals library), $p(x) - \hat{T}_n(x)$ must therefore have at least $n$ roots. But if $p(x)$ also has leading coefficient 1, that is impossible, since a polynomial of degree $n - 1$ cannot have $n$ roots. A refinement of this argument, considering repeated roots, yields the equality part.

A theorem we have not yet proved in HOL, due to Chebyshev, states that for any continuous real function $f(x)$, there is a polynomial approximation $p(x)$ of given degree which is best in this sense (the above theorem is the special case $f(x) = 0$). This generalizes to *rational* approximations, which are quite widely used in evaluation of transcendental functions. An algorithm developed by Remez [11] allows arbitrarily close approximation of the coefficients — see Fike [2] for a presentation. We would like to integrate this algorithm into the theorem prover in order to generate the constants reliably and prove the error bounds. The generation of constants is considered in more detail below.

## 2   A Floating Point Format

A floating point $f$ number usually consists of three fields, which in actual implementations are allocated a certain bitfield in a floating point value. We have a sign $s = \pm 1$, an exponent $e$ and a mantissa (or 'significand') $m$; we'll interpret these as numbers, ignoring the details of how they are actually stored. The corresponding value is as follows:

$$val(f) = s2^e m$$

In our work we will assume $1/2 \leq m < 1$; in other words the top bit of $m$ is set. Such floating point values are said to be *normalized*. By using normalized numbers we make the maximum use of the bitfield available to the mantissa[3]. On the other hand when the exponent drops below the minimum representable value $e_{min}$, the number cannot be represented at all (whereas it could, albeit to lower accuracy, by relaxing the normalization condition). Of course we also cannot represent 0 directly as a normalized number, so in practice one allocates a normally unused value for the exponent to represent zero. In fact the IEEE standard includes both positive and negative zeros, as well as positive and negative infinities and NaN's (NaN = not a number) to represent exceptional conditions in a way which (usually) propagates through a calculation.

Taking into account all these features of the IEEE standard is not especially difficult, but is tedious and distracts from the main points. Accordingly, we will in what follows assume we are dealing with normalized numbers which are (therefore) nonzero. Our verifications will be parametrized by certain values like the number of bits used to store the mantissa.

## 3 A Square Root Algorithm

First we explain the verification of an *integer* square root algorithm and then show how to adapt it to our floating point situation. The algorithm is a binary version of a quite well-known technique for extracting square roots by hand, analogous to performing long division. The algorithm is iterative; at each step we feed in two more bits of the input (starting with the most significant end), and get one more bit of the output (ditto). More precisely, we maintain three numbers, $x_n$, $y_n$ and $r_n$, all zero for $n = 0$. At each stage we claim:

$$x_n = y_n^2 + r_n \wedge r_n \leq 2y_n$$

This means that $y_n^2 \leq x_n < (y_n + 1)^2$, the latter inequality because $x_n < y_n^2 + 2y_n + 1$ is equivalent to $x_n \leq y_n^2 + 2y_n$, i.e. $r_n \leq 2y_n$ (remember that we are in the domain of the natural numbers). That is, $y_n$ is the truncation of the square root of $x_n$ to the natural number below it, and $r_n$ is the remainder resulting from this truncation. Accordingly, when the full input has been fed in, $y_n$ will hold a good approximation to the square root of the intended value $x$. The iteration step is as follows, where $s_n$ is the value (considered as a 2-bit binary integer) of the next two bits shifted in. We always set $x_{n+1} = 4x_n + s_n$, which amounts to shifting in the next two input bits. Also:

- If $4y_n + 1 \leq 4r_n + s_n$ then $y_{n+1} = 2y_n + 1$ and $r_{n+1} = (4r_n + s_n) - (4y_n + 1)$
- Otherwise $y_{n+1} = 2y_n$ and $r_{n+1} = 4r_n + s_n$.

It is not hard to see that this step does preserve the property appropriately. Consider the two cases separately:

---

[3] Since the top mantissa bit is always 1 in a normalized number, it is redundant, and can be used, e.g. to store the sign. Actually the IEEE standard interprets a mantissa as $1 + m$, not $(1 + m)/2$.

1. If $4y_n + 1 \leq 4r_n + s_n$ then we have

$$
\begin{aligned}
y_{n+1}^2 + r_{n+1} &= (2y_n + 1)^2 + ((4r_n + s_n) - (4y_n + 1)) \\
&= 4y_n^2 + 4y_n + 1 + 4r_n + s_n - 4y_n - 1 \\
&= 4y_n^2 + 4r_n + s_n \\
&= 4(y_n^2 + r_n) + s_n \\
&= 4x_n + s_n
\end{aligned}
$$

and furthermore

$$
\begin{aligned}
r_{n+1} \leq 2y_{n+1} &\iff (4r_n + s_n) - (4y_n + 1) \leq 2(2y_n + 1) \\
&\iff 4r_n + s_n \leq 8y_n + 3 \\
&\iff 4r_n + s_n \leq 4(2y_n) + 3
\end{aligned}
$$

But this is true because $r_n \leq 2y_n$ and $s_n \leq 3$ (since $s_n$ is only 2 bits wide).

2. Otherwise we have $4y_n + 1 > 4r_n + s_n$, so

$$
\begin{aligned}
y_{n+1}^2 + r_{n+1} &= (2y_n)^2 + (4r_n + s_n) \\
&= 4y_n^2 + 4r_n + s_n \\
&= 4(y_n^2 + r_n) + s_n \\
&= 4x_n + s_n
\end{aligned}
$$

and furthermore, using the above hypothesis:

$$
\begin{aligned}
r_{n+1} \leq 2y_{n+1} &\iff 4r_n + s_n \leq 2(2y_n) \\
&\iff 4r_n + s_n \leq 4y_n \\
&\iff 4r_n + s_n < 4y_n + 1
\end{aligned}
$$

since we are dealing with natural numbers; but this is true by hypothesis.

This shows that the basic algorithm works as claimed. The corresponding proof can be rendered in HOL without much difficulty. Dealing with the idea of shifting bits of $x$ in is a bit tedious, but not difficult. Let's now look at how to extend it to a floating point algorithm. To take the square root of $s2^e m$, we can first of all fail if $s = -1$. Otherwise we have a positive number. The square root is of course $2^{e/2} \sqrt{m}$. That is, we need to halve the exponent and take the square root of the mantissa. But the exponent $e$ might be odd. If that is so, we evaluate instead $2^{(e+1)/2} \sqrt{m/2}$. So we can assume hereafter that $e$ is even and $m$ is in the range $1/4 \leq m < 1$. We will perform all the arithmetic on the mantissa using $N$ bits of accuracy, which will in general be a little more than the number of bits in the mantissa (more on this below). In particular the right shift above does not lose bits.

We simply scale the mantissa $m$ and consider $2^{2N}m$. Of course we don't actually need to store $2N$ bits; we can simply feed bits of $m$ in till we run out, and thereafter feed in zeros. Now we apply the integer square root algorithm with $x = 2^{2N}m$. After $N$ iterations we have:

$$y_N^2 \leq 2^{2N}m < (y_N + 1)^2$$

so:

$$y_N \leq 2^N\sqrt{m} < y_N + 1$$

and therefore:

$$1 \leq \frac{\sqrt{2^e m}}{2^{e/2}(y_N/2^N)} \leq 1 + \frac{1}{y_N}$$

Therefore if we take $2^{e/2}(y_N/2^N)$ as the result, the relative error is bounded by $1/y$. Now $1/4 \leq m < 1$, so $1/2 \leq \sqrt{m} < 1$. Therefore $y_N \leq 2^N\sqrt{m} < 2^N$, and furthermore $y_N \geq 2^N\sqrt{m} \geq 2^{N-1}$. Consequently if we take the top bits of $y_N$ for the mantissa, the result will be correctly normalized, and also the relative error $1/y$ is bounded by $2^{1-N}$. Chopping the top $n$ bits out of the result gives an additional maximum relative error of $2^{-n}$. However a more refined final part to the procedure, taking note of the current remainder, would allow correct rounding.

# 4   A CORDIC Natural Logarithm algorithm

The CORDIC technique, invented by Volder [13] and developed further by Walther [14], provides a simple scheme for calculating a wide range of transcendental functions. It is an iterative algorithm which at stage $n$ only requires multiplications by $1 \pm 2^{-n}$, which can be done efficiently by a shift and add. Here is a CORDIC algorithm to find the natural logarithm of $x$, which we shall suppose to be in the range $1/2 \leq x < 1$, evidently the case for the mantissa of a normalized floating point number. We start off with $x_0 = x$ and $y_0 = 0$. At each stage:

- If $x_n(1 + 2^{-(n+1)}) < 1$ then $x_{n+1} = x_n(1 + 2^{-(n+1)})$ and $y_{n+1} = y_n + ln(1 + 2^{-(n+1)})$.
- Otherwise $x_{n+1} = x_n$ and $y_{n+1} = y_n$.

Now it is easy to establish by induction on $n$ that at stage $n$ we have $x_n < 1$ but $x_n(1 + 2^{-n}) \geq 1$, and that $y_n = ln(x_n) - ln(x)$. For example, if we know $x_n(1 + 2^{-n}) \geq 1$, then either:

- We have $x_{n+1} = x_n(1 + 2^{-(n+1)}) < 1$, and

$$x_{n+1}(1 + 2^{-(n+1)}) = x_n(1 + 2^{-(n+1)})^2$$
$$= x_n(1 + 2^{-n} + 2^{-2(n+1)})$$
$$\geq x_n(1 + 2^{-n})$$
$$\geq 1$$

- We have $x_n(1 + 2^{-(n+1)}) \geq 1$ and $x_{n+1} = x_n$, so $x_{n+1}(1 + 2^{-(n+1)}) \geq 1$ as required.

Now after $n$ iterations we find, since $1/(1 + 2^{-n}) \leq x_n < 1$, that $-2^{-n} \leq ln(x_n) < 0$. To get this we use the following theorem, which will be applied frequently in what follows:

$$\forall x.\, 0 \leq x \implies ln(1 + x) \leq x$$

This is not actually proved in the current version of the reals library, but is simple to derive from theorems like $0 \leq x \implies 1 + x \leq e^x$ which are. The same goes for other mathematical results we use below.

Now, we have $|ln(x_n)| \leq 2^{-n}$, so $|y_n - ln(x)| \leq 2^{-n}$. Therefore by a suitable number of iterations we can approximate $ln(x)$ as closely as we please. The only operations required are addition and shift. Of course it is necessary to have a suitable table of constants $ln(1 + 2^{-n})$ precomputed. We return to this matter below.

## Details of verification

The above was based on an abstract mathematical description of the algorithm. However a real implementation will introduce additional errors, because the shift-right will lose bits, and the stored logarithm values will be inexact. In our verification we must take note of these facts. We suppose:

- The $x_n$ values are stored as an $N_1$-bit binary fraction, and the shift-right operation truncates all values. Hence:

$$[x_n(1 + 2^{-(n+1)})] = x_n(1 + 2^{-(n+1)}) - \epsilon_n$$

where $0 \leq \epsilon_n \leq 2^{-N_1}$.
- The logarithms are stored as $N_2$-bit binary fractions, rounded down. They are all in the range $0 < ln(z) < 1/2$, so the truncated values have the following property:

$$[ln(1 + 2^{-n})] = ln(1 + 2^{-n}) - \delta_n$$

where $0 \leq \delta_n \leq 2^{-N_2}$.

The comparison is done on the basis of $[x_n(1 + 2^{-(n+1)})]$. Hence the step phase of the algorithm becomes:

- If $x_n(1 + 2^{-(n+1)}) - \epsilon_n < 1$ then $x_{n+1} = x_n(1 + 2^{-(n+1)}) - \epsilon_n$ and $y_{n+1} = y_n + ln(1 + 2^{-(n+1)}) - \delta_n$.
- Otherwise $x_{n+1} = x_n$ and $y_{n+1} = y_n$.

Clearly we still have $x_n < 1$. However the other properties become more complicated. A detailed error analysis has been conducted in HOL (it does not claim to be especially refined; indeed it is rather pessimistic). We find that it is convenient to assume that $n + 2 \leq N_1$. On that basis we find that:

$$x_n(1 + 2^{-n}) \geq (1 - n2^{-N_1})$$

Certainly this is true for $n = 0$ (it just says $x \geq 1/2$). So suppose it is true for $n$; we will prove it for $n + 1$. There are two cases to consider.

1. If $x_n(1 + 2^{-(n+1)}) - \epsilon_n < 1$, then we have

$$
\begin{aligned}
x_{n+1}(1 + 2^{-(n+1)}) &= (x_n(1 + 2^{-(n+1)}) - \epsilon_n)(1 + 2^{-(n+1)}) \\
&= x_n(1 + 2^{-n}) + x_n 2^{-2(n+1)} - \epsilon_n(1 + 2^{-(n+1)}) \\
&\geq (1 - n2^{-N_1}) + (1/2)2^{-2(n+1)} - 2^{-N_1}(1 + 2^{-(n+1)}) \\
&= (1 - (n+1)2^{-N_1}) + (1/2)2^{-2(n+1)} - 2^{-N_1}2^{-(n+1)}
\end{aligned}
$$

It suffices, then, to show that $2^{-N_1}2^{-(n+1)} \leq (1/2)2^{-2(n+1)}$. But this is equivalent to $n + 2 \leq N_1$, true by hypothesis.

2. Otherwise we have $x_n(1 + 2^{-(n+1)}) - \epsilon_n \geq 1$ and $x_{n+1} = x_n$, so

$$
\begin{aligned}
x_{n+1}(1 + 2^{-(n+1)}) &= x_n(1 + 2^{-(n+1)}) \\
&\geq x_n(1 + 2^{-(n+1)}) - \epsilon_n \\
&\geq 1 \\
&\geq 1 - (n+1)2^{-N_1}
\end{aligned}
$$

As far as the accuracy of the corresponding logarithm we accumulate is concerned, we claim:

$$|y_n - (ln(x_n) - ln(x))| \leq n(4\,2^{-N_1} + 2^{-N_2})$$

Again this is trivially true for $n = 0$, so consider the two cases.

1. If $x_n(1 + 2^{-(n+1)}) - \epsilon_n < 1$, then we have for the overall error $E = |y_{n+1} - (ln(x_{n+1}) - ln(x))|$:

$$
\begin{aligned}
E &= |(y_n + ln(1 + 2^{-(n+1)}) - \delta_n) - (ln(x_n(1 + 2^{-(n+1)}) - \epsilon_n) - ln(x))| \\
&= |(y_n - (ln(x_n) - ln(x))) + ln(1 + 2^{-(n+1)}) + ln(x_n) \\
&\quad - ln(x_n(1 + 2^{-(n+1)}) - \epsilon_n) - \delta_n|
\end{aligned}
$$

$$\leq |y_n - (ln(x_n) - ln(x))| + |ln\left(\frac{x_n(1 + 2^{-(n+1)})}{x_n(1 + 2^{-(n+1)}) - \epsilon_n}\right)| + |\delta_n|$$

$$\leq n(4\,2^{-N_1} + 2^{-N_2}) + |ln\left(1 + \frac{\epsilon_n}{x_n(1 + 2^{-(n+1)}) - \epsilon_n}\right)| + 2^{-N_2}$$

$$\leq n(4\,2^{-N_1}) + (n+1)2^{-N_2} + \frac{\epsilon_n}{x_n(1 + 2^{-(n+1)}) - \epsilon_n}$$

$$\leq n(4\,2^{-N_1}) + (n+1)2^{-N_2} + \frac{2^{-N_1}}{(1/2) - 2^{-N_1}}$$

$$\leq n(4\,2^{-N_1}) + (n+1)2^{-N_2} + \frac{2^{-N_1}}{(1/2) - (1/4)}$$

$$= (n+1)(4\,2^{-N_1} + 2^{-N_2})$$

2. Otherwise $x_{n+1} = x_n$ and $y_{n+1} = y_n$, so since $n(4\,2^{-N_1} + 2^{-N_2}) \leq (n+1)(4\,2^{-N_1} + 2^{-N_2})$ the result follows.

Finally we can analyze the overall error bound after $n$ iterations. Since $|y_n + ln(x)| \leq |y_n - (ln(x_n) - ln(x))| + |ln(x_n)|$, we just need to find a bound for $|ln(x_n)|$. Since

$$1 \geq x_n \geq \frac{1 - n2^{-N_1}}{1 + 2^{-n}}$$

we have:

$$|ln(x_n)| \leq ln\left(\frac{1 + 2^{-n}}{1 - n2^{-N_1}}\right)$$

$$\leq ln\left(\frac{(1 + 2^{-n})(1 + 2n2^{-N_1})}{(1 - n2^{-N_1})(1 + 2n2^{-N_1})}\right)$$

$$= ln\left(\frac{(1 + 2^{-n})(1 + 2n2^{-N_1})}{1 + n2^{-N_1} - 2n^2 2^{-2N_1}}\right)$$

$$= ln(1 + 2^{-n}) + ln(1 + 2n2^{-N_1}) - ln(1 + n2^{-N_1} - 2n^2 2^{-2N_1})$$

$$\leq 2^{-n} + 2n2^{-N_1} - ln(1 + n2^{-N_1} - 2n^2 2^{-2N_1})$$

If we introduce the additional assumption that $n \leq 2^{N_1 - 1}$, hardly a stringent requirement since in practice $N_1$ will be at least 20, then we find that $ln(1 + n2^{-N_1} - 2n^2 2^{-2N_1}) \geq ln(1) = 0$, so we have $|ln(x_n)| \leq 2^{-n} + 2n2^{-N_1}$. We get our grand total error bound:

$$|y_n + ln(x)| \leq n(6\,2^{-N_1} + 2^{-N_2}) + 2^{-n}$$

It's rather complicated, but if we have a desired accuracy (say $N$ bits), it's easy to find appropriate $N_1$, $N_2$ and $n$ to make sure it is met. Here is the final HOL form of the correctness theorem:

```
|- !x n.
    (inv(& 2)) <= x /\
    x < (& 1) /\
    (n num_add 2) num_le N1 /\
    (& n) <= (&(2 EXP (N1 num_sub 1))) ==>
    (abs((CORDIC_Y x n) + (ln x))) <=
    (((& n) * (((& 6) * (inv(&(2 EXP N1)))) + (inv(&(2 EXP N2))))) +
    (inv(&(2 EXP n))))
```

The step to a full floating point algorithm is quite interesting. Since $ln(2^e m) = e\, ln(2) + ln(m)$, we just need to multiply $e$ by the prestored constant $ln(2)$ and subtract $y_n$. This result has an absolute error only a little worse than that indicated for the mantissa. However it is then necessary to renormalize the result in general. Assuming the result is stored to $N_2$ places, then provided the magnitude of the answer is at least 1/4, no great additional loss of accuracy is concerned. However if $x$ was very close to 1, the logarithm will be very small. When renormalized, the *relative* error will be blown up.

What is to be done about this? One school of thought is to ignore it, since the input $x$ is presumably not exact, and since $ln(1 + x) \approx x$, the inaccuracy in $ln(x)$ is commensurate with that of $x$. Such a statement is yet another reasonable specification for a floating point operation in terms of real numbers! An analogous situation arises when taking $sin(z)$ for very large $z$. Usually one will first perform a range reduction to $z'$ where $|z'| < \pi/4$ and $(z' - z)/(\pi/4) \in \mathbb{Z}$. If $z$ is very large, it takes considerable effort to do this accurately, and one might feel it isn't worth it since the input is probably inexact anyway.

Alternatively, one could choose $N_2$ quite a bit bigger than the number of bits required in the mantissa, say $N_3$ larger. Now if the result is at least $2^{-N_3}$, we are OK. Otherwise an alternative algorithm can be kicked in, which is adequate for $x$ below $2^{-N_3}$. For example, one might evaluate the first one or two terms of the Taylor series.

### Generating the constants

For the logarithm algorithm, we need a precomputed table of constants $ln(1 + 2^{-n})$. Since this will not be the last time we require some mathematical functions evaluated in HOL to a certain accuracy, it is desirable to have a general package that will do it. The first stage is to have a reasonably efficient means of doing natural number and integer arithmetic. The HOL numeral library [6] meets this requirement, although it is still of course far slower than using the native machine bignums. Whether the numeral library is fast enough in practice can only be shown by case studies.

Now in different applications, the accuracy required will be different. Therefore a real number calculation package should be capable of accommodating any reasonable requirements on precision. The most elegant and flexible way of doing this follows the current body of research in 'exact real arithmetic', of which the work of Menissé [8] probably represents the state of the art. These methods normally use function closures to represent real numbers, the idea being that when given a required precision, they return a rational number (or integer with appropriate scaling understood) which approximates it to the required degree. The present author is involved in implementing such a package which not only returns the approximation but returns a theorem in HOL saying that the approximation is within the required bounds.

As yet the work described above is not complete for the transcendental functions. Therefore to generate the constants for the logarithm we use a slightly ad-hoc technique. The current reals library does not include a proof of McLaurin's theorem, but a proof survives from a previous version. This was exhumed and specialized to the case of the logarithm. The theorem (verbatim) is as follows; note that $Sum(0,n)f = f(0) + \cdots + f(n-1)$.

```
|- !n x.
     (& 0) < x /\ 0 num_lt n ==>
     (?y.
       (& 0) < y /\
       y < x /\
       (ln((& 1) + x) =
       (Sum
         (0,n)
         (\m.
           ((m = 0) =>
           & 0 |
           (((--(& 1)) pow (SUC m)) / (& m)) * (x pow m)))) +
       ((((--(& 1)) pow (SUC n)) / (& n)) *
       ((inv(((& 1) + y) pow n)) * (x pow n)))))
```

Now it is a fairly straightforward matter to evaluate the required logarithms by scaling everything up by a suitable power of 2 (depending on the eventual accuracy required) and exploiting this theorem. Once $n$ becomes reasonably large, only a few terms are required. The numbers can be handled by the numeral library. The only remaining difficulty is calculating $ln(2)$ (required to multiply the exponent by), since the above series converges far too slowly (as $1/n$) when $x = 1$. The easiest solution is to derive a similar Taylor expansion for $ln(1 - x)$ and use $ln(2) = -ln(1 - 1/2)$.

## Conclusions

We have illustrated how the current version of the HOL system contains ample mathematical infrastructure to verify floating point algorithms and derive precise error bounds. We believe that such proofs, precisely because they are rather messy and intricate, are difficult for humans to get right, and because they demand substantial mathematical infrastructure, difficult to tackle with tools like model checkers which are often useful in hardware verification. There have been plenty of hand proofs of correctness of floating point algorithms, but we are not aware of many mechanized proofs. The closest work to that described here is by O'Leary et. al. [9].

The two examples chosen were quite simple, and we ignored precise details of the floating point format (e.g. representation of zero); this was partly through laziness and partly because it adds nothing to the exposition. We also neglected the details of how the algorithms are represented in hardware (or software). This is deliberate, since we want to focus on algorithms rather than the details of implementation. However work is in progress on using a simple imperative language to specify the algorithms, based on a few assumed integer arithmetic operations. This would allow direct verification of algorithms close to the pseudocode often used to describe them. Before more substantial examples are undertaken, we want to finish this, as well as the package for performing real number calculations inside HOL. But the gap between these examples and real-world implementations is quantitative, not qualitative. We hope this paper illustrates that there are no substantial technical obstacles to complete mechanically checked verification of floating point algorithms.

## Acknowledgements

I am grateful to my supervisor, Mike Gordon, and to the EPSRC and the Newton Trust for financial support. Tim Leonard provided valuable encouragement and enabled me to talk to some real hardware and software designers.

## References

1. Alan Baker. *Transcendental Number Theory*. Cambridge University Press, 1975.
2. C. T. Fike. *Computer Evaluation of Mathematical Functions*. Series in Automatic Computation. Prentice-Hall, 1968.
3. David Goldberg. What every computer scientist should know about floating point arithmetic. *ACM Computing Surveys*, 23:5–48, 1991.
4. John Harrison. Constructing the real numbers in HOL. *Formal Methods in System Design*, 5:35–59, 1994.
5. IEEE. Standard for binary floating point arithmetic. ANSI/IEEE Standard 754-1985, The Institute of Electrical and Electronics Engineers, Inc., 345 East 47th Street, New York, NY 10017, USA, 1985.
6. Tim Leonard. The HOL numeral library. Distributed with HOL system, 1993.
7. F. Lindemann. Über die Zahl $\pi$. *Mathematische Annalen*, 120:213–225, 1882.
8. Valérie Ménissier-Morain. *Arithmétique exacte, conception, algorithmique et performances d'une implémentation informatique en précision arbitraire*. Thèse, Université Paris 7, December 1994.
9. John O'Leary, Miriam Leeser, Jason Hickey, and Mark Aagaard. Non-restoring integer square root: A case study in design by principled optimization. In T. Kropf and R. Kumar, editors, *Proceedings of the Second International Conference on Theorem Provers in Circuit Design (TPCD94): Theory, Practice and Experience*, volume 901 of *Lecture Notes in Computer Science*, pages 52–71, Bad Herrenalb (Black Forest), Germany, 1994. Springer-Verlag.

10. Vaughan R. Pratt. Anatomy of the pentium bug. In ?, editor, *Proceedings of the 5th International Joint Conference on the theory and practice of software development (TAPSOFT'95)*, volume 915 of *Lecture Notes in Computer Science*, pages 97–107, Aarhus, Denmark, 1995. Springer-Verlag.
11. E. Ya. Remez. Sur le calcul effectif des polynomes d'approximation. *Comptes Rendus de l'Académie des Sciences*, pages 337–340, 1934.
12. A. M. Turing. On computable numbers, with an application to the Entscheidungsproblem. *Proceedings of the London Mathematical Society (2)*, 42:230–265, 1936.
13. J. Volder. The CORDIC trigonometric computing technique. *IRE Transactions on Electronic Computers*, 8:330–334, 1959.
14. J. S. Walther. A unified algorithm for elementary functions. In *Proceedings of the AFIPS Spring Joint Computer Conference*, pages 379–385, 1971.

# Inductive definitions: automation and application

John Harrison

University of Cambridge Computer Laboratory
New Museums Site
Pembroke Street
Cambridge
CB2 3QG
England
jrh@cl.cam.ac.uk

**Abstract.** This paper demonstrates the great practical utility of inductive definitions in HOL. We describe a new package we have implemented for automating inductive definitions, based on the Knaster-Tarski fixpoint theorem. As an example, we use it to give a simple proof of the well-founded recursion theorem. We then describe how to generate free recursive types starting just from the Axiom of Infinity. This contrasts with the existing HOL development where several specific free recursive types are developed first.

## 1 Inductive definitions

Inductive definitions are very common in mathematics, especially in the definition of formal languages used in mathematical logic and programming language semantics. Camilleri and Melham [6] give some illustrative examples. Examples crop up in other parts of mathematics too, e.g. the definition of the Bòrel hierarchy of subsets of $\mathbb{R}$. A detailed discussion, from an advanced point of view, is given by Aczel [2].

Inductive definitions define a set $S$ by means of a set of *rules* of the form 'if ... then $t \in S$', where the hypothesis of the rule may make assertions about membership in $S$. These rules are customarily written with a horizontal line separating the hypotheses (if any) from the conclusion. For example, the set of even numbers $E$ might be defined as a subset of the reals by:

$$\frac{}{0 \in E}$$

$$\frac{n \in E}{(n+2) \in E}$$

Read literally, such a definition merely places some constraints on the set $E$, asserting its 'closure' under the rules, and does not, in general, determine it uniquely. For example, the set of even numbers satisfies the above, but so does the set of natural numbers, the set of integers, the set of rational numbers and even the the whole set of real numbers! But implicit in writing a definition like this is that $E$ is the *least* set which is closed under the rules. It is when understood in this sense that the above defines the even numbers.

This convention, however, needs to be justified by showing that there *is* a least set closed under the rules. A good try is to consider the set of *all* sets which are closed under the rules, and take their intersection. If only we knew that this intersection was closed under the rules, then it would certainly be the least such set. But in general we don't know that, as the following example illustrates:

$$\frac{n \notin E}{n \in E}$$

There are no sets at all closed under this rule! However it turns out that a simple syntactic restriction on the rules is enough to guarantee that the intersection is closed under the rules. Crudely speaking, the hypotheses must make only 'positive' assertions about membership in $S$. To state this precisely, observe that we can collect together all the rules in a single assertion of the form:

$$\forall x. \, P[S, x] \Rightarrow x \in S$$

The following example for the even numbers should be a suitable paradigm to indicate how:

$$\forall n. \, (n = 0 \lor \exists m. \, n = m + 2 \land m \in E) \Rightarrow n \in E$$

If we make the abbreviation $f(S) = \{x \mid P[S, x]\}$ the assertion can be written $f(S) \subseteq S$. Our earlier plan was to take the intersection of all subsets $S$ which have this property, and hope that the intersection too is closed under the rules. A sufficient condition for this is given in the following fixpoint theorem due to Knaster [14] and Tarski [22] (which holds for an arbitrary complete lattice): if $f : \wp(X) \rightarrow \wp(X)$ is monotone, i.e. for any $S \subseteq X$ and $T \subseteq X$

$$S \subseteq T \Rightarrow f(S) \subseteq f(T)$$

then if we define

$$F = \bigcap \{S \subseteq X \mid f(S) \subseteq S\}$$

not only is $f(F) \subseteq F$ but $F$ is actually a fixpoint of $f$, i.e. $f(F) = F$. To see this, define $B = \{S \subseteq X \mid f(S) \subseteq S\}$. Observe that for any $S \in B$ we have $F \subseteq S$, so by monotonicity $f(F) \subseteq f(S)$; but since $f(S) \subseteq S$ we also have $f(F) \subseteq S$. This is true for *any* $S \in B$ so we must also have $f(F) \subseteq F$. Now monotonicity gives $f(f(F)) \subseteq f(F)$, so $f(F) \in B$. This shows that $F \subseteq f(F)$ and putting everything together, $f(F) = F$. Obviously $F$ is the least set closed under the rules, for if $F'$ is another, we have $F' \in B$ and so $F \subseteq F'$.

The fixpoint property $f(F) = F$ yields what we will call a 'cases theorem' because it lists the ways an element of the inductively defined set can arise. In our example, we have:

$$\forall n.\, n \in E \iff (n = 0 \lor \exists m.\, n = m + 2 \land m \in E)$$

This is a bonus from the fact that we actually have a *fixpoint*. It also yields what we will call an 'induction theorem' asserting that $F$ is contained in every set closed under the rules. This justifies proofs by 'structural induction' or 'rule induction' (the terminology is from [25]).

$$\forall E' \subseteq \mathrm{N}.\, (0 \in E' \land (\forall n \in \mathrm{N}.\, n \in E' \Rightarrow (n + 2) \in E')) \Rightarrow E \subseteq E'$$

It is easy to derive from the cases theorem by a bit of simple logic that the inductively defined set *is* closed under the rules. We will refer to this as the 'closure theorem'.

$$0 \in E \land (\forall n \in \mathrm{N}.\, n \in E \Rightarrow (n + 2) \in E)$$

## 2  Automating inductive definitions

There are two existing packages for inductive definitions in HOL. One, by Melham [16], which does not use the Knaster-Tarski theorem, is limited to finitary hypotheses, excluding those of the form $(\forall x. \ldots) \Rightarrow t \in S$. On the other hand, it provides a very attractive interface and good proof support, and has been extended to support mutual induction by Roxas [21]. The other package was developed by Andersen and Petersen [4] as a byproduct of much more general theories. It is based on Tarski's theorem, and thus more general than Melham's package, but provides no substantial automation — in particular the user must prove the monotonicity theorem manually. Mutual induction has to be dealt with by hand too. More recently, Paulson [18] has produced a package for the Isabelle prover, also based on Knaster-Tarski. Our aim was to combine the best features of the available packages. The implementation can be divided into several stages which we discuss separately:

1. Transition from clausal definition to 'cases' form
2. Proof of monotonicity
3. Application of Tarski's theorem
4. Recovery of 'clausal' form of rules and induction theorem.

It is convenient to broaden the idea of defining a set (or equivalently, unary relation) and allow inductive definitions of arbitrary arities. We shall say that $f$ is monotone iff for each pair of $n$-ary relations $R$ and $R'$:

$$(\forall x_1, \ldots, x_n.\, R x_1 \ldots x_n \Rightarrow R' x_1 \ldots x_n)$$
$$\Rightarrow (\forall x_1, \ldots, x_n.\, f(R) x_1 \ldots x_n \Rightarrow f(R') x_1 \ldots x_n)$$

The Knaster-Tarski theorem generalizes in the obvious way. Making this generalization has two advantages:

- $n$-ary relations do not have to be reduced to an uncurried form before they can be defined.
- It fits in well with the automated proofs of monotonicity (see below).

**Transition from clausal to casewise form**

Each rule is of the form:

$$\forall x_1, \ldots, x_n. \, E[R, x_1, \ldots, x_n] \Rightarrow R \, t_1 \, \ldots \, t_k$$

except that the antecedant may be absent (for regularity, the basic package assumes all rules have an antecedant; a wrapper function then inserts $\top$ as an antecedant before calling the main function then modifies the resulting theorems). This gives rise to a casewise version:

$$(\exists x_1, \ldots, x_n. \, (z_1 = t_1) \wedge \ldots \wedge (z_k = t_k) \wedge E[R, x_1, \ldots, x_n]) \Rightarrow R \, z_1 \, \ldots \, z_k$$

Each rule $1, \ldots, p$ is transformed into something of this form, i.e.

$$E_i[R, z_1, \ldots, z_k] \Rightarrow R \, z_1 \, \ldots \, z_k$$

For the sake of readability of the cases theorem, we avoid introducing local variables when the relevant $t_i$ is itself a variable. However more could be done in other cases, e.g. if $t_i$ is a pair of variables. Now, the resulting composite rule is:

$$E_1[R, z_1, \ldots, z_k] \vee \ldots \vee E_p[R, z_1, \ldots, z_k] \Rightarrow R \, z_1 \, \ldots \, z_k$$

**Proof of Monotonicity**

The proof of monotonicity is managed by a function which makes a recursive traversal over the term. Monotonicity has good compositionality properties over the 'positive' logical operators like conjunction:

$$(A \Rightarrow A') \wedge (B \Rightarrow B') \Rightarrow ((A \wedge B) \Rightarrow (A' \wedge B'))$$

and the universal quantifier (the dollar emphasizes its role as a higher order function):

$$(\forall x. \, P \, x \Rightarrow Q \, x) \Rightarrow (\$\forall \, P \Rightarrow \$\forall \, Q)$$

We have an assignable variable of rules embodying such principles which the function tries to use, so users can add monotonicity rules for special functions they may need. They are stored as rules, not theorems, since some cannot be stored as a single theorem because they take various numbers of arguments (e.g. the rule for UNCURRY). However most of the rules just apply MATCH_MP to a theorem of the above form. We store some theorems which indicate *antitone* instances, and everything works provided they balance out (e.g. $\neg P \Rightarrow Q$). The monotonicity function works by composing the above rules, using the fact that formulas not involving the relation are trivially monotone, and dealing with abstractions as follows:

$$(\forall x.\, P\, x \Rightarrow Q\, x) \Rightarrow (\forall x\, y.\, (\lambda y.\, P)\, y\, x \Rightarrow (\lambda y.\, Q)\, y\, x)$$

We see here that to get a smooth recursion over the term structure, it is convenient to allow the variadic notion of 'monotonicity'. The function proves monotonicity completely automatically for almost every conceivable instance. For example, the reader may confirm that it deals with paired quantifiers correctly.

## Applying Knaster-Tarski

There is little to say about this stage, except that the theorem generalizes to the variadic case in a completely obvious way. The variadic form cannot be stored as a single HOL theorem, but it is simple to implement a derived rule, essentially performing the proof above (alternatively one could store an 'uncurried' form and 'curry' it appropriately each time).

## Recovering the clausal forms

The 'cases' theorem pops out of the previous stage in its final form. But for the closure and induction theorems we need to reverse the initial transformation. This is tedious, but perfectly straightforward.

## Extension to schemas and mutual induction

Sometimes (e.g. when defining the transitive closure of a relation) there are additional arguments. These can be handled by ignoring them during the main phase of the function and reintroducing them by Skolemization when an existence theorem has been derived. More constructively, the definition as an intersection can be retained explicitly and abstracted over the additional schematic arguments.

Mutually inductive relations can be handled by the basic package using additional 'tag' arguments (e.g. distinct natural numbers) to define a family of relations. However in our implementation we actually generalize the above Tarskian procedure to multiple relations. If we are defining a family of relations $R_1, \ldots, R_k$, then we carry through hypotheses of monotonicity for all the relations together. The rules involving the same relation in their hypothesis are collected together and the definition is modified simply by generalizing over *all* the relations together (i.e. intersecting over the all the relations involved). Essentially the same reasoning as in the one-relation case now goes through.

## A foundational remark

All of the above can be carried out in an extremely simple, intuitionistic subsystem of the HOL logic. No extensive logical or theory support is required. This is valuable since, as this paper sets out to show, many theory developments are greatly eased by having a system for inductive definitions available early on.

# 3 Well-founded relations

A binary relation $\prec$ on a set $X$ is said to be *well-founded* iff every nonempty subset of $X$ has a $\prec$-minimal member, i.e.

$$WF(\prec) = \forall S \subseteq X.\, S \neq \emptyset \Rightarrow \exists m \in S.\, \forall y \in S.\, y \not\prec m$$

Note that despite the suggestive notation and the use of the term 'minimal', we are making no special assumptions about the relation $\prec$. Evidently if $\prec$ is well-founded it must be antisymmetric, (if $x \prec y$ and $y \prec x$ then $\{x, y\}$ would fail to have a minimal member), and so irreflexive (set $x = y$ here). But we do not assume that $\prec$ is transitive, still less connected (it is easy to see that the latter implies the former). This means that $y \not\prec m$ is not, in general, the same as $m \preceq y$.

There are several important consequences of well-foundedness, such as the validity of proofs by induction and definitions by recursion on $\prec$. In fact it is not widely appreciated that all the following are *equivalent*.

1. $\prec$ is well-founded
2. There are no infinite descending $\omega$-chains
3. Well-founded induction holds for $\prec$.
4. Every admissible recursion is satisfied by *no more than one* function.
5. Every admissible recursion is satisfied by *exactly one* function.

An infinite descending $\omega$-chain is a function $s : \mathbf{N} \to X$ such that $\forall n \in \mathbf{N}.\, s_{n+1} \prec s_n$. The equivalence involving descending $\omega$-chains differs in two respects from the other equivalences. One is that to establish its equivalence involves the Axiom of (Countable Dependent) Choice. To the shamelessly nonconstructive HOL user this is of little import; more significant is the fact that it presupposes the existence of the natural numbers, which is something we want to avoid at this stage.

Most of the above are easy to prove, and the proofs are well-known. We want to focus on the recursion theorem, since we can give a simple proof using an inductive definition, and moreover one which makes essential use of the more general hypotheses our system allows. The recursion theorem justifies the existence of a unique function $f : X \to Y$ such that $\forall x \in X.\, f(x) = \psi(f, x)$, provided each $\psi(f, x)$ depends only on the values of $f(z)$ for $z \prec x$. This condition on a recursion equation is what we called 'admissibility' above. Formally, we can define admissibility of $\psi : (X \to Y) \times X \to Y$ with respect to a relation $\prec$ as follows:

$$Adm_\prec(\psi) = \forall f \; g. \, (\forall x \in X. \, (\forall z \prec x. \, f(z) = g(z)) \Rightarrow (\psi(f,x) = \psi(g,x))$$

We claimed that

$$\forall \prec . \, WF(\prec) \Rightarrow \forall \psi. \, Adm_\prec(\psi) \Rightarrow \exists! f : X \to Y. \, \forall x \in X. \, f(x) = \psi(f,x)$$

To prove this, we make the following inductive definition of a binary relation $R$ on $X$.

$$\frac{\forall z \prec x. \, (z, g(z)) \in R}{(x, \psi(g,x)) \in R}$$

The cases theorem is:

$$\forall x \; y. \, (x,y) \in R \iff \exists g. \, (\forall z \prec x. \, (z, g(z)) \in R) \land y = \psi(g,x)$$

We claim that $R$ is the graph of the desired recursive function. First we prove by induction on $\prec$ that it is the graph of some function, i.e. $\forall x \in X. \, \exists! y \in Y. \, (x,y) \in R$. Supposing this is true for all $z \prec x$, then we can certainly define a function $g : X \to Y$ so that $\forall z \prec x. \, (z, g(z)) \in R$ (for $z \prec x$ choose for $g(z)$ the unique $y$ with $(z,y) \in R$, and for other $z$ some arbitrary value). Now $(x, \psi(g,x)) \in R$, so we have proved existence. For uniqueness, suppose $(x,y) \in R$ and $(x,y') \in R$. Then by the cases theorem there are functions $g$ and $h$ with:

$$(\forall z \prec x. \, (z, g(z)) \in R) \land y = \psi(g,x)$$

and

$$(\forall z \prec x. \, (z, h(z)) \in R) \land y' = \psi(h,x)$$

But now by the uniqueness part of the inductive hypothesis, $\forall z \prec x. \, g(z) = h(z)$, and by the admissibility assumption $\psi(g,x) = \psi(h,x)$. Consequently $y = y'$ as required.

Let $f$ be the function with graph $R$. By the induction rule, we know that $(\forall z \prec x. (z, f(z)) \in R) \Rightarrow (x, \psi(f,x)) \in R$. By definition of $f$, we know that $\forall x \in X. (x, f(x)) \in R$; but this certainly implies the antecedant of this implication. So $\forall x \in X. \, (x, \psi(f,x)) \in R$; but the fact that $R$ is single-valued, already proven, shows that $\forall x \in X. \, f(x) = \psi(f,x)$.

Uniqueness of the function $f$ is a simple induction. More surprising is the converse, that just the *uniqueness* part of the recursion theorem implies that $\prec$ is well-founded. To see this, suppose that

$$\forall \psi. \, Adm_\prec(\psi) \Rightarrow \forall f \; g. \, (\forall x. \, f(x) = \psi(f,x)) \land (\forall x. \, g(x) = \psi(g,x)) \Rightarrow (f = g)$$

We claim well-founded induction holds. Suppose that for any property $P$ we have $\forall x. \, (\forall z \prec x. \, P(z)) \Rightarrow P(x)$. In the above, set:

$$\psi(f, x) = P(x) \lor \forall z \prec x.\, f(z)$$
$$f = P$$
$$g = \lambda x.\ \top$$

in the above theorem. Evidently $\psi$ is admissible, and furthermore $\forall x.(\lambda x.\top)(x) = P(x) \lor \forall z \prec x.(\lambda x.\top)(x)$ is trivially true. But also $\forall x.P(x) = P(x) \lor \forall z \prec x.P(x)$ is trivially true, since it is logically equivalent to $\forall x.\ (\forall z \prec x.\ P(z)) \Rightarrow P(x)$, true by hypothesis. Thus $P = \lambda x.\ \top$ and so $\forall x \in X.\ P(x)$ as required.

These equivalences have all been proved easily in HOL. Actually, the ones involving the recursion theorem can't be stated as single HOL theorems, because they require a local type quantification (over the set $Y$). To state them as equivalences directly one would need to extend the logic as proposed by Melham [17]. However, as with most other similar troublesome theorems (completeness for first order logic, pointwise continuity in terms of nets etc.) they can easily be split into two parts:

```
|- !L.
    WF L ==>
    (!H.
        (!f g x. (!z. L(z,x) ==> (f z = g z)) ==> (H f x = H g x)) ==>
        (?! f:*->**. !x. f x = H f x))

|- !L.
    (!H.
        (!f g x. (!z. L(z,x) ==> (f z = g z)) ==> (H f x = H g x)) ==>
        (?! f:*->bool. !x. f x = H f x)) ==>
    WF L
```

Note that we are considering relations on the whole type, not a specific subset. However it is clear from the definition that either expanding or contracting the domain of the well-founded relation still yields a well-founded relation.

## 4  Automating recursive definitions

The well-founded recursion theorem is sometimes quite practically useful. The only derived definitional mechanisms built into the HOL core are for *primitive* recursive functions. Primitive recursion in higher order logic is stronger than in first order logic. For example, the following function ($\psi_1(a, b) = ab$, $\psi_2(a, b) = a^b$ and so on):

$$\psi_0(a, b) = a + b$$
$$\psi_{n+1}(a, 0) = \begin{cases} 0 \text{ if } n = 0 \\ 1 \text{ if } n = 1 \\ a \text{ otherwise} \end{cases}$$
$$\psi_{n+1}(a, b + 1) = \psi_n(a, \psi_{n+1}(a, b))$$

was introduced by Ackermann [1] [1] precisely to show that some functions which are not first order primitive recursive can nevertheless be defined by higher order primitive recursion (using an iteration functional like ML's **funpow**). Nevertheless for more general recursions like that involved in quicksort, it is attractive to be able to use well-founded recursion in its full generality. The well-founded recursion theorem is an excellent starting point; the fact that we have not assumed $\prec$ to be antisymmetric means it subsumes the use of a separate 'measure' function. However as yet there is no support to define functions without manually instantiating the theorem. Making this process easier has been explored by Ploegaerts et. al. [20] and by van der Voort [23]. A rather different scheme for defining recursive functions in HOL via a theory of CPOs, is described by Agerholm [3]. Recently, Slind has been developing a very useful suite of automated tools for recursive definitions.

## 5 Free recursive types

Suppose we have any set $X$, together with an element $z \in X$ and a function $s : X \to X$. We can carve out a subset $N$ by the following inductive definition:

$$\overline{z \in N}$$

$$\frac{n \in N}{s(n) \in N}$$

From the inductive definition, we immediately find:

$$z \in N$$

$$\forall n \in N. \, n \in N \Rightarrow s(n) \in N$$

$$\forall P. \, P(z) \wedge (\forall n \in N. \, P(n) \Rightarrow P(s(n))) \Rightarrow \forall n \in N. \, P(n)$$

These are three of the Peano axioms for the natural numbers, with $z$ taking the place of zero and $s$ of successor. We have thus succeeded in inductively defining a set which shares these three important properties of N. However the natural numbers have the additional property that they are *freely* generated by the constructors 0 and $SUC$, i.e. two values produced from 0 and $SUC$ are the same only if they were constructed in exactly the same way, i.e.

$$\overbrace{SUC(\ldots(SUC(0))\ldots)}^{m \text{ times}} = \overbrace{SUC(\ldots(SUC(0))\ldots)}^{n \text{ times}}$$

if and only if $m = n$. This follows from the other two Peano axioms:

---

[1] The better-known 2-argument function often called Ackermann's function is actually due to Péter [19].

- The constructors are distinct, i.e. $SUC(n) \neq 0$.
- The constructors are injective, i.e. $(SUC(m) = SUC(n)) \Rightarrow (m = n)$.

Now if $m < n$ in the above equation, repeatedly using injectivity yields:

$$\overbrace{SUC(\ldots(SUC(0))\ldots)}^{n-m \; times} = 0$$

which contradicts distinctness. If we knew that $z$ and $s$ had these two properties, the set $N$ would obey all the (second order) Peano axioms and hence be isomorphic to N. The detailed nature of $x$, $s$ and $X$ are irrelevant. In fact the Axiom of Infinity in HOL uses the Dedekind definition of an infinite set: namely there exists a function from the set to itself which is injective but not surjective. We can take this function (eventually restricted to N) as $SUC$, and any point not in its range as 0. (This is how the type :num is currently developed in HOL, though it does not explicitly use an inductive definition.)

From the Peano axioms, a theorem justifying primitive recursion may be derived[2]. For any set $X$, element $a \in X$ and function $\psi : X \times N \rightarrow X$, there is a unique function $f : N \rightarrow X$ such that $f(0) = a$ and $\forall n \in N.\ f(SUC(n)) = \psi(f(n), n)$. Once again the easiest proof by far is to build up the graph of the function by an inductive definition:

$$\overline{(0, a) \in f}$$

$$\frac{(n, y) \in f}{(SUC(n), \psi(y, n)) \in f}$$

The details are much as in the well-founded case and are left to the reader. And just as in the well-founded case, the recursion theorem implies all the Peano axioms (the fact that $0 \in N$ and $n \in N \Rightarrow SUC(n) \in N$ are implicit in the types of the functions concerned). The uniqueness part proves that induction holds, and now the existence part proves freeness. For example, the recursion theorem allows a function $Z : N \rightarrow bool$ to be defined with $Z(0) = \top$ and $Z(SUC(n)) = \bot$, showing that the constructors are distinct. Likewise we can define a 'predecessor' function and so show injectivity.

---

[2] Often called an 'initiality' theorem since it asserts that N is an initial object in the category of (0,$SUC$)-algebras (or equivalently a free $SUC$-algebra on one generator), i.e. there is a unique homomorphism from N into any other (0,$SUC$)-algebra. Properly speaking, a statement of initiality does not involve the extra argument $n$ to $\psi$, but since we are talking about the category of *all* algebras with the same signature, the two are equivalent: set $O_A = (0, a)$ and $SUC_A(n, y) = (SUC(n), \psi(y, n))$. For more on these matters see [8].

# 6  Automating recursive types

Given the inductive definitions package, we are in a strong position to automate recursive types as described above. We use inductive definitions first to carve out a subset of an existing type to yield the new one, and also to justify the recursion theorem. However not all inductive definitions are as simple as the one for natural numbers. There may be many constructors, which may be mutually recursive and even nested with each other and previously defined constructors. They are perhaps best illustrated by the following example, which is meant to be a type of terms for an embedded syntax of first order logic, with the variables parametrized by natural numbers and the function symbols by integers.

```
Term = Var num | Fun int (Term list)
```

Now we can avoid having to deal with existing constructors like `list` explicitly by defining an isomorphic type by mutual recursion and then modifying the theorems at the end, as pointed out by Gunter [9]. Furthermore, nested instances of constructors can be unwound by introducing locally some additional types.

```
Term = Var num | Fun int Termlist
Termlist = Nil | Cons Term Termlist
```

First we collect together all the basic types $\alpha_1, \ldots, \alpha_n$ (here $num$ and $int$). All we need to generate some distinct injective constructors is a set $u$ with injective functions:

- $i_\alpha : \alpha \to u$ to inject the basic types.
- $t : ind \to u$ to inject distinct tag elements to distinguish the constructors, of which there may be arbitrarily many.
- $P : u \to u \to u$ to implement multivariate constructors by iteratively pairing up the arguments.

This is pretty easy to arrange, by perfectly straightforward cardinality reasoning. Without any nontrivial use of arithmetical properties, we can show that $X = ind \to bool$ admits an injective function $bool \to X \to X$; thus $Y = X \to bool$ admits one $Y \to Y \to Y$. Now we may take $Y \to \alpha_1 \to \cdots \to \alpha_n \to bool$ as the set $u$. For our example, we can define the constructors on $u$ as follows:

$$Var\ n = P\ (t\ 0)\ (i_{num}(n))$$
$$Fun\ i\ l = P\ (t\ 1)\ (P\ (i_{int}(i))\ l)$$
$$Nil = P\ (t\ 2)\ arb$$
$$Cons\ x\ l = P\ (t\ 3)\ (P\ x\ l)$$

Clearly these are distinct and injective, because $P$ and all the $i_X$ are. We may now define sets to represent $Term$ and $Termlist$ by mutual induction. The new types may then be declared (since $Termlist$ and $(Term)list$ are isomorphic, we can freely transform the former into the latter with a little effort). The recursion theorem may then be proved, again using an inductive definition.

And now, if we want to consider an extension to infinitely branching trees as Gunter [10] does, finding a corresponding universe is almost as easy. It suffices to create a new universe $v$ with an injection $(u \to v) \to v$. It's easy to see that $u \to u$ suffices: given a function $f : u \to (u \to u)$, form the function $(UNCURRY\ f) \circ P^{-1}$. In cardinal terms, $|u \to v| = |u \to (u \to u)| = |u \to u|^{|u|} = (|u|^{|u|})^{|u|} = |u|^{|u|^2} = |u|^{|u|} = |u \to u| = |v|$, using the fact that we already know $|u|^2 = |u|$. This contrasts with the explicit construction in Gunter's work. On the other hand, it may be that to store a form of the recursion theorem which allows explicit instantiation, it may be useful to carve out an inductive subset of the constructed space right from the beginning.

## Conclusions

We have developed a powerful and easy-to-use package for automating inductive definitions, and demonstrated the power of such definitions in important theory developments. As yet the proposed recursive types package has not been implemented, but we believe the idea holds considerable promise. The merit of our approach is that finding large enough underlying sets is done very generally using abstract set theory, so we avoid creating specific recursive types in a piecemeal way. It is intended to experiment with this more 'rational' theory development in the gtt system [11]. It also makes easier certain other changes, e.g. introducing a binary system for numeral constants along the lines of the existing numeral library [15].

## Acknowledgements

I am grateful to my supervisor, Mike Gordon, and to the EPSRC and the Newton Trust for financial support. Tom Melham's original package provided the inspiration, and I also learned a lot about free recursive types from Tom's work and conversation. (The proof that well-founded recursion implies well-foundedness was inspired by Tom's analogous HOL proofs for free recursive types.) The idea of creating recursive types using an inductive definition came from Larry Paulson's work. The possibility of storing a general form of the recursion theorem is due to Vernon Austel. The idea of using inductive definitions to prove recursion theorems came from the proof of the recursion theorem for N in [12]. My interest in well-founded relations in general was sparked by Konrad Slind. I have also benefited greatly from conversations with Elsa Gunter. Malcolm Newey impressed on me the aesthetic appeal of leaving numbers till late in the theory development. Comments from the anonymous referees were very helpful.

# References

1. Wilhelm Ackermann. Zum Hilbertschen Aufbau der reellen Zahlen. *Mathematische Annalen*, 99:118–133, 1928. English translation, 'On Hilbert's construction of the real numbers', in [24], pp. 493–507.
2. Peter Aczel. An introduction to inductive definitions. In J. Barwise and H.J. Keisler, editors, *Handbook of mathematical logic*, volume 90 of *Studies in Logic and the Foundations of Mathematics*, pages 739–782. North-Holland, 1991.
3. Sten Agerholm. A HOL basis for reasoning about functional programs. BRICS Report Series RS-94-44, Centre of the Danish National Research Foundation, Department of Computer Science, University of Aarhus, Ny Munkegade, DK-8000 Aarhus C, Denmark, 1994.
4. Flemming Andersen and Kim Dam Petersen. Recursive boolean functions in HOL. In Archer et al. [5], pages 367–377.
5. Myla Archer, Jeffrey J. Joyce, Karl N. Levitt, and Phillip J. Windley, editors. *Proceedings of the 1991 International Workshop on the HOL theorem proving system and its Applications*, University of California at Davis, Davis CA, USA, 1991. IEEE Computer Society Press.
6. Juanito Camilleri and Tom Melham. Reasoning with inductively defined relations in the HOL theorem prover. Technical Report 265, University of Cambridge Computer Laboratory, New Museums Site, Pembroke Street, Cambridge, CB2 3QG, UK, 1992.
7. Luc J. M. Claesen and Michael J. C. Gordon, editors. *Proceedings of the IFIP TC10/WG10.2 International Workshop on Higher Order Logic Theorem Proving and its Applications*, volume A-20 of *IFIP Transactions A: Computer Science and Technology*, IMEC, Leuven, Belgium, 1992. North-Holland.
8. J. A. Goguen, J. W. Thatcher, and E. G. Wagner. An initial algebra approach to the specification, correctness and implementation of abstract data types. In Raymond T. Yeh, editor, *Current Trends in Programming Methodologies, volume IV*, pages 80–149. Prentice-Hall, 1978.
9. Elsa L. Gunter. Why we can't have SML style datatype declarations in HOL. In Claesen and Gordon [7], pages 561–568.
10. Elsa L. Gunter. A broader class of trees for recursive type definitions for HOL. In Joyce and Seger [13], pages 141–154.
11. John Harrison and Konrad Slind. A reference version of HOL. Presented in poster session of 1994 HOL Users Meeting and only published in participants' supplementary proceedings. Available on the Web from http://www.dcs.glasgow.ac.uk/~hug94/sproc.html, 1994.
12. Nathan Jacobson. *Basic Algebra I*. W. H. Freeman, 2nd edition, 1989.
13. Jeffrey J. Joyce and Carl Seger, editors. *Proceedings of the 1993 International Workshop on the HOL theorem proving system and its applications*, volume 780 of *Lecture Notes in Computer Science*, UBC, Vancouver, Canada, 1993. Springer-Verlag.
14. B. Knaster. Un théorème sur les fonctions·d'ensembles. *Annales de la Société Polonaise de Mathématique*, 6:133–134, 1927. Volume published in 1928.
15. Tim Leonard. The HOL numeral library. Distributed with HOL system, 1993.
16. Thomas F. Melham. A package for inductive relation definitions in HOL. In Archer et al. [5], pages 350–357.
17. Thomas F. Melham. The HOL logic extended with quantification over type variables. In Claesen and Gordon [7], pages 3–18.

18. Lawrence C. Paulson. A fixedpoint approach to implementing (co)inductive definitions. Technical Report 320, University of Cambridge Computer Laboratory, New Museums Site, Pembroke Street, Cambridge, CB2 3QG, UK, 1993.
19. R. Péter. Konstruktion nichtrekursiver Funktionen. *Mathematische Annalen*, 111:42–60, 1935.
20. Wim Ploegaerts, Luc Claesen, and Hugo De Man. Defining recursive functions in HOL. In Archer et al. [5], pages 358–366.
21. Rachel E. O. Roxas. A HOL package for reasoning about relations defined by mutual induction. In Joyce and Seger [13], pages 129–140.
22. Alfred Tarski. A lattice-theoretical fixpoint theorem and its applications. *Pacific Journal of Mathematics*, 5:285–309, 1955.
23. Mark van der Voort. Introducing well-founded function definitions in HOL. In Claesen and Gordon [7], pages 117–132.
24. Jean van Heijenoort, editor. *From Frege to Gödel: A Source Book in Mathematical Logic 1879–1931*. Harvard University Press, 1967.
25. Glynn Winskel. *The formal semantics of programming languages: an introduction*. Foundations of computing. MIT Press, 1993.

# A formulation of TLA in Isabelle

Sara Kalvala    (sk@cl.cam.ac.uk)

Computer Laboratory, University of Cambridge,
New Museums Site, Cambridge, CB2 3QG UK

**Abstract.** The Temporal Logic of Actions is a formalism for reasoning about concurrent and reactive systems. In this paper I present a formulation of TLA in the Isabelle theorem prover, in which I make extensive use of facilities in the Isabelle system for embedding different logics, particularly the syntax for easy axiomatisation of logics and the existence of parsing and printing facilities. I show how these aspects of Isabelle have facilitated the embedding, and describe an example proof done within the implementation.

## 1 Introduction

When applying logical formalisms to reason about computational systems, it is often found useful to have intermediate formalisms and abstractions that make the problem tractable. One may describe a system through a programming language or a hardware description language, or use a mathematical calculus. The Temporal Logic of Actions [8] provides one useful abstraction level. TLA consists of two components: a logic of *actions*, where an action represents a relation between old state and new state, and a *temporal* logic for reasoning about (potentially infinite) sequences of states, arising from the execution of an algorithm.

TLA is quite similar to other formalisms such as Unity [2] and State Transition Systems [11]. A convenient feature of TLA is the use of lifted predicates to specify the relation between two consecutive states, as opposed to guarded assignments and explicit state transitions. There is then a single logic in which both specifications and processes can be described, much in the same way as logics such as HOL have been applied to hardware verification [5].

As in the case of other formalisms for temporal reasoning, some attempts have already been made in the theorem proving community to mechanize TLA. Engberg and Grønning have collaborated with Lamport and produced TLP, a purpose built tool that translates TLA specifications into input for the Larch Prover and a model-checking tool [4]. Von Wright and Långbacka have mechanized TLA in HOL [12, 13]. The work described in this paper follows the notion that some features of Isabelle would make it particularly attractive for implementing TLA; for example, the ease of implementing concrete syntax, the choice of different object logics onto which the TLA calculus could be grafted, and the use of higher-order unification and the natural-deduction style proof rules.

The next two sections describe TLA and the Isabelle system; this is followed by an exposition of the mechanisation strategy and a description of the largest example of a verification task accomplished using the implementation so far. The work is compared with the other existing tools, and the paper concludes with an assessment of the work and possible follow-ups to the research done.

# 2 The Temporal Logic of Actions

TLA has been described originally by Lamport [8]. What follows is a brief summary.

Truth values in TLA are lifted: a formula is *valid* iff it is true of every *behaviour*. A behaviour is an infinite sequence of states, where a *state* is an assignment of values to *flexible* variables. A flexible variable is one that can have different values in different states. An important subclass of formulae in TLA are *actions*, which relate values from one state to the next. An action is true or false on a pair of states. These semantic notions can be expressed as[1]:

$$\models F \quad\quad \triangleq \forall \sigma \in \mathbf{St}^\infty : \sigma[\![F]\!]$$
$$\langle s_0, s_1, \dots \rangle[\![A]\!] \triangleq s_0[\![A]\!]s_1$$

To facilitate the unlifting of TLA formulae into ordinary logic, TLA makes use of the *prime* operator. The meaning of an unprimed variable is its value at the first state, while the meaning of a primed variable is its value at the second state. As a syntactic convention, priming also applies to predicates and expressions, to represent the priming of all variables in the expression.

Another component of TLA syntax is the *stuttering* operator on actions. A stuttering on action $\mathcal{A}$ under the vector of variables $f$ occurs when either the action $\mathcal{A}$ occurs or the variables in $f$ remain the same (while either some other independent action occurs or the system remains idle). The stuttering operator and its converse the *angle* operator are given by:

$$[\mathcal{A}]_f \triangleq \mathcal{A} \vee (f' = f)$$
$$\langle \mathcal{A} \rangle_f \triangleq \mathcal{A} \wedge (f' \neq f)$$

The use of stuttering allows descriptions at different levels of granularity, where one state change at one level can be implemented by several state changes at the lower level.

---

[1] The representation of all definitions in this section is copied directly from [8]

Temporal aspects of the specification make use of the standard operators 'always' ($\Box$), 'eventually' ($\Diamond$) and 'leadsto' ($\leadsto$), defined as follows:

$$\langle s_0, s_1, \ldots \rangle[\Box F] \triangleq \forall n \in \mathrm{Nat} : \langle s_n, s_{n+1}, \ldots \rangle[F]$$
$$\langle s_0, s_1, \ldots \rangle[\Diamond F] \triangleq \exists n \in \mathrm{Nat} : \langle s_n, s_{n+1}, \ldots \rangle[F]$$
$$F \leadsto G \triangleq \Box(F \Rightarrow \Diamond G)$$

Reasoning about *fairness* is an important aspect when modeling concurrency. Fairness is concerned with progress properties, facts that ensure that no process is consistently neglected. (Notions of justice and fairness are well described in [9].) In TLA, two types of fairness properties are declared: a process is said to satisfy the *weak fairness* condition if at all times either it is eventually executed or it eventually becomes disabled, and it is said to satisfy the *strong fairness* condition if at all times either it will eventually be executed or eventually it will be disabled *at all later states*. The definitions for these two conditions are given as:

$$\mathrm{WF}_f(\mathcal{A}) \triangleq \Box\Diamond\langle \mathcal{A}\rangle_f \vee \Box\Diamond\neg \mathit{Enabled}\ \langle \mathcal{A}\rangle_f$$
$$\mathrm{SF}_f(\mathcal{A}) \triangleq \Box\Diamond\langle \mathcal{A}\rangle_f \vee \Diamond\Box\neg \mathit{Enabled}\ \langle \mathcal{A}\rangle_f$$

where the predicate *Enabled* $\langle \mathcal{A}\rangle_f$ is a predicate that specifies the possibility of executing $\langle \mathcal{A}\rangle_f$ from a particular state.

In the TLA methodology, systems are usually represented as a conjunction of an initial condition, an action that is continually repeated under stuttering, and a set of fairness conditions that are assumed to hold of the system. The same language is used to represent both specifications and implementations. As there is no assignment statement nor sequentiality operators, it becomes possible to use standard proof apparatus to mechanize it.

The temporal operators used are part of the standard repertoire for temporal logic, and standard proof systems for temporal logic, consisting of a series of tautologies that can be derived from the definitions, can be used in much the same way as in other mechanizations of temporal logic [9].

As in the case of the temporal operators, it is not always necessary to expand the definitions of the fairness conditions. Lamport describes a proof system for TLA, which links the action reasoning component and the temporal reasoning component with ordinary logic. Some of the rules are straightforward and easy to understand—such as the rule for invariance, which states that if a predicate remains true through zero or one excution of an action $\mathcal{N}$, and if the system starts with it holding and the only action being executed is $\mathcal{N}$, then it will always hold:

$$\mathrm{INV1.}\quad \frac{I \wedge [\mathcal{N}]_f \Rightarrow I'}{I \wedge \Box[\mathcal{N}]_f \Rightarrow \Box I}$$

Some rules are, however, more difficult to understand simply by examination, particularly the ones dealing with fairness. These rules incorporate a lot of information on how fairness conditions can be derived. An example of such a rule, for which an informal justification is provided by Lamport [8], is:

SF2.
$$\frac{\begin{array}{l} \langle \mathcal{N} \wedge \mathcal{B} \rangle_f \Rightarrow \langle \mathcal{M} \rangle_g \\ P \wedge P' \wedge \langle \mathcal{N} \wedge \mathcal{A} \rangle_f \Rightarrow \mathcal{B} \\ P \wedge Enabled \langle \mathcal{M} \rangle_g \Rightarrow Enabled \langle \mathcal{A} \rangle_f \\ \Box [\mathcal{N} \wedge \neg \mathcal{B}]_f \wedge SF_f(\mathcal{A}) \wedge \Box F \\ \qquad \wedge \Box \Diamond Enabled \langle \mathcal{M} \rangle_g \Rightarrow \Diamond \Box P \end{array}}{\Box [\mathcal{N}]_f \wedge SF_f(\mathcal{A}) \wedge \Box F \Rightarrow SF_g(\mathcal{M})}$$

Other rules are similar, and can be thought of as representing algorithms rather than more primitive forms of inference. One feature of a proof system specified in such a way is that many of the messy and complex steps of a proof are isolated into a few rules. However, such a system provides very little flexibility, and has very few desirable proof search properties. Lamport leaves the formal derivation of these rules as an exercise to the reader; some of the rules have been proved in HOL [13].

## 3   The Isabelle Theorem Prover

Isabelle is a general theorem-proving system [10]. A tactic-based, interactive proof assistant in the tradition of LCF, it has powerful facilities for embedding logics, and has been used for mechanizing different logics. It has also been proved useful for doing large proofs, having many tools that allow the automation of difficult and tedious details. Thus it is particularly suitable for both implementing TLA as well as actually using it to develop proofs in TLA.

Isabelle can be thought of as a *framework* for logics: it consists of a *meta*-logic (specifically intuitionistic higher-order logic) which manipulates theorems in the *object* logic, in a manner transparent to the user. Inference rules of the object logics are theorems of the meta-logic rather than programmed functions, as in the case of, say, HOL. Higher-order unification is used to allow meta-logic rules to manipulate theorems and rules of the object logic.

New object logics as well as extensions of logics are implemented in the form of *theories*, consisting in the declaration of axioms and a *signature*, which contains information such as types and classes, constants declared, and syntax rules. A theory is specified in the form of a stylized but easily readable 'theory file'. Theories follow a dag-like lineage, with each theory specifying its parents, and inheriting the signatures of all antecedent theories. Formal logics are developed from the antecedent 'Pure'; user theories for specifying particular problems typically make use of many theories that extend a logic in a useful way.

Proofs are developed in much the same way as in other LCF-style provers: a goal is specified and tactics expand the goal using inference rules. While in HOL different tactics correspond to the different inference rules of the logic, in Isabelle there are only a small handful of tactics, the basic set of which implement resolution in various forms, and their power arise from the different theorems they are given as arguments. Another difference with regards to HOL is that a proof is not a tree structure of subgoals but rather a sequence of lists of subgoals. In brief, this is because the higher-order unification behind Isabelle can generate multiple unifiers, and this representation allows backtracking over the different unifiers. Further details of how Isabelle works can be found elsewhere [10, 6]. One aspect of dealing with Isabelle is the (sometimes confusing) distinction between object and meta logic, this will be explained next.

Zermelo-Fraenkel Set Theory is one particular object logic that has been implemented in Isabelle. The ZF logic is constructed on top of First-Order Logic (FOL). There are two primary types in ZF, the type o of truth values inherited from FOL and the type i of sets. The usual statement that ZF is typeless is a simplification of the principle of different objects in ZF being constructed as sets, and therefore all have the same type i. The interpretation of ZF into the meta-logic is dependent upon combining sets into o-valued predicates, which can then be coerced into meta-level propositions.

Users of ZF must be aware of the distinction between object level functions and meta-level functions. In ZF, functions are usually represented as sets of pairs, so that both objects as well as functions on them have type i. On the other hand, meta-level functions on sets have type i => i. Each representation of functions has its advantage, so the implementation makes the use of ZF-function abstraction (lam x. y) and application (y`x) as well as abstraction (% x. y) and application (y(x)) at the meta-level.

Another facet of ZF which might puzzle readers not familiar with it is that several tools of ZF, particularly the inductive definitions package, manipulates set-valued objects only. To get around this restriction when dealing with truth values, the usual trick is to code 'true' as 1, and 'false' as 0 (as both 0 and 1 are represented as sets), and at some point equate the whole expression to 1 to translate back into type o. Thus, in the text below, the term 'boolean-like' is used to define these mock boolean values. Equality, conjunction, and disjunction are represented as (=, &, and |) in the type o, and as (eq, and, and or) in the type i, respectively.

# 4 Embedding TLA in Isabelle

TLA is a second-level logic, in the sense that it extends a base logic that provides all the basic operations over unlifted values. The first decision to have been made towards building the embedding of TLA was the choice of object logic to use. Higher-Order Logic and Zermelo-Fraenkel Set Theory were the primary candidates, as both are well developed logics with many tools and extensions built on them, particularly with the formalization of inductive datatypes. The decision to use ZF was made to keep to a close correspondence to the initial description of TLA. However, the formulation described here can be adequately carried over to HOL.

This embedding of TLA keeps very closely to the description of the semantics in Section 2. The heart of the embedding is the specification of the datatypes representing actions and temporal formulae. Three levels of objects are used:

- lifted objects, corresponding to the flexible variables (primed and unprimed) and functions on them.
- lifted predicates, which may be called actions, where lifted objects are combined to return boolean-like values; this corresponds not only to the usual equality but also set membership. As the datatype is inductively defined, lifted predicates can be combined with boolean-like operators.
- temporal formulae, where action predicates are the basic objects, extended by the temporal operators and boolean-like operators.

The coding of these datatypes is reasonably straightforward. The declaration displayed in Figure 1 shows the datatype for lifted formulae, which includes some parsing and pretty-printing information. It may be observed that the syntactic convention for flexible variables is slightly different here than in Lamport's notation: unprimed and primed variables (say $a$ and $a'$) are represented by prepending them or appending them to a hash symbol (in this case a# and #a), respectively. This was felt to correspond closer to the usual reading of "value of X before/after the action". Rigid variables–ones that have the same value independent of the state at which they are observed–are denoted by being prepended by a comma.

The datatype also includes lifted abstraction (not shown) and application (represented by the constant . '. to parallel the ' operator for application in ZF) which allows users to define their own functions (of arbitrary arity) and thus extend the datatype.

It may also be noted that there is different syntax for operators within actions and for temporal actions (for example, .&. represents conjunction in actions, and && conjunction in formulae). Though the primary reason for this is to disambiguate and speed up parsing, it is also a good idea to make it easy for readers to know at which level of the specification they are looking at. This function is achieved an extension of TLA called TLA+, by using modules to structure specifications, and separate the definitions for simple predicates, actions and temporals [7]. Here however they are not separated, and they can coexist in the same definition.

The interpretation of the lifted datatype is in terms of a pair of states and of the

```
datatype
  "lifted" =
    before ("x: nat")                          ("_#" [100] 100)
  | after  ("x: nat")                          ("#_" [100] 100)
  | constnt("x:nat")                           (";_" [100] 100)
  | ".'." ("f:liffun", "x:(lifted)list")  (infixr 40)

datatype
  "actions" =
    ".=." ("x: lifted", "y:lifted")     (infixr 40)
  | ".:." ("x: lifted", "y:lifted")     (infixr 40)
  | ".&." ("x:actions", "y:actions")    (infixr 40)
  | ".|." ("x:actions", "y:actions")    (infixr 40)
  | acneg ("x:actions")               (".~._" [100] 100)

datatype
  "temporals" =
    pred ("x:actions")
  | primed ("x:actions")
  | "&&" ("x:temporals","y:temporals")  (infixr 50)
  | "||" ("x:temporals","y:temporals")  (infixr 50)
  | "-o" ("x:temporals","y:temporals")  (infixr 50)
  | "##" ("x:temporals","y:temporals")  (infixr 50)
  | stut ("x:actions","y:list(nat)")    (" [[ _ ]] _ ")
  | angle ("x:actions","y:list(nat)")   (" << _ >> _ ")
  | box ("x:temporals")                 ("[=] _")
  | diam ("x:temporals")                ("<=>_")
  | enabled ("x:actions","y:list(nat)")
```

**Fig. 1.** Inductive datatype for temporal formulae

temporal datatype in terms of a behaviour. This unravelling of TLA formulae is given by recursive functions, which generates the equalities shown in Figure 2.

Finally, the validity of temporal formulae makes use of unlifting the formula with a behaviour, using the definition:

```
validity  "|= y   == ALL b:behaviour. b |- y"
```

where

```
temporal_lemmas
[ "temp ([[P]]y) =
    (lam b:behaviour.
      ((act(P)'(hd2(b))) or (nochange(y)'(hd2(b)))))",
  "temp (<<P>>y) =
    (lam b:behaviour.
      ((act(P)'(hd2(b))) and not (nochange(y)'(hd2(b)))))"
  "temp (P && Q) =
    (lam b:behaviour. ((temp (P)'b) and (temp (Q)'b))",
...]
```

**Fig. 2.** Unraveling the inductive datatype

```
satisfiability    "b |- y   == (temp(y)'b) = 1"
```

The turnstyle above is just a syntactic device to isolate the particular behaviour over which a formula is lifted (it is not an indication of the use of sequents).

One detail in which the implementation differs from the given semantics, transparent to the use of the logic, is the representation of states and behaviours. Lamport describes behaviours as infinite sequences of states, where states are functions that given the name of a variable return the value of that variable in that state. However, the formulation of these sequences of states is rather complicated when done from first principles. A decision was made to represent states simply as natural numbers (rather than as functions from variables to values) and behaviours as lists of natural numbers. The value $s[\![a' = a + 1]\!]t$ is therefore implemented by the predicate value(a,t) = value(a,s) + 1, where the function value is an uninterpreted constant. Pretty printing makes it possible to hide the representation of the function value and the state, as this information is not required during proofs. This modification makes it simpler to use ZF, with few changes in the interpretation of TLA formulae. Potentially finite lists are made infinite in ZF because it is always possible to take the head of an empty list (which returns 0). The desired relation between behaviours and pairs of states to be preserved is

```
"B : behaviour ==> hd2(B) : state_pair"
```

where the function hd2(x) is defined to be the pair <hd(x),hd(tl(x))>. That is, if B is a behaviour, it is always possible to take the first two elements off of it, which will correspond to a valid representation of a pair of states. This property is still preserved in the representation used.

Actions and temporal formulae are unravelled by simplification. As illustrated in Figure 3, the output at all stages remains readable. The first expression at Level 0 shows a goal as it is entered, in a lifted form. The expression at Level 2 is the goal after simplifying some of the temporal operators while the expression at Level 4 is the result of simplifying over the action operators.

```
    Level 0
    |= (pred(x# .:. Nat) &&
        [[ (#x .=. Succ(x#)) .&. #y .=. y# ]] [x] ) -o
      pred(#x .:. Nat)

    ...
    Level 2
      1. !!xa.
            xa : behaviour ==>
            xa |- pred(x# .:. Nat) &
            xa |- [[ (#x .=. Succ(x#)) .&. #y .=. y# ]] [x]  -->
            xa |- pred(#x .:. Nat)

    ...
    Level 4
      1. !!xa.
            [| xa : behaviour;
               hd2(xa) : state_pair |] ==>
            x# : nat &
            ((#x = succ(x#) & (#y = y#)) | #x = x#)  -->
            x# : nat
```

**Fig. 3.** Interacting with the TLA implementation

While the temporal operators have been defined in terms of their semantics, their use in practice is instead governed by an axiomatic proof system. To illustrate how this system looks, one may look at the Isabelle incarnation of the SF2 rule from Page 4:

```
SF2
"[| Be:behaviour;
    Be|- (<<B .&. N>>f) -o (<<M>>g);
    Be|- (pred(P) && primed(P) && (<<A .&. N>>f)) -o pred(B);
    Be|- (pred(P) && enabled(M,g)) -o enabled(A,f);
    Be|- (((([=]([[.~. B .&. N]]f)) && SF(A,f) && ([=]F)
       && ([=](<=> (enabled(M,g)))))) -o (<=>([=](pred(P)))));
```

```
  Be|- ([=]([[N]]f)) && SF(A,f) && ([=]F) |]
==> |= SF(M,g)"
```

## 5  The 'increment' Example

The test of the mechanisation of an application-oriented logic is in how well it supports proofs of at least a reasonable size. Thus, TLA in Isabelle has been used to re-prove the correctness of an increment algorithm, which has already been proved in both TLP [4] and HOL [13].The proofs concerning this algorithm are explained informally by Lamport [8], and in much more detail by Engberg [3].

The specification of the algorithm is illustrated in Figure 4. Formula $\Phi$ describes the problem in an abstract level: there are two variables, $x$ and $y$, which are asynchronously updated, in such a way that while one is being updated the other one remains constant, There is also a fairness condition, stating that at any state both the possible actions must occur infinitely often (unless disabled).

Formula $\Psi$ represents an algorithm conjectured to implement the specification. The property that $y$ remains constant when $x$ is being updated and vice-versa is assured by means of a semaphore. The control between all the different actions is implemented by the use of two program counters ($pc_1$ and $pc_2$). The strong fairness conditions on both $\mathcal{N}_1$ and $\mathcal{N}_2$ ensure that each sequence occurs infinitely often or is eventually forever disabled.

Many parts of the verification of the algorithms concern invariants and the step simulation; these are fairly straightforward, and have been easy to execute in Isabelle. A typical proof is the step simulation, which is the proof of correspondence modulo the fairness conditions, and which is illustrated in Figure 5.

More complicated proofs arise during the derivation of fairness conditions of $\mathcal{M}$ from the execution of $\mathcal{N}$ under the fairness conditions of $\mathcal{N}$. However, much of the complexity arises from the details of the problem itself, and not with the use of Isabelle. This can be observed by examining the proof of derivation of one of the fairness conditions of $\mathcal{M}$ from the description of $Phi$, shown in Figure 6. The complication is in deriving the auxiliary lemmas; once they are done, it is enough to integrate them into the proof state, performing some rewriting steps, and then resolving with the rule for strong fairness SF2 (already shown on Page 4). The use of the classical reasoner takes care of instantiating most variables, and the simplifier finishes the proof. The proofs of the lemmas used have been similarly facilitated by the use of the solver and the simplifier. It is encouraging that starting from the description of the complete proof in such a different system such as TLP, tailoring it to Isabelle was a straightforward task.

$Init_\Phi \;\triangleq\; (x = 0) \wedge (y = 0)$

$\mathcal{M}_1 \;\triangleq\; (x' = x + 1) \wedge (y' = y) \qquad \mathcal{M}_2 \;\triangleq\; (y' = y + 1) \wedge (x' = x)$

$\mathcal{M} \;\triangleq\; \mathcal{M}_1 \vee \mathcal{M}_2$

$\Phi \;\triangleq\; Init_\Phi \wedge \Box[\mathcal{M}]_{\langle x, y \rangle} \wedge \mathrm{WF}_{\langle x, y \rangle}(\mathcal{M}_1) \wedge \mathrm{WF}_{\langle x, y \rangle}(\mathcal{M}_2)$

$Init_\Psi \;\triangleq\; \wedge\; (pc_1 = \text{``a''}) \wedge (pc_2 = \text{``a''})$
$\phantom{Init_\Psi \;\triangleq\;} \wedge\; (x = 0) \wedge (y = 0)$
$\phantom{Init_\Psi \;\triangleq\;} \wedge\; sem = 1$

$\alpha_1 \;\triangleq\; \wedge\; (pc_1 = \text{``a''}) \wedge (0 < sem) \qquad \alpha_2 \;\triangleq\; \wedge\; (pc_2 = \text{``a''}) \wedge (0 < sem)$
$\phantom{\alpha_1 \;\triangleq\;} \wedge\; pc'_1 = \text{``b''} \qquad\qquad\qquad\qquad\quad \wedge\; pc'_2 = \text{``b''}$
$\phantom{\alpha_1 \;\triangleq\;} \wedge\; sem' = sem - 1 \qquad\qquad\qquad\quad\; \wedge\; sem' = sem - 1$
$\phantom{\alpha_1 \;\triangleq\;} \wedge\; Unchanged\; \langle x, y, pc_2 \rangle \qquad\qquad \wedge\; Unchanged\; \langle x, y, pc_1 \rangle$

$\beta_1 \;\triangleq\; \wedge\; pc_1 = \text{``b''} \qquad\qquad\qquad\quad\; \beta_2 \;\triangleq\; \wedge\; pc_2 = \text{``b''}$
$\phantom{\beta_1 \;\triangleq\;} \wedge\; pc'_1 = \text{``g''} \qquad\qquad\qquad\qquad\qquad\;\; \wedge\; pc'_2 = \text{``g''}$
$\phantom{\beta_1 \;\triangleq\;} \wedge\; x' = x + 1 \qquad\qquad\qquad\qquad\qquad\; \wedge\; y' = y + 1$
$\phantom{\beta_1 \;\triangleq\;} \wedge\; Unchanged\; \langle y, sem, pc_2 \rangle \qquad\qquad \wedge\; Unchanged\; \langle x, sem, pc_1 \rangle$

$\gamma_1 \;\triangleq\; \wedge\; pc_1 = \text{``g''} \qquad\qquad\qquad\quad\; \gamma_2 \;\triangleq\; \wedge\; pc'_2 = \text{``a''}$
$\phantom{\gamma_1 \;\triangleq\;} \wedge\; pc'_1 = \text{``a''} \qquad\qquad\qquad\qquad\qquad\;\; \wedge\; pc_2 = \text{``g''}$
$\phantom{\gamma_1 \;\triangleq\;} \wedge\; sem' = sem + 1 \qquad\qquad\qquad\quad\; \wedge\; sem' = sem + 1$
$\phantom{\gamma_1 \;\triangleq\;} \wedge\; Unchanged\; \langle x, y, pc_2 \rangle \qquad\qquad \wedge\; Unchanged\; \langle x, y, pc_1 \rangle$

$\mathcal{N}_1 \;\triangleq\; \alpha_1 \vee \beta_1 \vee \gamma_1 \qquad\qquad\qquad \mathcal{N}_2 \;\triangleq\; \alpha_2 \vee \beta_2 \vee \gamma_2$

$\mathcal{N} \;\triangleq\; \mathcal{N}_1 \vee \mathcal{N}_2$

$w \;\triangleq\; \langle x, y, sem, pc_1, pc_2 \rangle$

$\Psi \;\triangleq\; Init_\Psi \wedge \Box[\mathcal{N}]_w \wedge \mathrm{SF}_w(\mathcal{N}_1) \wedge \mathrm{SF}_w(\mathcal{N}_2)$

**Fig. 4.** The TLA formulae describing the 'increment' algorithm

## 6 Discussion

Isabelle is a relatively new system. The goal of this work was to examine the feasibility of using Isabelle for significant verification tasks; the results of the work seem to indicate that Isabelle is certainly appropriate for such use.

Lamport stresses the importance of rigorous reasoning as 'the only way to avoid subtle errors in concurrent algorithms'; one way to maintain such rigour is with the use of theorem provers. There have been two previous attempts at mechanizing TLA in the literature. The TLP tool is a purpose-built system, consisting of

```
val sim_init = prove_goal thy
"|= pred(InitPsi) -o pred(InitPhi)"
(fn _ => [temporal_tac increment_defs 1,
          rtac ballI 1 THEN forward_tac [hd2_type] 1,
          action_tac [] 1]);

val n_implies_m = prove_goal thy
"|= (([[N]]w) -o ([[M]]v))"
(fn _ => [temporal_tac increment_defs 1,
          rtac ballI 1 THEN forward_tac [hd2_type] 1,
          action_tac [] 1, rtac impI 1,
          step_tac (empty_cs addSEs [conjE,disjE]) 1,
          ALLGOALS (asm_full_simp_tac arith_ss)]);

val step_simulation = prove_goalw thy [hold_def]
"|= ([=]([[N]]w)) -o ([=]([[M]]v))"
(fn _ => [rtac ballI 1, cut_facts_tac [unlift n_implies_m] 1,
  temporal_tac [] 1, etac (unlift ImplBox1) 1, atac 1]);
```

**Fig. 5.** Proof of step simulation

```
val fairness_M1 = prove_goal thy "|- Psi -o SF(M1,v)"
(fn _ =>
[temporal_tac [] 1, rtac ballI 1, rtac impI 1,
 cut_facts_tac (map unlift [M1SF2v, M1SF2vi]) 1,
 temporal_tac [] 1,
 forward_tac [unlift BoxElim1] 1 THEN atac 1,
 cut_facts_tac (map unlift [M1SF2i,M1SF2ii,M1SF2iii, M1SF2iv]) 1,
 temporal_tac [] 1,
 eres_inst_tac [("F","SF(N2,w)")] (unlift SF2) 1,
 TRYALL (fast_tac ZF_cs) THEN ALLGOALS (temporal_tac []),
 ALLGOALS (temporal_tac [])]);
```

**Fig. 6.** Proof of fairness

an Emacs front-end, an SML translator, and several back-ends, which at present include the Larch Prover and a BDD procedure for temporal tautologies. The translator program translates TLA expressions into input for the various back-ends. TLP has been used extensively [3].

Another mechanisation of TLA, this time in HOL, is due to von Wright and Långbacka. Instead of axiomatizing the proof rules given by Lamport, they have decided to derive them from the interpretation of temporal logic. Their representation of the semantics of state is slightly different from the one described by Lamport; they represent state pairs through tuples of variables, with two tuples representing the state before and after an action, and primed and unprimed variables as distinct unrelated variables, as in:

$$\backslash \; (x,y,z) \; (x',y',z') . \; (x' = x + 1) \; /\backslash \; (y' = y + 1)$$

where \ is the representation of lambda-abstraction in HOL.

Though they haven't provided too much syntactic sugaring, they have been able to prove the same 'increment' example, and are using concepts from TLA in their work on refinements.

Both the implementations described above have deviated from Lamport's semantics, in that they encode action formulae and temporal formulae separately, in the case of HOL using distinct types. This is partly due to the complexities involved, but partly because it was assumed that un-lifting all objects would make terms unreadable and hard to manipulate. However, this was not observed in the mechanisation in Isabelle, and it seems that it has been possible to achieve the same user-friendliness and power without sacrificing the elegant semantics. Also, this representation has allowed more flexibility in the way of specifying actions and temporal formulae, and has also allowed the direct expression of priming and the **enabled** predicate. This solved some of the problems in the previous implementations: Engberg has had to formulate the INV1 rule in three versions, and this isn't needed here. Also, it has been possible to generate **enabled** predicates automatically, rather than relying on users having to write them.

With the results obtained so far, it seems appropriate to enhance the implementation further. The first step will be to add quantification and refinement mappings [8]. These will allow mechanisation to be applied to more complex examples. It is also envisaged that a version of TLA running on Isabelle's HOL theory will be completed soon.

Another path for further work is in designing improvements to the proof system for TLA. The axiomatization of TLA described by Lamport does work, but unfortunately it is not very amenable for automated proof search in Isabelle, which relies instead on natural deduction -style formulations. There are several different temporal logics that can be used, and one possibility is to explore some of these, and try to find a temporal logic that can be added to the action reasoning inherent to TLA. Abadi describes an alternative axiomatisation of a previous variation of TLA based on modal logic [1]; it will be interesting to experiment with such a system and see if it enhances the proof search capabilities. Isabelle is particularly well-suited to rapidly prototype variations of logics.

Another part of the work will involve incorporating external reasoners to the temporal reasoning package. In particular, *model checkers* are automatic tools which are useful to prove *some* aspects of reactive and concurrent systems, and it would be advantageous to use them in conjunction with the prover, as long as this can be done in a safe and consistent way. More generally, this will allow implementors to explore ways to incorporate other decision procedures to Isabelle.

There are still some problems with this work, such as the lack of a theory of lexemes in Isabelle and more support for reasoning about numbers. Nevertheless, these gaps in Isabelle's facilities should be solved soon as Isabelle reaches maturity. However, it is hoped that even as it is this mechanisation of TLA in Isabelle may be of use.

# Acknowledgements

The idea of implementing TLA in Isabelle arose after being invited to attend a meeting between Leslie Lamport, Urban Engberg, and Peter Grønning. Urban Engberg has been particularly helpful by answering questions and discussing issues in mechanization.

# References

1. M. Abadi. An axiomatization of Lamport's Temporal Logic of Actions. In *CONCUR '90*, volume 458 of *Lecture Notes in Computer Science*. Springer Verlag, 1990. updated version available by ftp.
2. K. M. Chandy and J. Misra. *Parallel Program Design: A Foundation*. Addison Wesley Publishing Company, Inc., Reading, Massachusetts, 1988.
3. U. Engberg. *Reasoning in the Temporal Logic of Actions*. PhD thesis, Aarhus University, 1994.
4. U. Engberg, P. Gronning, and L. Lamport. Mechanical verification of concurrent systems with TLA. In *Proceedings of the Fourth International Workshop on Computer-Aided Verification*, 1992.
5. M. Gordon. Why higher-order logic is a good formalism for specifying and verifying hardware. In G. Milne and P. Subrahmanyam, editors, *Formal Aspects of VLSI Design*. Elsevier Science, 1986.
6. S. Kalvala. *A Gentle Introduction to Isabelle*. Isabelle distribution, available from ftp.cl.cam.ac.uk, 1994.
7. L. Lamport. Hybrid systems in TLA+. In Grossman et al., editors, *Hybrid Systems*, volume 736 of *LNCS*. Springer Verlag, 1993.

8. L. Lamport. The temporal logic of actions. *ACM Transactions on Programming Languages and Systems*, 16(3), 1994.

9. Z. Manna and A. Pnueli. *The Temporal Logic of Reactive and Concurrent Systems: Specification*. Springer-Verlag, 1992.

10. L. Paulson. *Isabelle: A generic theorem prover*, volume 828 of *Lecture Notes in Computer Science*. Springer-Verlag, 1994.

11. A. U. Shankar. An introduction to assertional reasoning for concurrent systems. *ACM Computing Surveys*, 25(3), 1993.

12. J. von Wright. Mechanizing the temporal logic of actions in HOL. In M. Archer, J. J. Joyce, K. N. Levitt, and P. J. Windley, editors, *Proceedings of the 1991 International Workshop on the HOL Theorem Proving System and its Applications*. IEEE Computer Society Press, 1992.

13. J. von Wright and T. Langbacka. Using a theorem prover for reasoning about concurrent algorithms. In *Proceedings of the Fourth International Workshop on Computer-Aided Verification*, 1992.

# Formal Verification of Serial Pipeline Multipliers*

Jang Dae Kim and Shiu-Kai Chin

Department of Electrical and Computer Engineering
Syracuse University, Syracuse, NY 13244-4100

**Abstract.** Serial data-path circuits are often more difficult to analyze than their parallel counterparts. The major reason is data and operations are spread over time. Here the strength of formal verification is demonstrated with verification of classical serial pipeline multipliers. The designer's informal notions of how to interpret the design are formally captured in well-defined functions and standard mathematical notation. A linear-time temporal logic is found to be useful for analyzing such circuits; temporal operators are succinct in expressing certain operating conditions that are otherwise verbose, and temporal laws and operators enable us to work more efficiently in a higher level of reasoning.

## 1 Introduction

Bit-serial data-path circuits are used widely in digital signal processing (DSP) such as digital filters [1]. Because of the bit-serial nature of their operation, it is often difficult to analyze and verify their correctness. Part of the reason is that data objects are spread over time as are the operations on them. Simulation can validate some sample data, but exhaustive simulation is not possible.

Lyon has described a set of bit-serial multipliers in [2]. Lu has formally verified a serial-parallel multiplier [3] although the circuit is not an industry-standard one. We verified classical serial pipeline multipliers [1] [2] which are *parameterized* on the size of the multiplier word as in [3] [4].

Bit-serial data path circuits pose some unique problems for formal verification. State transition analysis which has been the primary tool for sequential circuit analysis is not adequate for verifying such data path circuits because state information is an implementation detail and its analysis does not answer the intended behavior (specification) directly.

We expressed the specifications of serial pipeline multipliers without any reference to internal state, internal signals or even the initial state; the specification is described entirely with input and output streams. Temporal logic helps to express the specification more succinctly and clearly. It also makes the proof more

---

* This work was supported by the NY State Center for Advanced Technology in Computer Applications and Software Engineering (CASE) at Syracuse University. The authors are grateful to Anand Chavan for his help with the GDT tools for getting layouts and simulations for this work.

efficient by bringing the level of reasoning to a more intuitive and natural level with temporal properties.

This paper is organized as follows: Section 2 describes a linear-time temporal logic we have embedded in HOL and use in the subsequent sections. A serial adder is described in Section 3 which is a core part of the serial multipliers in the subsequent sections. Section 4 describes a multiplier cell which uses the serial adder of Sect.3. Section 5 describes a basic pipeline multiplier. Finally, an enhanced pipeline multiplier is described in Sect.6.

## 2 Embedding a Temporal Logic in HOL

The temporal logic we use is described by Manna and Pnueli in [5]. Temporal formulas are represented in HOL as terms of type ": $num \rightarrow bool$" where type ": $num$" (the natural numbers) represents time. Temporal operators are defined in HOL according to their semantic definitions in [5]. For example, the definitions of temporal connectives corresponding to Boolean conjunction and negation appear below. These temporal connectives are prefixed with a symbol * to distinguish them from Boolean connectives.

$\vdash_{def}$ ∀w1 w2 t. (w1 *∧ w2) t = w1 t ∧ w2 t
$\vdash_{def}$ ∀w t. (*¬ w) t = ¬w t

The temporal formula (w1 *∧ w2) evaluates to 'T' at present time t iff both sub-formulas w1 and w2 evaluate to 'T' at present time t.

Modal operators which are unique in temporal logic are divided into 'future' operators and 'past' operators. Among the future operators are ○ (Next) and □ (Henceforth) as defined below:

$\vdash_{def}$ ∀w t. ○ w t = w(t+1)
$\vdash_{def}$ ∀w t. □ w t = (∀t1. t ≤ t1 ⊃ w t1)

Temporal formula ○w evaluates to 'T' at present time t iff the sub-formula w evaluates to 'T' at next time t+1. And temporal formula □w evaluates to 'T' at present time t iff the sub-formula w evaluates to 'T' at present time as well as at all time thereafter.

Past operators include *S (Since) and ⊖ (Previous) whose definitions are as follows:

$\vdash_{def}$ ∀w1 w2 t. (w1 *S w2) t = (∃k. k ≤ t ∧ w2 k ∧
(∀i. k < i ∧ i ≤ t ⊃ w1 i))
$\vdash_{def}$ ∀w t. ⊖ w t = 0 < t ∧ w(t − 1)

Temporal formula (w1 *S w2) evaluates to 'T' at present time t iff there exists a time k no later than time t that the sub-formula w2 evaluates to 'T' and the sub-formula w1 evaluates to 'T' at all time i after k through time t. The formula ⊖ w evaluates to 'T' at present time t iff there exists a previous time (that is, 0 < t) and the sub-formula w evaluates to 'T' at time (t − 1) (which

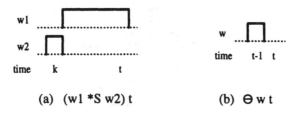

(a)  (w1 *S w2) t          (b)  ⊖ w t

**Fig. 1.** Waveforms corresponding to the definitions of *Since* and *Previous*

is immediately before time $t$). Figure 1 illustrates example waveform diagrams corresponding to the two definitions.

From the semantic definitions of the temporal operators, we were able to *prove* all the 'axioms' of the formal deductive system described in [5]. Thus the 'axioms' are nothing but derived theorems in our extensional approach. Temporal inference rules are also implemented as derived inference rules in HOL. Other important temporal laws are also proved. The one often employed in the serial multiplier verification is the expansion law for the operator *S (*Since*):

$$\vdash\ (p\ *S\ q) = (q\ *V\ (p\ *\wedge\ \ominus\ (p\ *S\ q)))$$

The law says that temporal formula $(p\ *S\ q)$ evaluates to '$T$' at present time iff either the sub-formula $q$ evaluates to '$T$' at present time or both the sub-formula $p$ evaluates to '$T$' at present time and the formula $(p\ *S\ q)$ evaluates to '$T$' at the previous time immediately before the present time.

## 3  Bit-Serial Adder

The bit-serial adder is a core component of serial multipliers. An implementation of the serial adder *SA* is shown in Fig.2 which consists of a delay flip-flop *DFF* and a full adder *FAR* with a *reset* input. When the reset input of the full adder is high ('$T$'), the carry output *cout* is low ('$F$') regardless the values on other inputs. The *sum* output is unaffected by the *reset* input. The truth table for *FAR* is shown in Table 1 and the corresponding HOL definition of *FAR* appears below.

```
⊢def  ∀in1 in2 cin reset sum cout.
     FAR in1 in2 cin reset sum cout = ∀t.
     (sum t = in1 t ∧ in2 t ∧ cin t ∨
          ¬in1 t ∧ ¬in2 t ∧ cin t ∨
          ¬in1 t ∧ in2 t ∧ ¬cin t ∨
          in1 t ∧ ¬in2 t ∧ ¬cin t) ∧
     (cout t = ¬reset t ∧
               (in1 t ∧ in2 t ∨ in1 t ∧ cin t ∨
               in2 t ∧ cin t))
```

The HOL definition of the delay flip-flop *DFF* appears below, which says that the output value at next time is always same as the present input value.

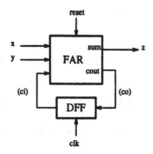

**Fig. 2.** An implementation of bit-serial adder

**Table 1.** Truth table for a full adder with reset $FAR$

| $in1(t)$ | $in2(t)$ | $cin(t)$ | $sum(t)$ | $cout(t)$ |
|:---:|:---:|:---:|:---:|:---:|
| $F$ | $F$ | $F$ | $F$ | $F$ |
| $F$ | $F$ | $T$ | $T$ | $F$ |
| $F$ | $T$ | $F$ | $T$ | $F$ |
| $F$ | $T$ | $T$ | $F$ | $\neg reset(t)$ |
| $T$ | $F$ | $F$ | $T$ | $F$ |
| $T$ | $F$ | $T$ | $F$ | $\neg reset(t)$ |
| $T$ | $T$ | $F$ | $F$ | $\neg reset(t)$ |
| $T$ | $T$ | $T$ | $T$ | $\neg reset(t)$ |

$\vdash_{def}$ ∀in out. DFF in out = ∀t. (out (t+1) = in t)

The clock signal *clk* is implicit in our model of time; the discrete time represents the advancement of system clock.

As the component definitions are given, the implementation description of $SA$ is defined in the standard way as described by Gordon in [6]. The implementation of the serial adder $SA$ is just the logical conjunction of the two components, $FAR$ and $DFF$, with internal signals ($co$ and $ci$) being hidden by existential quantification as shown below.

$\vdash_{def}$ ∀x y z reset.
    SA x y z reset = ∃ co ci. DFF co ci ∧ FAR x y ci reset z co

The implementation description of $SA$ is done in the standard way as described by Gordon in [6]. All the signals are boolean streams, i.e. functions from *time* to *bool* where *time* is represented by natural numbers (type *num* in HOL).

In order to derive the intended behavior (that is, *addition* of numbers) from this structural implementation description, some abstraction functions are needed to interpret the Boolean streams as numbers. This interpretation is part of the design concept. We formally capture such an interpretation with well-defined functions and standard mathematical notation as follows.

The control signal *reset* initiates a new computation as illustrated in Fig.3. The *reset* signal must be high (*'T'*) some time in the past and then it stays low (*'F'*) through the time $(t-1)$. The value of *reset* at time $t$ is unimportant. This condition is expressed by " $\ominus\;((*\neg reset)*S(reset))t$ " which reads "*since the 'reset' was last 'true', it has been 'false' through time* $(t-1)$." Without the temporal operators, it becomes rather verbose as " $0<t\wedge\;(\exists t1.t1\leq(t-1)\wedge reset(t1)\wedge\;(\forall t2.t1<t2\wedge\;t2\leq(t-1)\supset\neg reset(t2)))$ ."

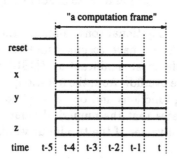

**Fig. 3.** Timing diagram for serial adder *SA*

If the *reset* signal is controlled properly, we can assemble boolean words (i.e. lists of boolean values) of the two operands and the sum from the corresponding signals of the circuit. The boolean words are assembled in two steps. First, the word size parameter is extracted from the *reset* signal using a recursive function called *elap* as defined below. *elap* counts the number of clock periods after the *reset* was last high ($T$).

$$\vdash_{def}\;\;(\forall x.\;elap\;x\;0\;=\;0)\;\wedge$$
$$(\forall x\;t.\;elap\;x(SUC\;t)\;=\;(x\;t\;\rightarrow\;0\;|\;SUC(elap\;x\;t)))$$

The base case is unimportant since we never use this function at time 0. (Note: $SUC$ increments numbers by one.) For the *reset* signal shown in Fig.3, (*elap reset t*) returns 4.

As the word size parameter is set, another recursively defined function *word* assembles boolean words (i.e. list of boolean values) from the corresponding signals (that are functions from *time* to *bool*) as appears below. (The *elap* function is not needed if the word size is not signal dependent.)

$$\vdash_{def}\;\;(\forall x\;t.\;word\;0\;x\;t\;=\;[x\;t])\;\;\;\wedge$$
$$(\forall n\;x\;t.\;word(SUC\;n)x\;t\;=\;CONS(x\;t)(word\;n\;x(PRE\;t)))$$

(Note: $CONS$ is a list operator that pushes elements onto the front of lists and $PRE$ decrements numbers by one.) For the signals in Fig.3, (*word 4 x t*) evaluates to $[F;x(t-1);x(t-2);x(t-3);x(t-4)]$, (*word 4 y t*) to $[F;y(t-1);y(t-2);y(t-3);y(t-4)]$, and (*word 4 z t*) to $[z(t);z(t-1);z(t-2);z(t-3);z(t-4)]$.

The boolean words are then evaluated to natural numbers by using a standard evaluation function *BLVAL* (which is the same as the function *BNVAL* in WORD library created by Wong [7]) whose definition in HOL is given below.

$\vdash_{def}$ (BLVAL [] = 0)   ∧
      (∀b bl. BLVAL(CONS b bl) = ((BV b) * (2 EXP (LENGTH bl)))
                               + (BLVAL bl))

*BV* maps a boolean value '*T*' to 1 and '*F*' to 0. For example, $(BLVAL\ [T; F; T; F])$ is decimal 10.

There are additional operational constraints for the circuit to operate properly as a serial adder. As we have assumed uniform word-length for the two operands and the sum, the most-significant bit (MSB) of the two operands must be $F$ in order not to have overflow from the addition; that is, $\neg x(t)$ and $\neg y(t)$.

With these mapping functions and operational constraints, the correctness theorem given below states that the numerical value of output word is indeed addition of the numerical values of the two input words.

$\vdash$  SA z y z reset ⊃
    ∀t. ⊖ (*¬reset *S reset) t ∧ ¬x t ∧ ¬y t ⊃
        let m = elap reset t in
        let x_word = word m x t
        and y_word = word m y t
        and z_word = word m z t in
         (BLVAL z_word = BLVAL x_word + BLVAL y_word)

Details of the proof for this theorem are not described here because of the space limitations. We mention that after induction on time $t$, temporal laws such as the expansion law for the *Since* operator are used for rewriting. Such rewriting makes the case analysis on the *reset* signal straightforward.

## 4   Basic Multiplier-Cell

Figure 4 shows the basic multiplier cell *MCELL* which is employed in the array multipliers in the following sections. The cell contains *DFF2* - a D-type flip-flop with complementary outputs, *WFF* - a write-controlled D-type flip-flop that latches its input value if the control signal is asserted and keeps old value otherwise, and two *and*-gates. The truth table of *WFF* is shown in Table 2. HOL definitions for *DFF2*, *WFF* and *and*-gate appear below.

$\vdash_{def}$ ∀in q qb.
    DFF2 in q qb = ∀t. (q(t+1) = in t) ∧ (qb(t+1) = ¬in t)

$\vdash_{def}$ ∀in w out.
    WFF in w out = ∀t. ( w t ⊃ (out(t+1) = in t)) ∧
                    ( ¬w t ⊃ (out(t+1) = out t))

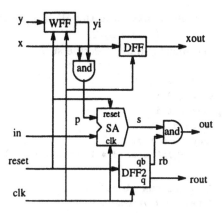

**Fig. 4.** Basic multiplier-cell *MCELL*

**Table 2.** Truth table for a latch *WFF*

| $w(t)$ | $out(t+1)$ |
|--------|------------|
| T | $in(t)$ |
| F | $out(t)$ |

$\vdash_{def}$ ∀a b c. and a b c = ∀t. (c t = a t ∧ b t)

The implementation description of *MCELL* appears below. It is a logical conjunction of *WFF*, two *and*-gates, serial adder *SA*, and *DFF2* with internal signals being hidden by existential quantification.

$\vdash_{def}$ ∀x y in reset xout out rout.
 MCELL x y in reset xout out rout = ∃ yi p s rb.
 WFF y reset yi ∧ and x yi p ∧ DFF x xout ∧
 SA p in s reset ∧ DFF2 reset rout rb ∧ and s rb out

The two *and*-gates take different roles in the circuit; the one connected to *WFF* computes the bit-product, and the other one connected to *DFF2* truncates the least-significant bit (LSB) of serial adder *SA* output. Note that the signal *yi* of *WFF* doesn't change from the last time that the *reset* was falling. Thus the first *and*-gate with *WFF* computes the bit-product of *x_word* (that is (*word m x t*) where *m* is (*elap reset t*)), and *yi-bit* (which is $y(t-(m+1))$ where $m = elap$ *reset t*).

The *out_word* from the second *and*-gate is taken with a one time-unit delay relative to other words, that is, *out_word* is (*word m out* $(t+1)$) where $m =$ (*elap reset t*). As the *reset* is high at time $t$ by assumption, the *out* at time $t+1$ (which is MSB of *out_word*) is forced to low ('*F*'). Since the word length is the same for *s_word* (which is *word m s t*) and *out_word*, *out_word* is the one-bit shift right of *s_word* with '$F$' being shifted into the MSB of *out_word*. Arithmetically this relation is integer division ($DIV$) by 2.

The following two lemmas formally describe such behavior.

```
⊢  WFF y reset yi ∧ and x yi p ⊃
   ∀t. ⊖ (*¬reset *S reset) t ⊃
       let m = elap reset t in
       let x_word = word m x t
       and p_word = word m p t in
       (BLVAL p_word = BLVAL x_word * BV(y(t−(m+1))))
```

```
⊢  DFF2 reset rout rb ∧ and sum rb out ⊃
   ∀t. ⊖ (*¬reset *S reset) t ∧ reset t ⊃
       let m = elap reset t in
       let s_word = word m s t
       and out_word = word m out (t+1) in
       (BLVAL out_word = (BLVAL s_word) DIV 2)
```

These lemmas and the correctness theorem of $SA$ provide the logical "bridge" to relate the input and output words of the multiplier cell $MCELL$ through major internal signals with their associated words. The correctness theorem proved for the $MCELL$ appears below. It adopts the same abstraction functions used in the serial adder $SA$.

```
⊢  MCELL x y in reset xout out rout ⊃
   ∀t. ⊖ (*¬reset *S reset) t ∧ ¬x t ∧ ¬in t ∧ reset t ⊃
       let m = elap reset t in
       let x_word = word m x t
       and in_word = word m in t
       and out_word = word m out (t+1) in
       (BLVAL out_word = (((BLVAL x_word) * (BV(y(t−(m+1)))))
                         + (BLVAL in_word)) DIV 2)
```

The theorem states that if *reset* was once '$T$' in the past and the present values of signals $x$ and *in* are '$F$' and the present value of *reset* is '$T$', then the numerical value of output word taken with a one time unit delay is the same as the numerical value of multiplicand word multiplied by the bit-value of signal $y$ at the time when the *reset* was last '$T$', then added to the numerical value of offset word of signal *in*, and then divided by 2. (See also Fig.5.) Note that the latency of one time unit is the consequence of taking the *out_word* one time unit later, relative to other words, in order to get the effect of one-bit shift right or integer division by 2 as explained for the *and*-gate connected to $DFF2$.

In addition, the $MCELL$ has a property shown in the following lemma, which is useful in the verification of array multipliers in the following sections.

```
⊢  MCELL x y in reset xout out rout ⊃
   ∀t. ⊖ (*¬reset *S reset) t ∧ ¬x t ∧ ¬in t ∧ reset t ⊃
       ⊖ (*¬rout *S rout) (t+1) ∧ ¬xout(t+1) ∧ ¬out(t+1) ∧ rout(t+1)
```

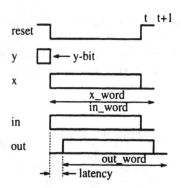

**Fig. 5.** Timing diagram for basic multiplier-cell MCELL

The lemma says that all the operational conditions of input signals appearing in the correctness theorem above are transferred to the corresponding output signals with one time-unit delay.

## 5 Basic Pipeline Multiplier

The serial pipeline multipliers described in this section and in the next section are suggested in [2]. The two multipliers are identical except the optional delays between stages for improved electrical delay characteristics.

The basic pipeline multiplier (*BPM*) is an iterative array of the multiplier cell *MCELL*. The number of stages is same as the length of multiplier word. A block diagram of a 5-stage multiplier is shown in Fig.6. Note that the signal $y$ is global which is connected to all cells. The recursive definition of the multiplier *BPM* in HOL is given below.

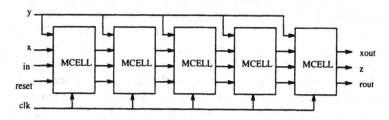

**Fig. 6.** Basic pipeline multiplier *BPM* of 5 stages

$\vdash_{def}$ (∀x y in reset xout z rout.
BPM 0 x y in reset xout z rout = MCELL x y in reset xout z rout) ∧
(∀n x y in reset xout z rout.
BPM(SUC n)x y in reset xout z rout = ∃ xi zi ri.
BPM n x y in reset xi zi ri ∧
MCELL xi y zi ri xout z rout)

A cell generator is obtained by using our HOL2L compiler [8] with the implementation description. A layout is obtained by invoking the cell generator and by using the AutoCells [10] automated placement and routing tool in the Mentor Graphics GDT design system. Figure 7 shows such a layout that the generator created for 5 stages.

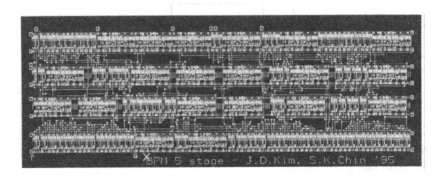

**Fig. 7.** Layout of basic pipeline multiplier BPM of 5 stages

The multiplier word arrives serially on the $y$-port beginning right before the '$T$-$F$' transition of the *reset* signal with least-significant bit entering first. The multiplicand and offset words arrive on the $x$ and *in* ports, respectively, beginning right after the '$T$-$F$' transition of the *reset* signal, with the least-significant bits entering first, and the most-significant bits entering last at which time the *reset* is '$T$' again. (The *reset* pulse serves as a mark which separates adjacent computational frames.)

The operation is pipelined; each stage (*MCELL*) performs the *bit-multiplication – addition – division-by-2* as described in the previous section and contributes to the overall function of integer division (*DIV*) by ($2^{n+1}$ and the overall delay of $(n + 1)$ time units for a $(n + 1)$ stage multiplier. Note that the *out_word* of each *MCELL* is the *offset_word* to the next cell. The first cell computes the partial-product with the LSB of the *y_word*, then added to the *offset_word* on the signal *in*, and then divided by 2. The second cell performs a similar operation one time unit later; all the input signals to this cell are delayed by one time unit. The second LSB of *y_word* is provided to this cell. Eventually the *z_word* of the last stage is delayed by $(n + 1)$ time units relative to the input words to the first cell.

The correctness theorem proved for this $(n + 1)$-stage multiplier appears below. The variable $n$ in the theorem is the multiplier size (word length) parameter to BPM. The actual number of stages (and the multiplier word length) is $(n + 1)$ since parameter 0 corresponds to a single stage multiplier which is just *MCELL* itself. The theorem states that "if operational conditions (which are the same as that for *MCELL*) are satisfied, then the numerical value of output word which is taken $(n+1)$-time units later than the inputs is same as the numerical value of multiplicand-word multiplied by the numerical value of multiplier word then added to the numerical value of offset word of signal *in* and then divided by $2^{n+1}$, where $n+1$ is the word length of the multiplier word."

```
∀a b. a DIV (2 EXP (b+1)) = (a DIV (2 EXP b)) DIV 2
⊢  ∀n. BPM n x y in reset xout z rout ⊃
   ∀t. ⊖ (*¬reset *S reset) t ∧ ¬x ∧ ¬in t ∧ reset t ⊃
      let m = elap reset t in
      let x_word = word m x t
      and in_word = word m in t
      and y_word = word n y (t+n−(m+1))
      and z_word = word m z (t+n+1) in
        (BLVAL z_word = ((BLVAL x_word * BLVAL y_word) + BLVAL in_word)
                       DIV (2 EXP (n+1)))
```

A lemma on integer division is included in the theorem explicitly as a hypothesis, which says "$a$ divided by $2^{b+1}$," is same as "$a$ divided by $2^b$ and then divided again by 2." The offset word applied to *in* port can be a small number (value 1) so that the truncation of the least-significant $(n+1)$ bits from the multiplication – addition – division operation effects an actual rounding if the intended function is just multiplication and division with rounding.

The proof proceeds with induction on $n$, the size parameter of the multiplier. The proof uses the following lemma which tells that the operating conditions which appear in the correctness theorem are transferred to the output side with $n+1$ time-unit delay.

```
⊢  BPM n x y in reset xout z rout ⊃
   ∀t. ⊖ (*¬reset *S reset) t ∧ ¬x t ∧ ¬in t ∧ reset t ⊃
      ⊖ (*¬rout *S rout)(n+1) ∧ ¬xout(n+1) ∧ ¬z(n+1) ∧ rout(n+1)
```

This lemma enables us to use the *MCELL* correctness theorem in the proof of subgoal corresponding to inductive case.

A sample simulation result is shown in Figure 8. For this simulation, the netlist is extracted from the layout for the 5-stage multiplier. The simulation is for the case where the multiplier input is decimal 17 (binary 10001), the multiplicand is decimal 111 (binary 01101111), and the offset is decimal 1 (binary 0000001). With this value of the offset word, the result is as shown in Fig.8 decimal 59 (binary 00111011) with a latency time of 5 clock ticks. Note that the result would have been decimal 58 (binary 00111010) if the value of the offset word were 0 because $(17 * 111)/2^5 = 58.96875$.

# 6   Enhanced Pipeline Multiplier

The basic pipeline multiplier *BPM* suffers a long propagation (combinational) delay time when the multiplicand word (*x_word*) is long because a signal may have to go through many *FAR-and* stages in a single clock cycle. This limits the clock rate of the multiplier as the total propagation delay time (combinational) from the inputs to the first cell to the outputs of the last cell.

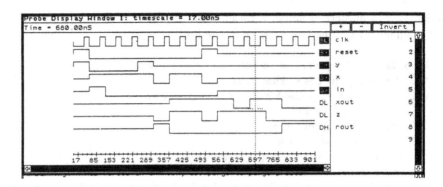

**Fig. 8.** Simulation of 5-stage basic pipeline multiplier *BPM*

The enhanced pipeline multiplier *EPM* breaks up the *FAR-and* ripple chain by inserting delay elements (*DFF*) between the multiplier cells *MCELL*, thereby maintaining the clock rate as high as that of a single *MCELL*. Although the latency is nearly doubled in terms of the clock cycles, each clock period can be decreased by a factor of $n$ where $n$ is the number of stages in the multiplier. Consequently the enhanced pipeline multiplier has *decreased* latency in real time and increased throughput, both by a factor of $n$ where $n$ is the number of stages in the multiplier. Figure 9 illustrates a 5-stage multiplier with the system clock signal connection being suppressed. All the connections are local except the system clock and power rails.

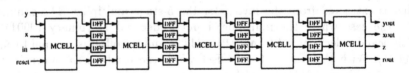

**Fig. 9.** Enhanced pipeline multiplier *EPM* of 5 stages

The formal implementation description of the enhanced pipeline multiplier *EPM* in HOL appears below. It generates an iterative array of *MCELL*s and *DFF*s as illustrated in Fig.9. Note that delay stage is not included for the base case; it is inserted only when a new stage is added.

```
⊢_def  (∀x y in reset yout xout z rout.
        EPM 0 x y in reset yout xout z rout =
        (∀t. yout t = y t) ∧
        (MCELL x y in reset xout z rout)) ∧
      (∀n x y in reset yout xout z rout.
        EPM(SUC n)x y in reset yout xout z rout =
        ∃ xi zi ri yd xd zd rd.
          EPM n x y in reset yd xi zi ri ∧
          DFF yd yout ∧ DFF xi xd ∧ DFF zi zd ∧ DFF ri rd ∧
          MCELL xd yout zd rd xout z rout)
```

From this implementation description, a cell generator and layout are obtained in the same way as in the case of *BPM*. Figure 10 shows a layout that the generator created for 5 stages.

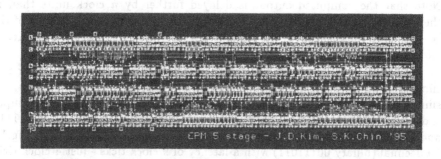

**Fig. 10.** Layout of enhanced pipeline multiplier *EPM* of 5 stages

The operation of *EPM* is the same as *BPM* except timing. Because of the additional *DFF*s between stages, all the signals between stages including the $y$ signal are delayed by one clock cycle. Thus, the $z\_word$ is delayed by a total $(2n + 1)$ time units for an $(n + 1)$ stage multiplier. It is $n$ more time units than in *BPM* of the same size, which is caused by the insertion of *DFF*s between *MCELL* stages.

The correctness theorem proved for this multiplier appears below. This theorem is parameterized on the multiplier size parameter $n$ (the actual number of stages and multiplier word length are $(n + 1)$ since parameter 0 corresponds to a single stage multiplier which is just an *MCELL*); that is, the enhanced pipeline multiplier is formally verified for all multiplier word sizes as well as the multiplicand word size $m$ which is determined by the signal *reset*.

∀a b. a DIV (2 EXP (b+1)) = (a DIV (2 EXP b)) DIV 2
⊢  ∀n. EPM n x y in reset yout xout z rout ⊃
    ∀t. ⊖ (*¬reset *S reset) t ∧
        ¬x t ∧ ¬in t ∧ reset t ⊃
        let m = elap reset t in
        let x_word = word m x t
        and in_word = word n in t
        and y_word = word n y (t+n−(m+1))
        and z_word = word m z (t+n+n+1) in
        let x_num = BLVAL x_word
        and in_num = BLVAL in_word
        and y_num = BLVAL y_word
        and z_num = BLVAL z_word in
        (z_num = ((x_num * y_num) + in_num)
                 DIV (2 EXP (n+1)))

As before, the division lemma is included as an explicit hypothesis. The proof is not significantly harder than the basic pipeline multiplier. In fact, the proof structure is almost the same as that of *BPM* and similar lemmas are employed. Note that the computed output is delayed further by $n$ clock units than in the basic pipeline one. In total, the output word appears on the output of the $(n + 1)$-stage multiplier $(2n + 1)$ clock cycles later than the input words ($x\_word$ and $in\_word$).

As before, a sample simulation is shown in Figure 11. The netlist for this simulation is extracted from the layout for the 5-stage multiplier. Again, the simulation is for the same case as in the simulation of *BPM*; the multiplier input is decimal 17 (binary 10001), the multiplicand is decimal 111 (binary 01101111), and the offset is decimal 1 (binary 0000001). The result is as shown in Fig.11, 59 decimal (binary 00111011) with a latency of 9 clock ticks - just 4 clock ticks more than in *BPM*.

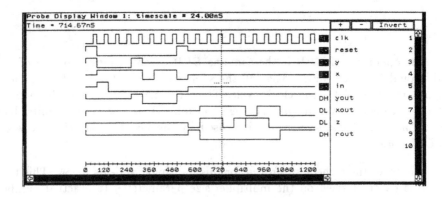

**Fig. 11.** Simulation of 5-stage enhanced pipeline multiplier *EPM*

# 7 Conclusion

The bit-serial multiplier with pipelining and truncation is intricate because the result is not exact. Bit-serial circuits have suffered from informal justification of their claimed behavior. Such difficulty is overcome in formal verification by using a set of well-defined abstraction functions. The formal abstraction functions enable us to describe the intended behavior in standard mathematical notations.

The verified bit-serial pipeline multipliers are *parameterized* on the length of the multiplier word; they can be instantiated to arbitrary size. The actual operation is *multiplication-addition-division*. The multipliers are verified for *all* sizes of the multiplier word (i.e. coefficient) as well as the multiplicand and offset words, unlike Boolean Decision Diagram (BDD)-based verification approaches in which the word sizes are typically fixed. Truncation of least-significant bits is captured in the standard mathematical notation of integer division. If an appropriate number is supplied as the offset word, the *multiplication-addition-division* will produce rounding.

The linear-time temporal logic that we have embedded in HOL has helped to express the behavior succinctly. It also makes the the proof more efficient by allowing us to reason with its temporal laws.

The properties proved support embedding into larger systems and reuse. The HOL definitions have been compiled to parameterized cell generators and layouts [8] to augment the Mentor Graphics GDT VLSI design library [9]. The arithmetic properties support design at the register-transfer level and instruction-set architecture levels.

# References

1. Henry S. McDonald, Leland B. Jackson, James F. Kaiser, "An approach to the implementation of digital filters," *IEEE Trans. on Audio and Electroacoustics*, AU-16(3):413–421, Sept 1968.
2. R. F. Lyon, "Two's complement pipeline multipliers," *IEEE Transactions on Communications*, pages 418–425, April 1976.
3. Shiu-Kai Chin, Juin-Yeu Lu, "The mechanical verification and synthesis of parameterized serial/parallel multiplier," Technical Report 9140, CASE Center, Syracuse University, 1991.
4. Shiu-Kai Chin, "Verified Functions for Generating Signed-Binary Arithmetic Hardware," *IEEE Trans. Computer-Aided Design*, pages 1529–1558, December 1992.
5. Amir Pnueli, Zohar Manna, *The Temporal Logic of Reactive and Concurrent Systems*, Springer-Verlag, 1992.
6. Michael J.C. Gordon, "Why higher-order logic is a good formalism for specifying and verifying hardware," In G. J. Milne and P. A. Subrahmanyam, editors, *Formal Aspects of VLSI Design*, pages 153–177. Elsevier Scientific Publishers, 1986.
7. Wai Wong, "Modelling bit vectors in HOL: the word library," *Proc. of 6th Intl. HOL Users Group Workshop 1993*, Vancouver, B.C, Canada, August 1993, Springer-Verlag, New York, 1994.

8. J. Y. Lu, S. K. Chin, "Linking HOL to a VLSI CAD system," *Higher Order Logic Theorem Proving and Its Applications*, Lecture Notes in Computer Science 780, Springer-Verlag, Berlin Heidelberg 1994.
9. Mentor Graphics Inc., *GDT Led, Lx Standard Cell, Explorer Lsim V.5.3 users manuals*, San Jose, CA, 1990.
10. Mentor Graphics Inc., *Explorer AutoCells Users Guide*, San Jose, CA, 1990.

# TkWinHOL:
# A Tool for Window Inference in HOL

Thomas Långbacka, Rimvydas Rukšenas, Joakim von Wright

Åbo Akademi University
Department of Computer Science
Lemminkäinengatan 14–18
SF–20520 Åbo, Finland

**Abstract.** Window inference is a method for contextual rewriting and refinement, supported by the HOL Window Inference Library. This paper describes a user-friendly interface for window inference. The interface permits the user to select subexpressions by pointing and clicking and to select transformations from menus. The correctness of each transformation step is proved automatically by the HOL system. The interface can be tailored to particular user-defined theories. One such extension, for program refinement, is described.

## 1 Introduction

Though the original purpose of the HOL system [10] was as a tool for hardware verification, it has become popular also as a basis for reasoning about software (see for example [1, 7, 8]). However, theories built for supporting the software development process are normally difficult to use, especially if one does not have any previous detailed knowledge of the HOL system. In order to make such theories available to a general audience, it is essential that users have support in the form of a user interface, either a general HOL interface such as [19] or a specialised tool such as [2]. This observation has motivated us to design a tool that supports development of provably correct programs within the framework of the Refinement Calculus [3, 4, 6].

Our work is based on a number of layers. The starting point is the HOL system and its Window Inference Library [11, 12]. Window inference (originally described in [17]) supports the notion of contextual refinement which is an important ingredient in the refinement calculus as well. We also use the HOL theory for the Refinement Calculus [21]. On top of these packages we have designed a graphical user interface that itself is designed in a layered fashion. The basic user interface is called TkWinHOL, and it offers users an X-windows based graphical interface to the window inference system.

TkWinHOL is built using the Tcl/Tk programming system [15], which is a general purpose environment for building user interfaces. Tcl/Tk applications are normally memory conserving, which is useful when building interfaces to large systems such as HOL. TkWinHOL is purposely designed to be easy to extend and modify. As an extension to the TkWinHOL system, we are developing a tool for the Refinement Calculus.

## 2  Background

Our original motivation for developing a tool, such as the one presented in this paper, was to provide an environment supporting development of provably correct programs within the framework of the Refinement Calculus [3,4]. This calculus is a formalization of the stepwise refinement method for program construction, based on the weakest precondition calculus of Dijkstra [9]. The basic idea is to start from a high level specification and refine it through a sequence of correctness preserving steps into an executable program. The basic notion in the Refinement Calculus is the refinement relation between program statements. Statement $S'$ is said to be a correct refinement of statement $S$ if it preserves the total correctness of the latter. The refinement relation is a preorder (i.e., it is reflexive and transitive). This fact justifies the stepwise refinement method. Another important aspect of the Refinement Calculus is its top-down approach to program development; a part of a larger program can be refined independently of its context.

A starting point for our work was the Centipede environment [5]. Centipede was a mouse and menu driven environment that supported stepwise development of programs within the Refinement Calculus. However, Centipede did not support mechanical verification of refinement steps. Thus, we wanted to develop a similar tool with this added capability.

The choice of HOL as the proof engine underlying the system was simple, because of previous work on formalizing the Refinement Calculus in HOL, which has spanned several years [20,21]. Furthermore, the Window Inference Library of the HOL system provides support for doing refinement transformations easily in HOL. In window inference, terms (or subterms) of a logic are transformed, preserving some preorder. The user indicates what transformation is desired and provides information describing the path to the subterm that should be transformed. Computing paths is tedious for an ordinary user, but it can be automated within a user interface. In fact, using the mouse to select the data (in this case subterms) one wants to perform operations on is a well established principle in modern user interfaces. Thus window inference lends itself well as the basis for a modern user interface to HOL. Most of the features that one wants from such an interface are not specific to program refinement but are desirable for arbitrary transformational reasoning using window inference. TkWinHOL is an attempt to provide such a general user interface. The basic idea is to provide facilities for manipulating any object that is already present in the system just by pointing and clicking with the mouse. New functionality needed for program refinement or any other kind of transformational reasoning can then be obtained by extending the basic TkWinHOL interface.

There were initial attempts to use both the Centaur system (which is used also in Théry's chol system [19]) and the Synthesizer Generator [16]. Both these systems make it possible to build syntax directed editors (in fact the Centipede system also had a syntax directed editor as it's base) for constructing programs in some user defined language. Both systems also allow one to define transformation rules that operate on the programs (or more generally some syntactically correct term in a given language). However, the executables that these systems produce are very big. Keeping in mind that HOL90 itself is a big system this is not a nice feature. Furthermore, both systems have built-in assumptions that can be restrictive, e.g., regarding the ways in which syntax can be manipulated.

To implement the interface we have chosen the Tcl/Tk programming system. This choice has turned out to be a good one. Tcl/Tk is a general purpose system that is well suited for building user interfaces for existing UNIX applications (HOL in our case). We have managed to keep TkWinHOL small and efficient. And given the freedom provided by a general purpose system we have also been able to add certain 'unorthodox' features to the system (e.g., dealing with associative operators as explained in Section 4).

## 3 Window inference

In window inference, the user transforms a term so that a preorder (a reflexive transitive relation) is preserved. The user can also restrict attention to a subterm of the term that is being transformed. A transformation of the subterm is at the same time a transformation of the whole term, provided that certain context-dependent monotonicity conditions hold.

In the HOL Window Inference Library there is built-in support for window reasoning under equality, implication and reverse implication. Furthermore, the user can add other preorders. In HOL, window reasoning is handled as follows. Assume that $E[t]$ is a term with subterm $t$. Our starting point is a *window stack* containing the window

$$R \star E[t]$$

The *focus* of the window is the term $E[\ldots t \ldots]$ and the relation that we want to preserve (the *window relation*) is $R$ (the star simply separates these two). In order to set up such a stack, the user must first load the HOL Window Inference Library and then type the appropriate command for beginning a stack.

The theorem associated with this window (the *window theorem*) is

$$\vdash E[\ldots t \ldots] \, R \, E[\ldots t \ldots]$$

which holds because $R$ is reflexive.

*Opening subwindows.* We can *open a subwindow* on the subterm $t$. This is done by entering a command of the form DO(OPEN_WIN path), where path is a list of path elements that indicate how we navigate in the abstract syntax tree of $E$ in order to get to the subterm $t$. In view of the simple syntax of higher order logic, the possible path elements are RATOR, RAND and BODY (we assume that the reader is familiar with higher order logic and has some experience of the HOL system). The result of the opening command is the following window:

$$R' \star t$$

with the corresponding window theorem (again by reflexivity)

$$\vdash\ t\ R'\ t$$

Here $R'$ may be the same relation as $R$, but it can be different. For example, $R$ can be implication and $R'$ equality (this would be the case if the focus was $x + 1 - 1 \leq y$, the relation was implication and we opened a subwindow on $x + 1 - 1$). We now transform $t$ under the relation $R'$, using available transformation rules (e.g., rewriting, beta-conversions, matching with pre-proved theorems etc). Assume that the result of these transformations is $t'$. The new window theorem is then

$$\vdash\ t\ R'\ t'$$

(the system uses transitivity of $R'$ to draw this conclusion). We can now close the subwindow by entering the command DO CLOSE_WIN.

*Closing and window rules.* At this point, the window inference system uses a collection of *window rules* (ML functions that perform HOL proofs) to do the following inference:

$$\frac{\vdash\ t\ R'\ t'}{\vdash\ E[t]\ R\ E[t']}$$

Essentially, window rules embody monotonicity properties of contexts, with respect to the relations involved. The system chooses suitable window rules automatically; the choice is made when the subwindow is opened. The rules are stored anonymously in a database so the user does not have to know names or other details of these rules. The result of closing is the new window

$$R \star E[t']$$

and the window theorem shows the result of the the whole sequence of steps:

$$\vdash\ E[t]\ R\ E[t']$$

We can now store this theorem in a theory file or we can continue and transform $E[t']$ further.

*Assumptions.* The window inference system keeps tracks of what assumptions can be used in a transformation. Consider opening a subwindow on the right conjunct of the following window:

$$\Rightarrow\ \star\ x = 1 \wedge x > 0$$

The new window has x=1 as an *assumption*:

$$\begin{array}{l} !\ x = 1 \\ \Rightarrow\ \star\ x > 0 \end{array}$$

The assumption can be used to transform the focus to T (truth). After this we can close and rewrite; this transforms the original focus to x=1.

*Extensibility.* The Window Inference library lets the user add a new relation $R$ into the system, once $R$ has been proved reflexive and transitive. An example is the refinement relation on programs which is used in the example in Section 6. The user can also add window rules. In the refinement example, we have added one rule for each argument of each statement constructor (e.g., one rule for closing after opening a subwindow on the left component of a sequential composition, one rule for the body of a loop, etc.).

## 4 TkWinHOL

TkWinHOL is a graphical user interface to the Window Inference Library. It runs on UNIX based systems under X-windows. It requires the Tcl/Tk programming system and the Expect [14] extension to Tcl/Tk. The HOL environment supported is HOL90.

TkWinHOL is designed to simplify the work a user has to do while transforming an initial focus towards the desired goal. One of the main features of TkWinHOL is that it gives the user the possibility to use the mouse to point out subterms that need to be operated on. The user is spared from the task of typing in explicit path information which can be tedious if operating on a large focus. The system has other interesting features as well (see below) and we have tried to build in as much graphical support as possible.

Using Tcl/Tk for implementing TkWinHOL makes the system small yet powerful. It also makes the system 'open' in the sense that it can easily be modified and extended. Indeed, one of the basic ideas of TkWinHOL is that it should be possible to use as a basic building block for the more specialised system discussed in Section 6. Tcl/Tk has turned out to be a good choice also because there is work going on using Tcl/Tk for interfacing to other aspects of the HOL system that we have not considered (for example theorem retrieval) [18], and there is a great potential for cooperation.

### 4.1 The basic appearance of TkWinHOL

When TkWinHOL is started three windows appear on the screen. First there is a simple text editor window that offers a small set of menu commands for file and edit operations. The text editor window also offers a subset of the ordinary Emacs key sequences for moving the insertion point etc. However, the user does not have to use this editor (this is explained below).

*Executing HOL commands.* The second window (the HOL window) is similar to a terminal window (like Xterm). Commands can be typed directly into the HOL window and executed. Alternatively, users can execute any HOL command they wish by selecting (using the mouse) a text string containing the command and pressing an execute button that is associated with the HOL window. The text string is sent verbatim to the HOL process running in the background and the reply given by HOL is presented in the HOL window. The selected text can be

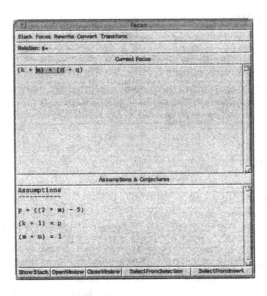

**Fig. 1.** A sample view of the focus window of TkWinHOL

in the text editor window but it can also be in a separate window not opened by the TkWinHOL application. This window can, e.g., be running Emacs provided the Emacs version used exports the selection.

*Window inference in TkWinHOL.* The third window is the heart of the application. This window (called the focus window) displays the status of the current window stack. As can be seen in Figure 1 the window is divided into parts. At the top of the window there is a menubar from which the user can select operations that transform the appearance of the the current focus. Below the menubar there is a field that presents the relation that is currently begin preserved, (in Figure 1 this relation is equality). The top subwindow presents a pretty printed version of the current focus. The pretty printer associated with the environment supplies not only the pretty-printed text, but also information about the paths to subterms and also the starting and ending positions on the screen of subterms. Using this information, TkWinHOL can access subterms. Without an interface, manipulating a subterm would mean typing the operation to be performed and supplying the path to the subterm in question, which can be a tedious task. In TkWinHOL, double-clicking the right mouse button at some position in the term will highlight the smallest enclosing subterm. After this, the user can select the desired operation from the menus (or a button in some cases); no path information is required. A typical operation is opening a window. The user selects, by clicking, the subterm to open a window on and then press the button labeled OpenWindow (which can be seen at the bottom of the focus window in Figure 1).

TkWinHOL then sends the corresponding command string to HOL and waits for a reply (which is either new pretty-printing information or an error message).

To illustrate how this works we can again refer to Figure 1. If the user clicks the k in the focus then only the k is selected (and highlighted). This is because k is a subterm on which it would make sense to open a subwindow. If the user clicks the leftmost + in the focus, then the selected subterm is (k + m). Similarly, if the user clicks the middle + the whole focus is selected.

*Handling associative operators.* In a sense TkWinHOL is, as for example chol [19], built around the structure of terms. However, this is not the only way to operate on the focus. Often there are situations where the structure of the focus is such that one cannot open a window on the part of the focus that one wants to, without modifying the focus first. Typically, such situations occur with associative operators where the standard association may not be what one wants. To simplify the user's work in such situations TkWinHOL provides a way to select (by dragging, not clicking) a segment of the string representing the focus, and then opening a window on that selection although it may not be accessible using a path in the ordinary sense.

This is made possible by a routine that automatically reassociates the focus according to what the user wants and then opens a window on the subterm the user has selected. There are restrictions on the use of this feature; the text segment selected by dragging has to be of the correct form[1] in order for reassociation to work.

We again refer to Figure 1 to illustrate how this works. As seen from the figure, the segment that has been selected (by dragging) is m) + (n. This is not a subterm in itself and cannot be selected by double-clicking as described above, but it is still possible to press the **OpenWindow** button which will result in a new window with the focus m + n.

*Working with assumptions.* The subwindow at the bottom of the focus window contains the assumptions (assumptions, conjectures and lemmas in window inference jargon) of the current stack. Operations that work on the assumptions are supported by TkWinHOL; users can click on the term that they want to operate on after which the term is highlighted. This term can then be used (as a theorem) as an argument to any command chosen from a menu. This saves the user from the trouble of typing the term (and additional type information).

*Supplying parameters to commands.* In the operations described above, TkWin-HOL constructs HOL commands that use only information that is present by default in the focus window. Sometimes, however, the user has to supply additional parameters to some menu command chosen (e.g. the command **REWRITE_WIN** in the **Rewrite** menu could be passed a list of names of theorems to be used for the

---

[1] This means that the selection must cover two subterms (having the same type) separated by an associative operator (disregarding parentheses denoting current association)

rewrite operation). Such parameters are passed via dialogue boxes or by selecting a string for example in the text editor window. Strings can also be selected in an emacs window if the user prefers to work in such a way. It is also always possible to modify the current window stack by issuing commands directly to HOL as described above. In this case the result of the issued command appears in the HOL window. The state of the focus window can be updated by pressing the ShowStack button.

*Proof scripts.* The commands that are issued via the focus window while transforming a given stack are recorded as a script. This is a simple process which records every command sent to HOL, separating them with semicolons. The complications that arise when trying to store a proof script from a session with the subgoal package (e.g. lists of tacticals to handle subgoals) do not exist in window inference.

For the moment being, the system does not record commands that change the current stack but are issued directly via the HOL window. This means that in order to record the whole proof process, all commands that actually operate on the stack must be issued via the focus window. Likewise, small lemmas etc. that are needed in the proof process have to be constructed via the focus window.

## 4.2 Implementing TkWinHOL

As mentioned in previous sections, the main motivation for implementing TkWinHOL is to use it as a building block for something larger; an environment supporting formal program derivation in the Refinement Calculus framework. This means that the interface layer must be reasonably small and efficient. Tcl/Tk applications are interpreted but that does not generally affect their efficiency substantially. In fact, the overhead produced by adding TkWinHOL on top of HOL90 is negligible. Tcl/Tk applications are also easy to extend since the language is easy to learn and additions are simply loaded into the application, so no recompilation or linking is needed.

To connect the interface to HOL we use an extension to the Tcl/Tk system called Expect. This extension provides a set of functions lacking from Tcl/Tk, like job control over several processes running in the background simultaneously, and communicating with standard command line based applications like HOL. Using Expect makes it easy to run, for example, a computer algebra system simultaneously with HOL under the same interface. There are situations where this possibility could prove useful [13].

TkWinHOL relies on a special purpose pretty printer implemented in SML. Besides pretty printing terms this pretty printer also supplies information about the subterms of the pretty printed term. This information is needed to implement the selection of subterms by clicking the mouse.

## 4.3 Customising TkWinHOL

When customising TkWinHOL for some specific HOL theory, there are basically three things one might want to do: modify the basic pretty printer supplied with

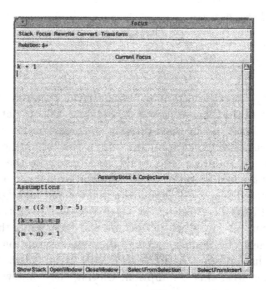

**Fig. 2.** An other sample view of the focus window of TkWinHOL

the environment, produce a parser suited for the theory used and add new choices to the menus or even entire new menus. A parser is useful in the situations where one has to enter terms as arguments to menu choices.

As mentioned above the format of the the pretty printer is fairly simple. It has to supply the TkWinHOL interface with information about the start and end point of subterms, using a specific row + column format and path information, in addition to the actual pretty printing. In general, it is not necessary to replace the basic pretty printer. It is only in the case that one wants the output to look 'nice' that a new pretty printer is needed. A parser is also needed only in case one is not satisfied with the 'HOL-syntax' of the theory one uses.

Adding new choices in the menus or altogether new menus to the menubar is done using Tcl/Tk programming.

## 5   Using TkWinHOL – an example

We shall now illustrate how TkWinHOL supports transformations in a simple and intuitively appealing way. In particular we want to show features dealing with selection and regrouping of expressions. Selection is always essential and automatic regrouping substantially simplifies derivations that handle associative operators.

Suppose that in a derivation we have opened a subwindow with the current focus and assumption list seen in the Figure 1. While preserving equality the focus

$$(k + m) + (n + q)$$

can be simplified to

$$p + q$$

just by exploiting context information present in the assumption subwindow (the lower subwindow of the focus window). In Figure 1, the only applicable assumption $(m + n) = 1$ can not be used directly since association of the variables m and n in the focus is different from the one in the assumption. Using window inference without an interface, one would need to rewrite the focus twice using the associativity theorem in order to get an appropriate association. Here, however, we can proceed just by dragging the pointer and highlighting the desired expression (see Figure 1). Pressing the OpenWindow button then automatically rearranges the current focus before opening and as a result the new focus is m + n. Now we select (by double-clicking) the first assumption $(m + n) = 1$ in the assumption list and use the REWRITE_WIN menu entry. TkWinHOL then rewrites the focus with this assumption. Closing the window with the button CloseWindow gives us the new focus

$$k + (1 + q)$$

Once again, by dragging the pointer we can highlight k + 1, then regroup the whole expression (with the OpenWindow button) and focus on

$$k + 1$$

Now double-clicking on the second assumption $(k + 1) = p$ selects it for the next rewriting (see Figure 2). Note that in this way we can easily choose the assumptions to be used in a transformation and avoid unnecessary (and possibly looping) rewriting which occurs when directly using functions from the Window Inference Library such as ASM_REWRITE_WIN (in our example we prevented variable p from being rewritten using the first assumption $p = ((2 * m) - 5)$).

Having closed the window we achieved our goal and transformed the initial expression into p + q. This fact is expressed by the window theorem:

$$|- (k + m) + (n + q) = p + q$$

which shows the result of the part of the derivation described here.

## 6 The Refinement Calculator

The Refinement Calculator, a customized version of TkWinHOL, is a tool for program refinement within the framework of the Refinement Calculus. The language used to express programs and specifications is essentially Dijkstra's language of guarded commands, with some extensions. In the Refinement Calculus, assertions can be used as statements, to indicate facts that are known to hold at certain points in the program text. There is also a nondeterministic assignment, which is used for specification purposes. Furthermore, blocks with local variables are included.

The Refinement Calculator extends TkWinHOL with a number of menu alternatives (in the Transform menu) for refinement transformations. For each transformation, HOL proves the correctness using the underlying theory of refinement. The example below illustrates a few basic transformations. More advanced transformations are also available, such as introduction of blocks with local variables, recursion introduction and data refinement.

The Refinement Calculator contains a parser (written by Michael Butler) which parses program statements entered by the user into the abstract syntax of the HOL system. It also contains an extension of the TkWinHOL pretty-printer to handle program statements.

## 6.1  How programs are handled

The HOL theory of refinement is a *shallow embedding*, meaning that the theory manipulates semantic objects, rather than the actual syntax of the Refinement Calculus. The semantics used is a *weakest precondition semantics*, where every program statement denotes a predicate transformer. A predicate transformer is a function of type $pred \rightarrow pred$ from predicates to predicates, where a predicate is a function of type $state \rightarrow bool$. The type of states is in the general theory a polymorphic type. For any given program, this type is instantiated to a product with one component for each program variable. If, for example, a program works on two natural number variables x and y and a boolean variable b, then the state space has type $num \times num \times bool$. The names of program variables are handled using a let-construction; in this case the term representing the program is of the form

```
let x = FST in let y = FST o SND in let b = SND o SND in ...
```

Thus program variables are *projection functions*; they indicate positions in the state tuple.

One main reason for using a shallow embedding is that this permits reuse of HOL data types in programs. When a program variable ranges over the natural numbers, we can use the HOL theory of natural numbers directly when doing transformations (e.g., to simplify a subexpression of a statement). Programs can also work on lists, sets, real numbers etc, relying on built-in theories and libraries. If a user adds a new data type to HOL, then that data type can immediately be used in programs. If we had decided on a deep embedding rather than a shallow one, then we would have had to embed the type structure in the theory of refinement. This would have fixed the available data types for program variables, making it impossible to introduce new types later without extending and rebuilding the whole HOL theory.

Even though the HOL theory only manipulates semantic objects, the Tk-WinHOL interface always presents programs and statements according the the syntax of the extended guarded command language. The correct syntactic form can always be computed from the semantic object and the list of current program variables. This list is given by the user for the initial specification from which a derivation starts, and it is then updated every time local variables are added.

Consider the following subset of the statement syntax:

$$S ::= \textbf{skip} \mid \{\textbf{b}\} \mid \textbf{x} := \textbf{e} \mid S_1; S_2 \mid \ldots$$

where **b** is a boolean term, **x** is a program variable and **e** is an expression of the same type as **x**. Given a tuple of current program variables u (containing **x**), each of these statements denotes a predicate transformer, as follows. The **skip** statement denotes the **skip** predicate transformer. The assertion $\{\textbf{b}\}$ denotes $\textbf{assert}(\lambda \textbf{u}.\textbf{b})$, the assignment $\textbf{x}:=\textbf{e}$ denotes $\textbf{assign}(\lambda \textbf{u}.\textbf{e})$, and the denotation of the sequential composition is the result of applying the operator **seq** to the denotations of $S_1$ and $S_2$.

The definitions in higher order logic of the semantic operators are as follows:

```
skip q = q
assert p q = λs. p s ∧ q s
assign f q = λs. q(f s)
seq S1 S2 q = S1(S2 q)
```

The right hand side shows the predicate which is the result when the predicate transformer in question is applied to the predicate q. For details on the weakest precondition semantics and its embedding in the HOL system, see [9, 20, 21].

Every time a transformation is applied to the current (syntactic) focus, the interface sends a command to ML which first performs the transformation on the semantic level (using window inference) and then computes (using the pretty-printer) the syntactic form of new focus. This new focus is returned to the interface and presented on the screen.

## 6.2 A small refinement example

We illustrate the Refinement Calculator by showing a derivation of a program that computes a factorial. We begin a program derivation by entering the initial specification

```
"program ex var x,y:num . {y=N}; x,y:=x',y'. x' = FACT N"
```

A special menu alternative for beginning derivations sends the string to a parser for the Refinement Calculus notation. The Window Inference system then begins a new stack and returns the program term which is pretty-printed in the focus window. The specification has name **ex**, its state has type num×num, the program variables are **x** and **y** and the specification contains an assertion and a nondeterministic assignment. The assertion gives the precondition y=N and the nondeterministic assignment says the final value of **x** is N! while the final value of **y** is not constrained. Here N is a specification constant, not a program variable (in fact, an initial specification need not mention any specification constants; they can be introduced during the derivation).

Our first transformation step is to add an initialisation of **x**. We open a subwindow (by pointing and clicking) on the nondeterministic assignment and select the refinement-specific menu alternative ADD ASSIGNMENT with argument "**x:=1**". After this, we close and then use a special menu alternative PUSH ASSERT to propagate the assertion through the assignment. The focus is then the following:

```
{ y = N };
x := 1;
{ (x = 1) /\ (y = N) };
x,y := x',y' . x' = (FACT N)
```

So far, the correctness of each refinement step has been proved automatically by the Refinement Calculator (i.e., by HOL in the background). The rule for adding an initialisation states that

> **x,y:=x',y'. b** *is refined by* **y:=e; x,y:=x',y'. b**
> *provided* **b** *does not mention* **y**.

The rule for pushing an assertion essentially says that

> **{p};S** *is refined by* **S;{sp(S,p)}**

where **sp(S,p)** is the *strongest postcondition* of **S** with respect to precondition **p**.

Now we want to replace the combination of an assertion and a nondeterministic assignment with a do-loop. We open a subwindow on this statement pair and select a LOOP INTRODUCTION menu alternative from the Transform menu and give the arguments that the interface prompts for:

> guard:                 "0<y"
> body:                  "x,y:=x*y,y-1"
> invariant:           "FACT N = x*(FACT y)"
> termination function: "y"

Figure 3 shows what the focus window looks like when the menu alternative for loop introduction is selected.

The Refinement Calculator replaces the focus with the statement

```
do 0<y -> x,y := x*y,y-1 od
```

and adds three proof obligations for the refinement. These proof obligations say that the invariant is true initially, that the invariant is preserved (and the termination function is decreased) and that the postcondition is established when the loop terminates. They can either be discharged (i.e., proved) immediately or later, also using the Refinement Calculator (there is a special menu alternative for setting up such a proof obligation as a focus, with the aim of transforming

**Fig. 3.** The Refinement Calculator

---

it to truth while preserving backward implication). In this simple example, one
is proved automatically, one requires only trivial arithmetic and one is a simple
total correctness assertion which can be proved using basic properties of the
factorial function.

Finally, we can close and use a special menu alternative to drop assertions
from the program text (again, correctness proofs of the transformations are
automatic). This leaves us with the final program for computing a factorial

```
x := 1;
do 0<y -> x,y := x*y, y-1 od
```

Once we have discharged the proof obligations, we can store the current window
theorem, which states that the final program is in fact a correct refinement of
the original specification.

# 7 Conclusions

We have described a graphical interface to the HOL Window Inference library.
The main inspiration for developing it was to provide a user friendly tool for
transformational reasoning. Using the interface, the user can indicate what transformations should be applied and what subterms they should be applied to. One
important aim was to automate certain simple but nevertheless tedious steps
needed in window inference, such as selecting a subexpression or regrouping an
expression into appropriate form. We have provided facilities which allow these
steps to be carried out in a simple way. An expression can be selected by pointing
to an appropriate part of it and clicking a mouse button. A part of an expression

can be regrouped using associativity properties of the operators just by selecting the segment of the text that one is interested in.

This simple way to select 'interesting' subexpressions in the focus window allows easy selection of the objects that are to be manipulated in some transformation (for instance, selecting an assumption to be used in rewriting). In the future, a multiple selection facility (allowing discontinuous selections) will make this feature even more powerful, e.g. when regrouping expressions that contain operators that are commutative and associative. Note that the user neither needs to enter HOL terms nor supply any extra information (types, numbers of assumptions, etc.) concerning the objects of interest.

We have also provided an example of how TkWinHOL can be used for more specific purposes: it serves as a basis for an environment supporting formal program derivation within the Refinement Calculus framework.

The choice of the Tcl/Tk programming system (with the Expect extension) for implementing the tool turned out to be a good one. It made our interface customisable and easily extensible with new functionality and new menu choices. The Refinement Calculator is a good illustration of this.

## Acknowledgements

We wish to thank Michael Butler who implemented the parser for the Refinement Calculator, as well as Jim Grundy who implemented the Window Inference Library and also commented on a draft version of this paper.

## References

1. F. Andersen. *A Theorem Prover for UNITY in Higher Order Logic*. PhD thesis, Technical University of Denmark, Lyngby, 1992.
2. Flemming Andersen, Kim Dam Petersen, and Jimmi S. Petterson. A Graphical Tool for Proving UNITY Progress. In T. F. Melham and J. Camilleri, editors, *Higher Order Logic Theorem Proving and Its Applications – 7th International Workshop. Valletta, Malta, September 1994*, volume 859 of *Lecture Notes in Computer Science*. Springer Verlag, 1994.
3. R.J.R. Back. *Correctness Preserving Program Refinements: Proof Theory and Applications*, volume 131 of *Mathematical Center Tracts*. Mathematical Centre, Amsterdam, 1980.
4. R.J.R. Back. A calculus of refinements for program derivations. *Acta Informatica*, 25:593–624, 1988.
5. R. J. R. Back, J. Hekanaho and K. Sere. Centipede — a Program Refinement Environment, Reports on Computer Science and Mathematics, Ser. A 139, Åbo Akademi University, 1992.
6. R.J.R. Back and J. von Wright. Refinement calculus, part I: Sequential programs. In *REX Workshop for Refinement of Distributed Systems*, volume 430 of *Lecture Notes in Computer Science*, Nijmegen, The Netherlands, 1989. Springer–Verlag.
7. R. J. R. Back and J. von Wright. Refinement concepts formalized in higher order logic. *Formal Aspects of Computing*, 2:247–272, 1990.

8. A. Camillieri. Mechanizing CSP trace theory in Higher Order Logic. *IEEE Transactions on Software Engineering*, 16(9):993–1004, 1990.

9. E.W. Dijkstra *A Discipline of Programming*. Prentice–Hall, 1976.

10. M.J.C. Gordon and T.F. Melham, editors. *Introduction to HOL*. Cambridge University Press, 1993.

11. J. Grundy. A window inference tool for refinement. In Jones et al, editor, *Proc. 5th Refinement Workshop*, London, Jan. 1992. Springer–Verlag.

12. J. Grundy. HOL90 window library manual. 1994.

13. John Harrison and Laurent Théry. Extending the HOL Theorem Prover with a Computer Algebra System to Reason about the Reals. In Jeffrey J. Joyce and Carl-Johan H. Seger, editors, *Higher Order Logic Theorem Proving and Its Applications – 6th International Workshop, HUG '93 Vancouver, B. C., Canada, August 1993*, volume 780 of *Lecture Notes in Computer Science*, pages 174–184. Springer Verlag, 1993.

14. Don Libes. *Exploring Expect: A Tcl-based Toolkit for Automating Interactive Programs*. O'Reilly & Associates, 1994.

15. John K. Ousterhout. *Tcl and the Tk Toolkit*. Addison–Wesley, 1994.

16. Thomas W. Reps and Tim Teitelbaum, editors. *The Synthesizer Generator. A System for Constructing Language-Based Editors*. Springer-Verlag, 1988.

17. P.J. Robinson and J. Staples. Formalising the hierarchical structure of practical mathematical reasoning. Techn. Rep. 138, Key Centre for Software Technology, University of Queensland, Australia, 1990.

18. Donald Syme. A New Interface for HOL - Ideas, Issues and Implementation. submitted to the HUG95.

19. Laurent Théry. A Proof Development System for the HOL Theorem Prover. In Jeffrey J. Joyce and Carl-Johan H. Seger, editors, *Higher Order Logic Theorem Proving and Its Applications – 6th International Workshop, HUG '93 Vancouver, B. C., Canada, August 1993*, volume 780 of *Lecture Notes in Computer Science*, pages 115–128. Springer Verlag, 1993.

20. J. von Wright, J. Hekanaho, P. Luostarinen and T. Långbacka. Mechanising some advanced refinement concepts. *Formal Methods in Systems Design*, 3:49–81, 1993.

21. J. von Wright. Program refinement by theorem prover. In *BCS FACS Sixth Refinement Workshop – Theory and Practise of Formal Software Development. 5th – 7th January, City University, London, UK.*, 1994.

# Formal Verification of Counterflow Pipeline Architecture

Paul N. Loewenstein

Sun Microsystems Computer Company

**Abstract.** Some properties of the Sproull counterflow pipeline architecture are formally verified using automata theory and higher order logic in the HOL theorem prover. The proof steps are presented. Despite the pipeline being a non-deterministic asynchronous system, the verification proceeded with minimal time and effort.

## 1 Introduction

Design of sequential and especially concurrent systems poses severe challenges to the designer. In the absence of a rigorous proof of correctness, design errors are almost impossible to avoid.

One approach to the modelling of sequential and concurrent systems is to use automata. Automaton models can be used to abstract away the notion of an infinite time-line, and replace it with a notion of "now" and "next". This can simplify proofs considerably.

This paper provides another example of the application of the HOL theory in [5], on top of the examples in [4, 6, 7]. This paper follows in the same tradition of taking advantage of the abstraction power of higher order logic to demonstrate more powerful results than can be achieved with purely automatic model checking [8, 1]. The price paid is increased human guidance of the verification process and the inability to use existing hardware description languages for input. This is more a statement of the poor abstraction capabilities of present-day hardware description languages than of any fundamental limitation of theorem proving.

## 2 Motivation

The primary reason for performing this work was to demonstrate the utility of theorem proving for demonstrating the correctness of the asynchronous pipeline architecture.

In particular, we wished to demonstrate that progress could be made with reasonable time and effort. To this end, we do not use any new fundamental verification theory for this work; the only innovation is for dealing with specifics of the pipeline architecture.

Ultimately we would like to link the high level descriptions used here with lower level descriptions used for implementation. However, this paper does not address this process.

## 2.1 Notation

Much of the definitions and theorems are included in this paper, but some have been omitted to save space. We have modified the notation somewhat to improve readability for readers who are not very familiar with HOL (for a detailed description of the HOL logic, see [2]).

HOL logic reads very much like ordinary predicate logic with a few differences. Variables and constants can range over functions or values, depending on their type. Types are left implicit in this paper; they are explained in the text unless they are obvious from context. Function application is indicated by juxtaposition: $f\,x$ denotes function $f$ applied to $x$. Often multiple argument functions are *curried*; $g\,x\,y$ denotes a function $g$ applied to $x$, the result (again a function) is then applied to $y$. Informally one can consider $g$ to be merely a multiple argument function. The $(x, y)$ notation is used for pairing (cartesian product). When pairs are used as function arguments, as in $f(x, y)$, the notation appears conventional. Sometimes, (as in this paper), currying and pairing are mixed, as in $g(u, v)(x, y)$.

If $x$ then $y$ else $z$ is represented in HOL as $x \rightarrow y \mid z$.

Proven theorems are indicated by the symbol $\vdash$. These have been derived using the mechanical theorem prover, HOL[3], and should be correct apart from errors in typographic transcription. Defined constants are printed in **sans serif** typeface.

We have adopted some special conventions for describing sequences and automata. Variables denoting sequences, or functions of time, are emboldened ($\boldsymbol{e}$), as opposed to other variables ($e$). When referring to next-state relations, unprimed variables ($s$) refer to the current state, and primed variables ($s'$) to the next state. The initial state predicate of an automaton with behaviour Name, is always Name_Q. The next-state relation is Name_N and any invariant is Name_I.

An operator Zip is used to join a pair of sequences of values to a sequence of pairs. This is needed to preserve the typing rules of higher order logic.

## 2.2 Rest of Paper

The paper starts by introducing the Sproull pipeline sufficiently to understand this paper. The automaton models are briefly described, together with the theorems we use for verifying the pipeline. We then go through the HOL definition of the pipeline in considerable detail, and describe some features of the proof of properties of the pipeline. The paper ends with results and suggestions for further work.

# 3 Counterflow Pipeline Processor Architecture

The Counterflow Pipeline Processor (CFPP) architecture is covered in much more detail in [9]. It is designed to be implemented using purely asynchronous circuits.

Its features include:
- Bi-directional pipeline in which partially executed instructions flow in the opposite direction to computed results.
- Local control and communication—no global stall signals.
- Regular structure—an ideal candidate for verification by induction.

In a CFPP, a pipeline connects an instruction fetch unit at the bottom with a register file at the top (Fig. 1). In this figure instructions flow up while results flow down. Earlier instructions are above later ones and remain in order as they move up the pipeline. Instructions move up the pipe and are executed in any stage equipped to execute it. Stages can be empty, as gaps can open up as instructions propagate at different speeds.

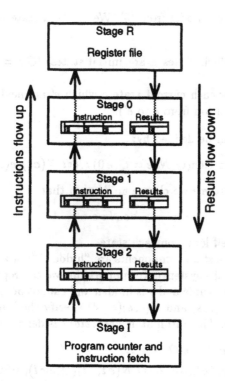

**Fig. 1.** Simplified schematic of a counterflow pipeline processor.

Each instruction, as it moves up the pipe, collects or *garners* values for each of its source operands, carrying them up as source bindings until it reaches the stage where it executes. When it executes, it carries the result(s) as result bindings until it is retired into the register file, where the results are written. The pipeline instruction stages in Fig. 1 have three source bindings and two result bindings. The proof in this paper assumes no particular value for the number of source and destination bindings.

When an instruction is executed, the instruction's destination bindings are passed into the result pipe to supply subsequent instructions, as well as continuing to flow up the instruction pipe until they reach the register file.

Instructions may modify result bindings flowing down. Each stage must prevent result bindings passing instructions (executed or not) whose destination matches that binding, and therefore modifies that binding. Subsequent instructions will therefore garner the most recent binding for any register.

In this paper we are only concerned with the correctness of the results produced by the execution of each instruction, we do not look at the rules in [9] which are concerned with guaranteeing forward progress.

## 4    The Automaton Model

We use the automaton model from [5]. We define a state transition system as follows:

- Let $Q$ specify which states $e$ are initial states. ($Q\,e = \mathsf{T}$ if $e$ is an initial state).
- Let $N$ specify for each current state $e$ which $e'$ are next states. ($N\,e\,e' = \mathsf{T}$ if there is a transition from $e$ to $e'$).

The *behaviour* can be defined by:

(1)    $\vdash \forall QNe.\,\mathsf{Run}(Q,N)e = Q(e\,0) \wedge (\forall t.N(e\,t)(e(\mathsf{Suc}\,t)))$

For asynchronous systems we choose an $N$ such that:

(2)    $\forall e.N\,e\,e$

so that there is a "self loop" on each state.

Often, some of the state components are "hidden" (not available for connecting to other parts of the system). To support this we can represent the state as a pair $(e,s)$, where $e$ is visible and $s$ is hidden and cannot be directly connected to other components. $Q(e,s)$ and $N(e,s)(e',s')$ specify the initial states and state transitions as before. The automaton's *behaviour* is defined by Trace:

(3)    $\vdash \forall QNe.\,\mathsf{Trace}(Q,N)e =$
$(\exists s.Q(e\,0,s\,0) \wedge (\forall t.N(e\,t,s\,t)(e(\mathsf{Suc}\,t),s(\mathsf{Suc}\,t))))$

### 4.1    Demonstrating Invariants

We shall be demonstrating invariants of the automaton describing the counter-flow pipeline. The necessary induction over time is performed in advance and encapsulated in a theorem.

We define an automaton with an invariant $P$ that we wish to prove:

(4)    $\vdash \forall QPNe.\,\mathsf{Trace\text{-}Inv}(Q,P,N)e =$
$(\exists s.Q(e\,0,s\,0) \wedge (\forall t.P(e\,t,s\,t)) \wedge (\forall t.N(e\,t,s\,t)(e(\mathsf{Suc}\,t),s(\mathsf{Suc}\,t))))$

We can then prove by induction on time:

(5)     $\vdash \forall Q P N.$

$$(\forall es.Q(e,s) \Rightarrow P(e,s)) \wedge$$
$$(\forall ese's'.N(e,s)(e',s') \wedge P(e,s) \Rightarrow P(e',s')) \Rightarrow$$
$$(\forall e. \, \mathsf{Trace}(Q,N)e = \mathsf{Trace\text{-}Inv}(Q,P,N)e)$$

## 4.2  Simulation Relations

To demonstrate that one state-transition system implements another we can find an invariant relation $R$ between the visible and the hidden state of the two systems:

(6)     $\vdash \forall Q_1 N_1 Q_2 N_2.$

$$(\exists R.$$
$$(\forall es_1.Q_1(e,s_1) \Rightarrow (\exists s_2.Q_2(e,s_2) \wedge R\,e\,s_1\,s_2)) \wedge$$
$$(\forall ee's_1 s_1' s_2.R\,e\,s_1\,s_2 \wedge N_1(e,s_1)(e',s_1') \Rightarrow$$
$$(\exists s_2'.R\,e'\,s_1'\,s_2' \wedge N_2(e,s_2)(e',s_2')))) \Rightarrow$$
$$(\forall e. \, \mathsf{Trace}(Q_1,N_1)e \Rightarrow \mathsf{Trace}(Q_2,N_2)e)$$

# 5  Modelling the Pipeline in HOL

## 5.1  Register file

We represent the set of registers using HOL type variable $\alpha$. Variables of type $\alpha$ therefore range over register *addresses*. To represent the range of values stored in each register, we use type $\beta$. The state of the entire register file is therefore represented with a variable of type $\alpha \rightarrow \beta$.

## 5.2  Instructions

An instruction can be represented as a function from the current state of the register file to the state it will assume after the instruction has executed. An instruction therefore has type $(\alpha \rightarrow \beta) \rightarrow (\alpha \rightarrow \beta)$.

Rather than use the "source" and "destination" fields of some specific instruction set, we use an abstract, architecture independent definition that takes account of the particular properties of each instruction.

We define a register address to be a *destination* of instruction $f$ if and only if the register's value can be changed by the instruction:

(7)     $\vdash \forall fa. \, \mathsf{Dest}\, f\, a = (\exists s.(f\, s\, a \neq s\, a))$

An address $a$ is a *source* of $f$ iff there exist two states that match each other except for component $a$, and which result in different values for the destination of $f$.

$$(8) \qquad \vdash \forall fa.\, \mathsf{Src}\, f\, a = (\exists ss'.$$
$$(\forall a'.(a = a') \lor (s\, a' = s'\, a')) \land$$
$$(\exists a'.\, \mathsf{Dest}\, f\, a' \land (f\, s\, a' \neq f\, s'\, a')))$$

We shall see in Sect. 5.7 that this definition of source leads to some proof management problems.

## 5.3   Latched and Bound values (results, sources)

The pipeline stages can either contain a value, or be empty. Similarly, a source address of an instruction can have a value bound to it, or not. We therefore define a type $:bind$ parameterised by type $\tau$ which has two possible values Absent corresponding to no bound value and $\mathsf{Present}(x{:}\tau)$ to denote a bound value of $x$.

We define $\mathsf{Empty}\, y$ to indicate that $y = \mathsf{Absent}$, and $\mathsf{Val}$ to extract $x$ from $\mathsf{Present}\, x$.

A set of (result or source) bindings can be modelled with a function from register address to Absent or $\mathsf{Present}(x{:}\beta)$. This has type $:\alpha{\rightarrow}(\beta)bind$.

We define $\mathsf{Bound}\, y\, a$ to indicate that $y\, a \neq \mathsf{Absent}$.

## 5.4   Instruction Pipe

Prior to execution, an instruction in the pipeline can be represented as an abstract instruction function as defined above, paired with a set of source bindings:

$$\mathsf{Closure}(f{:}(\alpha{\rightarrow}\beta){\rightarrow}(\alpha{\rightarrow}\beta))(s{:}\alpha{\rightarrow}(\beta)bind)$$

After execution, it is represented as a set of result bindings, which are the new values for the destination registers:

$$\mathsf{Result}(r{:}\alpha{\rightarrow}(\beta)bind)$$

Let the type representing the disjoint union of Closure and Result be $(\alpha, \beta)inst$.

An instruction *stage* in the pipeline can either contain an instruction, or be empty, so we can model contents of an instruction stage with a variable of type $:((\alpha, \beta)inst)bind$.

We introduce some useful definitions:

$\mathsf{Pending}\, i$ to indicate that the stage $i$ contains an unexecuted instruction.

$\mathsf{Executed}\, i$ to indicate that the stage $i$ contains an executed instruction.

$\mathsf{Src\text{-}Bound}\, i\, a$ to indicate that the address $a$ is a source of the (Pending) instruction in the stage $i$.

$\mathsf{Dest\text{-}Match}\, i\, a$ to indicate that the address $a$ is a destination of the instruction in the stage $i$ (either Executed or Pending).

$\mathsf{Inst}\, i$ returns the next-state function of the (Pending) instruction in stage $i$.

$\mathsf{Source}\, i$ returns the source bindings of the (Pending) instruction in stage $i$.

## 5.5 Result Pipe

We can model a result packet (set of bindings) using type $:\alpha \rightarrow (\beta)bind$. Were we to restrict ourselves to modelling a single result pipe, we would model a stage in that pipe using type $:(\alpha \rightarrow (\beta)bind)bind$. This is both unnecessarily tedious and restrictive. We can model each stage using type $:\alpha \rightarrow (\beta)bind$, representing a register *cache* attached to each stage. At this level of abstraction, this can also represent multiple result pipes, each propagating result packets independently.

## 5.6 Top Stage and Register File

We now move from the choice of HOL types, to the definition of predicates representing the behaviour of pipeline components. We construct a minimum, one-stage pipeline out of a register file and a single pipeline stage (Fig. 2). This minimum pipeline or "top stage" has two ports: $i$ for instructions in and $r$ for results out:

$$(9) \qquad \vdash \forall ir.\; \mathsf{Top}\, i\, r = \mathsf{Trace}(\mathsf{Top\text{-}Q}, \mathsf{Top\text{-}N})(i\, \mathsf{Zip}\, r)$$

We define the initial state predicate, Top-Q, so that $i$ and $r$ are empty, and the register file contents are undefined:

$$(10) \qquad \vdash \forall irs.\; \mathsf{Top\text{-}Q}((i,r),s) = \mathsf{Empty}\, i \wedge (\forall a.\; \mathsf{Empty}(ra))$$

The transition relation, Top-N, is defined in terms of other predicates. This not only makes the definition shorter and more readable, but it also allows us to easily direct the proof by choosing which definitions to expand.

$$(11) \qquad \vdash \forall irsi'r's'.\; \mathsf{Top\text{-}N}((i,r),s)((i',r'),s') =$$
$$\mathsf{Top\text{-}State\text{-}N}(i,s)(i',s') \wedge \mathsf{Top\text{-}Result\text{-}N}(i,r,s)r' \wedge$$
$$\mathsf{Inst\text{-}Res\text{-}N}\, i(i',r') \wedge \mathsf{Instruction\text{-}N}(i,r)i'$$

## 5.7 Retiring Instructions

In this paper we do not show that all instructions are executed before arriving at the register file. In order to decouple this property from what we prove here, we have the top stage execute all unexecuted instructions before writing to the register file:

$$(12) \qquad \vdash \forall isi's'.\; \mathsf{Top\text{-}State\text{-}N}(i,s)(i',s') =$$
$$(\forall a.s'\, a =$$
$$((\mathsf{Empty}\, i' \wedge \mathsf{Dest\text{-}Match}\, i\, a) \rightarrow$$
$$((\mathsf{Executed}\, i) \rightarrow (\mathsf{Val}(\mathsf{Result}\, i\, a))\,|$$
$$(\mathsf{Inst}\, i\, s\, a))\,|\, (s\, a)))$$

For a complete proof, once we have shown that all instructions are executed before reaching the register file, we no longer need that this feature.

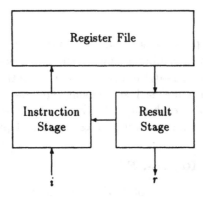

**Fig. 2.** Schematic of top stage of pipeline

**Top Stage Results** The result, $r$, propagating down the pipe can come from the register file or an executed instruction:

(13) $\vdash \forall irsr'.\; \text{Top-Result-N}(i, r, s)r' = (\forall a.$
$\qquad (r'\, a = r\, a)\; \vee$
$\qquad (\text{Bound}\; r'\, a \Rightarrow \text{Dest-Match}\; i\, a \wedge \text{Executed}\; i \wedge (r'\, a = \text{Result}\; i\, a)\; \vee$
$\qquad (r'\, a = \text{Present}(s\, a))))$

**Instruction Operations** The top stage uses the same generic instruction stage as every other stage in the pipeline. This is itself defined in terms of other predicates. The contents of an instruction stage can:

- Remain unchanged.
- Be empty (and possibly become full).
- Be possibly full and become empty.
- Garner a source operand.
- Execute.

(14) $\qquad \vdash \forall iri'.\; \text{Instruction-N}(i, r)i' =$
$\qquad\qquad (i' = i)\; \vee\; \text{Empty}\; i\; \vee\; \text{Empty}\; i'\; \vee\; \text{Garner}(i, r)i'\; \vee\; \text{Execution}\; i\, i'$

**Garnering a Source Operand** A source operand can be garnered from the result pipe if its address is a source of the instruction and the instruction is not yet executed.

(15)    ⊢∀iri'. Garner$(i, r)i'$ =

    Pending $i$ ∧ Pending $i'$ ∧ (Inst $i'$ = Inst $i$) ∧

    (∀a.

      Bound(Source $i'$)$a$ ⇒

      (Source $i'$ $a$ = Source $i$ $a$) ∨ Src(Inst $i'$)$a$ ∧ (Source $i'$ $a$ = $r$ $a$))

**Instruction Execution** An instruction can be executed if there are source operands for all source addresses of the instruction.

(16)    ⊢∀ii'. Execution $i$ $i'$ =

    Pending $i$ ∧ Executed $i'$ ∧

    (∀a. Src(Inst $i$)$a$ ⇒ Bound(Source $i$)$a$) ∧

    (Bound(Result $i'$) = Dest(Inst $i$)) ∧

    (∀s.

      (∀a. Src(Inst $i$)$a$ ⇒ ($s$ $a$ = Val(Source $i$ $a$))) ⇒

      (∀a. Bound(Result $i'$)$a$ ⇒ (Val(Result $i'$ $a$) = Inst $i$ $s$ $a$)))

Which states that the current state of the stage is **Pending**, the next state is **Executed**, that all source operands have been bound and that the set of result bindings corresponds to the destination of the instruction.

The remainder (∀s ... ) describes the value of the result bindings. This term deserves some explanation. To calculate the result bindings of an instruction we:

- Choose some arbitrary state that matches the source bindings of the instruction.
- Apply the instruction function to that state.
- Keep the resulting state bindings that correspond to the destination of the instructions.

This process would be represented by the term:

    (∃s.

    (∀a. Src(Inst $i$)$a$ ⇒ ($s$ $a$ = Val(Source $i$ $a$)))∧

    (∀a. Bound(Result $i'$)$a$ ⇒ (Val(Result $i'$ $a$) = Inst $i$ $s$ $a$)))

Given our definition 8 of the *source* of an instruction, it is possible to show that it makes no difference what state $s$ we choose, *provided that the number of registers is finite*. We make no such statement in this paper, so the result does not follow and we are obliged to use the ∀ form which states explicitly that which state $s$ we choose makes no difference.[1]

---

[1] For example, if we have a processor with an infinite number of registers, and an instruction that sets register 0 to 1 if there are an infinite number of non-zero registers, then we can no longer identify any particular set of registers as the source of the instruction.

**Instructions Passing Results** It is essential to prevent "stale" result bindings passing an instruction that will update that result binding.

Any results whose addresses match a passing-up instruction's destination must have values matching the executed instruction's result.

(17) $\qquad \vdash \forall i i' r'.\, \text{Inst-Res-N}\, i(i', r') = (\forall a.$

$\qquad\qquad \text{Empty}\, i' \wedge \text{Dest-Match}\, i\, a \wedge$

$\qquad\qquad \text{Bound}\, r'\, a \Rightarrow \text{Executed}\, i \wedge (r'\, a = \text{Result}\, i\, a))$

## 5.8 Intermediate Stage

The other instruction stages each have four ports:

- $i$ instruction in.
- $i_u$ instruction out.
- $r$ result out.
- $r_u$ result in.

and because they have no internal state, their behaviour is described using the Run predicate:

(18) $\vdash \forall i r i_u r_u.\, \text{Stage}\, i\, r\, i_u\, r_u = \text{Run}(\text{Stage-Q}, \text{Stage-N})(i\, \text{Zip}\, r\, \text{Zip}\, i_u\, \text{Zip}\, r_u)$

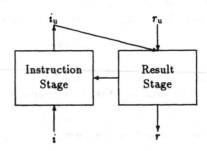

**Fig. 3.** Schematic of intermediate stage of pipeline

**Stage Initial States and Transition Relation** Each stage starts empty:

(19) $\qquad \vdash \forall i r i_u r_u.\, \text{Stage-Q}(i, r, i_u, r_u) = \text{Empty}\, i \wedge (\forall a.\, \text{Empty}(r\, a))$

As for the top stage, the transition relation defined using other definitions.

(20) $\qquad \vdash \forall i r i_u r_u i' r' i'_u r'_u.\, \text{Stage-N}(i, r, i_u, r_u)(i', r', i'_u, r'_u) =$

$\qquad\qquad \text{Instruction-N}(i, r)i' \wedge \text{Inst-Res-N}\, i(i', r') \wedge$

$\qquad\qquad \text{Pass-Up-N}(i, i_u)(i', i'_u) \wedge \text{Result-N}(i, r, i_u, r_u)(i', r')$

**Updating of Results** A result stage (Fig. 3) can take bindings from::

- The result of an executed instruction in the same stage or in the stage above.
- A result in the stage above, provided that it does not match the destination of the instruction above.

(21) ·     $\vdash \forall i r i_u r_u i' r'.\ \text{Result-N}(i, r, i_u, r_u)(i', r') =$

$\qquad (\forall a.(r'\, a = r\, a) \lor$

$\qquad\quad (\text{Bound } r'\, a \Rightarrow$

$\qquad\qquad \text{Executed } i \land \text{Dest-Match } i\, a \land (r'a = \text{Result } i\, a) \lor$

$\qquad\qquad \text{Bound } r_u\, a \land \neg(\text{Dest-Match } i_u\, a) \land (r'\, a = r_u\, a) \lor$

$\qquad\qquad \text{Executed } i_u \land \text{Dest-Match } i_u\, a \land (r'\, a = \text{Result } i_u\, a)))$

**Propagation of Instructions** In our model instructions pass up "instantaneously." The stage empties on the "same tick" as the stage above fills.

(22)     $\vdash \forall i i_u i' i'_u.\ \text{Pass-Up-N}(i, i_u)(i', i'_u) =$

$\qquad (\text{Empty } i_u \land \neg(\text{Empty } i'_u) \Rightarrow \text{Empty } i' \land (i'_u = i)) \land$

$\qquad (\neg(\text{Empty } i) \land \text{Empty } i' \Rightarrow \text{Empty } i_u \land (i'_u = i))$

## 5.9   Instruction Launch

Instructions are launched into the pipe as pending instructions with no source bindings.

(23)

$\quad \vdash \forall i i'.\ \text{Launch-N } i\, i' = \text{Empty } i \Rightarrow (\forall a. \neg(\text{Src-Bound } i'\, a)) \land \neg(\text{Executed } i')$

And the temporal behaviour of the instruction launch is:

(24)     $\vdash \forall i.\ \text{Launch } i = \text{Run}((\lambda x.\ \text{T}), \text{Launch-N})i$

## 5.10   Assembling the Pipeline

We can define a counterflow pipeline with no launch conditions recursively:

(25)   $\vdash (\forall i r.\ \text{Pipe}'\, 0\, i\, r = \text{Top } i\, r) \land$

$\qquad (\forall n i r.\ \text{Pipe}'(\text{Suc } n)i\, r = (\exists i_u r_u.\ \text{Stage } i\, r\, i_u\, r_u \land \text{Pipe}'\, n\, i_u\, r_u))$

To this recursive definition, we add the launch conditions:

(26)     $\vdash \forall n i r.\ \text{Pipe } n\, i\, r = \text{Launch } i \land \text{Pipe}'\, n\, i\, r$

and so we can use the automaton theory we also derive (by proof):

(27)     $\vdash \forall n i r.\ \text{Pipe } n\, i\, r = \text{Trace}(\text{Pipe-Q } n, \text{Pipe-N } n)(i\, \text{Zip } r)$

where the complete initial conditions Pipe-Q are given by:

(28)  $\vdash \forall n i r i_p r_p s. \text{Pipe-Q} \, n((i,r), i_p, r_p, s) =$

$\qquad (i = i_p \, n) \wedge (r = r_p \, n) \wedge$

$\qquad (\forall m.m < n \Rightarrow \text{Stage-Q}(i_p(\text{Suc } m), r_p(\text{Suc } m), i_p \, m, r \, p \, m)) \wedge$

$\qquad \text{Top-Q}((i_p \, 0, r_p \, 0), s)$

and where the complete pipeline transition relation Pipe-N is:

(29)  $\vdash \forall n i r i_p r_p s i' r' i'_p r'_p s'. \text{Pipe-N} \, n((i,r), i_p, r_p, s)((i',r'), i'_p, r'_p, s') =$

$\qquad (r' = r'_p \, n) \wedge (i' = i'_p \, n) \wedge$

$\qquad (\forall m.m < n \Rightarrow$

$\qquad\quad \text{Stage-N}(i_p(\text{Suc } m), r_p(\text{Suc } m), i_p \, m, r_p \, m)$

$\qquad\quad (i'_p(\text{Suc } m), r'_p(\text{Suc } m), i'_p \, m, r'_p \, m)) \wedge$

$\qquad \text{Top-N}((i_p \, 0, r_p \, 0), s)((i'_p \, 0, r'_p \, 0), s') \wedge$

$\qquad \text{Launch-N} \, i \, i'$

# 6  Proving Properties of Pipeline

Two properties of the pipeline were proven:

- That instructions see the correct register values that result from the execution of the earlier instructions.
- That (viewed from the launching end) a long pipeline implements a short pipeline.

The former property is a minimum statement that the counterflow pipeline behaves as a "conventional" computer, with the instructions appearing to update the register file as if they were executed atomically in program sequence.

The latter property is more of a curiosity, because most counterflow pipeline implementations have supplementary processing attached between stages in "sidings", so the instruction stages are visible to the rest of the system. This is in particular true of the memory system (see [9]).

Most of the work was demonstrating the former property. That the pipe in this paper has its only visible state at the launch end does not significantly affect this part of the proof.

## 6.1  Pipeline Effective State

We have to start by formally expressing what we are setting out to prove. A key concept is pipeline effective state. The effective state of a stage is the state of the register file after executing all instructions in the stages above. We define this recursively in terms of the state of the pipe:

(30) $\vdash (\forall i_p s a.\ \text{Eff-State } 0\ i_p\ s\ a = s\ a) \land$

$\qquad (\forall n i_p s a.\ \text{Eff-State}(\text{Suc } n) i_p\ s\ a =$

$\qquad\qquad ((\text{Dest-Match}(i_p\ n)a) \rightarrow$

$\qquad\qquad ((\text{Executed}(i_p\ n)) \rightarrow$

$\qquad\qquad (\text{Val}(\text{Result}(i_p\ n)a)) \mid (\text{Inst}(i_p\ n)(\text{Eff-State } n\ i_p\ s)a)) \mid$

$\qquad\qquad (\text{Eff-State } n\ i_p\ s\ a)))$

Given this, we can then express the invariant to be proven. We split this into four conjuncts; the first two conjuncts are trivial proof administration and the second two refer to supplementary definitions.

(31) $\vdash \forall n i r i_p r_p s.\ \text{Pipe-I } n((i,r), i_p, r_p, s) =$

$\qquad\qquad (i = i_p\ n) \land (r = r_p\ n) \land$

$\qquad\qquad (\forall m.m \le n \Rightarrow \text{Result-I } m(i_p, r_p, s)) \land$

$\qquad\qquad (\forall m.m \le n \Rightarrow \text{Instruction-I } m(i_p, s))$

The result invariant Result-I states that a result in the result pipe either agrees with the result of the executed instruction in the same stage, or it agrees with the effective state.

(32) $\vdash \forall n i_p r_p s.\ \text{Result-I } n(i_p, r_p, s) =$

$\qquad (\forall a.\ \text{Bound}(r_p\ n)a \Rightarrow$

$\qquad\qquad \text{Dest-Match}(i_p\ n)a \land \text{Executed}(i_p\ n) \land (r_p\ n\ a = \text{Result}(i_p\ n)a) \lor$

$\qquad (\text{Val}(r_p\ n\ a) = \text{Eff-State } n\ i_p\ s\ a))$

The instruction invariant Instruction-I states that the garnered source of an instruction always agrees with the effective state of the stage.

(33) $\vdash \forall n i_p s.\ \text{Instruction-I } n(i_p, s) =$

$\qquad (\forall a.\ \text{Src-Bound}(i_p\ n)a \Rightarrow (\text{Val}(\text{Source}(i_p\ n)a) = \text{Eff-State } n\ i_p\ s\ a))$

## 6.2 Remarks on Proof

The proof clearly has to proceed by induction on $n$, the stage index. However, blindly applying induction to the goal we are trying to prove fails because our goal being satisfied at stage $n$ is insufficient to demonstrate the property on stage $n + 1$. As often is the case in inductive proofs, we have to strengthen the goal before performing the induction. Choosing precisely what to demonstrate by induction is the key inspiration required.

The key property that can be demonstrated by induction is encapsulated in Eff-State-Instruction, which states that the effective state at stage $n + 1$ only changes when an instruction moves from stage $n+1$ to stage $n$, and is then the result of executing that instruction.

(34)    $\vdash \forall n i_\mathrm{p} s i'_\mathrm{p} s'$. Eff-State-Instruction $n(i_\mathrm{p}, s)(i'_\mathrm{p}, s') =$

   $(\forall a.$ Eff-State$(\text{Suc } n)i'_\mathrm{p} s' a =$

   $((\text{Empty}(i_\mathrm{p} n) \wedge \text{Dest-Match}(i'_\mathrm{p} n)a) \rightarrow$

   $((\text{Executed}(i'_\mathrm{p} n)) \rightarrow$

   $(\text{Val}(\text{Result}(i'_\mathrm{p} n)a)) \mid (\text{Inst}(i'_\mathrm{p} n)(\text{Eff-State}(\text{Suc } n)i_\mathrm{p} s)a)) \mid$

   $(\text{Eff-State}(\text{Suc } n)i_\mathrm{p} s a)))$

After performing the induction, the proof proceeds by straightforward if rather tedious case analysis, starting with a case split on the possible transitions of the instruction stage using the theorem:

(35)    $\vdash \forall i r i'$. Instruction-N$(i, r)i' \Rightarrow$

   $(i' = i) \vee \text{Empty } i \wedge \neg(\text{Empty } i') \vee \neg(\text{Empty } i) \wedge \text{Empty } i' \vee$

   Garner$(i, r)i' \vee$ Execution $i\, i'$

Here the decision *not* to fully expand the transition relation definitions enables us to keep the proof on track. Terms in the transition relation of a typical asynchronous system are disjunctive, leading HOL to perform excessive case splitting on the disjunctive assumptions. However, we keep the assumptions as unexpanded definitions, applying IMP_REC_TAC with the appropriate definition or theorem that we wish to use. This not only limits the automatic case splitting, but it also serves to keep the proof state compact and readable. Each subgoal usually fits in a single screen.

The unoptimized proof script took 1010 lines. Although this script is an SML program, it has no "free inputs" and does not have to be tested and debugged on multiple input data. It is therefore much quicker to develop than a typical "ordinary" program of similar length.

## 6.3  Implementation of Short Pipe by Long Pipe

The theorem that expresses that a short pipe implements a long pipe is:

(36)    $\vdash \forall m n i r.$ Pipe$(m + n)i\, r \Rightarrow$ Pipe $n\, i\, r$

which follows trivially by induction on $m$ from:

(37)    $\vdash \forall n i r.$ Pipe$(\text{Suc } n)i\, r \Rightarrow$ Pipe $n\, i\, r$

which we demonstrate by simulation relation.

As in most simulation relation proofs, the key is to find the simulation relation. For this example, it was not difficult to find one; it is almost certainly not the only one. The simulation relation between an $n + 1$-long pipe and a $n$-long pipe states:

- That the bottom $n$ stages match.
- The register file of the $n$-stage pipe contains the effective state of stage 1 of the $n + 1$-stage pipe.

(38)  $\vdash \forall n i r i_{p1} r_{p1} s_1 i_{p2} r_{p2} s_2.$
   $\quad$ Pipe-Pipe-R $n(i, r)(i_{p1}, r_{p1}, s_1)(i_{p2}, r_{p2}, s_2) =$
   $\quad (\forall m.m \leq n \Rightarrow (i_{p1}(\text{Suc } m) = i_{p2}\, m) \wedge (r_{p1}(\text{Suc } m) = r_{p2}\, m)) \wedge$
   $\quad (s_2 = \text{Eff-State}(\text{Suc } 0)i_{p1}\, s_1)$

The proof was somewhat less tedious than the invariant proof, and occupies 293 lines of SML.

# 7   Results

The verification was a success in that the theorems promised were delivered on time. The whole project took about 12 working days, including trials of various different ways of representing the pipeline as a state-transition system, and several days spent finding a workable induction scheme.

# 8   Conclusion

Using familiar techniques, rather than developing new ones, enables much faster progress in verifying real designs. Delaying the expansion of disjunctive terms was a major contributor to progress, as it made the proof state much easier to read, and prevented HOL from performing unnecessary goal splits.

## 8.1   Further work

There is much more to be done verifying the Sproull pipeline:

- Verifying progress—much of this can be done by invariance arguments similar to those used in this paper.
- Verifying the use of "sidings" (see [9]).
- Linking the low-level design to the abstract architecture.
- Verifying the properties of the asynchronous circuits used to implement the pipeline.

# Acknowledgements

I would like to thank Bob Sproull for giving me the opportunity to perform this work, and also for providing Fig. 1.

# References

1. David L. Dill, Andreas J. Drexler, Alan J. Hu, and C. Han Yang. Protocol verification as a hardware design aid. In *1992 IEEE International Conference on Computer Design: VLSI in Computers and Processors*, pages 522–525. IEEE Computer Society, 1992. Cambridge, MA, October 11-14.
2. Mike Gordon. HOL: A machine oriented formulation of higher-order logic. Technical Report 68, University of Cambridge Computer Laboratory, 1985.
3. Mike Gordon. HOL: A proof generating system for higher-order logic. In G. Birtwistle and P. A. Subrahmanyam, editors, *VLSI Specification, Verification and Synthesis*. Kluwer Academic Publishers, 1988.
4. Paul Loewenstein. The formal verification of state-machines using higher-order logic. In *IEEE International Conference on Computer Design*, pages 204–207. IEEE Computer Society Press, 1989.
5. Paul Loewenstein. A formal theory of simulations between infinite automata. *Formal Methods in System Design*, 3(1/2):117–149, August 1993.
6. Paul Loewenstein and David Dill. Formal verification of cache systems using refinement relations. In *IEEE International Conference on Computer Design*, pages 228–233. IEEE Computer Society Press, 1990.
7. Paul Loewenstein and David L. Dill. Verification of a multiprocessor cache protocol using simulation relations and higher-order logic. *Formal Methods in System Design*, 1:355–383, 1992.
8. K. L. McMillan. *Symbolic Model Checking*. Kluwer Academic Publishers, 1993.
9. Robert F. Sproull, Ivan E. Sutherland, and Charles E. Molnar. Counterflow pipeline processor architecture. *IEEE Design and Test of Computers*, 11(3), 1994.

# Deep Embedding VHDL *

Ralf Reetz

Sonderforschungsbereich 358, "Automatisierter Systementwurf"
Institut für Rechnerentwurf und Fehlertoleranz (Prof. D. Schmid)
Universität Karlsruhe, Zirkel 2, Postfach 6980, 76128 Karlsruhe, Germany
e-mail:reetz@informatik.uni-karlsruhe.de
WWW:http://goethe.ira.uka.de/people/reetz/reetz.html

**Abstract.** It is shown how a significant subset of VHDL has been *deep embedded* in HOL along with the four abstraction types of hardware: *behavioral, structural, data, temporal.*

First, a method for simplifying deep embedding of languages in HOL is presented: derivation trees as a representation of a syntax in HOL are automatically generated out of a given context–free grammar; a formal, functional compiler as a representation of the semantics in HOL is automatically generated out of a given set of attribution and translation rules.

Second, the formal base for a VHDL semantics in HOL consists of: a flow-graph model as a special form of a state transition system for describing *behavioral* VHDL, a hierarchical combination of state transition systems for describing *structural* VHDL and a formalization of scalar *datatypes* of VHDL.

Third, a VHDL semantics is presented, which enables *temporal* abstraction by preserving hierarchy.

## 1 Introduction

The overall aim of circuit design is the development of correct circuits and systems. This assumes the existence of an unequivocal specification of the intended behavior. Having created a circuit implementation, methods of hardware verification may then be used to formally prove the correctness of the implementation with regard to the formal specification.

Hardware proofs are usually performed in logics. However, the standardized means of describing hardware is the hardware description language VHDL [1]. This hinders the design of correct circuits in two ways: VHDL is known to have a partially blurred semantics (it is defined in plain English) and there is a major gap between VHDL as a design language and logics as a formal system to perform proofs in. To be able to carry out correctness proofs of VHDL descriptions, the semantics of VHDL has to be formalized in the logic.

A lot of work has already been done to formalize the semantics of VHDL. Criteria for a comparison of these works with the work presented here follow:

---

* This work was financed by the German research society under contract SFB 358.

| work | model | logic | method shallow deep | steps elaboration execution | temporal physical delta | beha- vioral | struc- tural | data minimum scalar array |
|---|---|---|---|---|---|---|---|---|
| [15] | none | Boyer–Moore | d | x | p | no | yes | m |
| [4] | none | EA-machines | s | x | p,d | yes | no | m |
| [9] | none | Prolog | d | x | p | yes | no | m |
| [5] | petri net | finite automaton | s | x | p,d | yes | no | s,a |
| [7] | none | FOCUS | s | x | d | yes | no | m |
| [16] | petri net | petri net | s | x | p,d | yes | yes | s,a |
| [13] | fgm | HOL | d | l,x | p,d | yes | no | m |
| here | fgm,hsts | HOL | d | l,x | p,d | yes | yes | s |

**Table 1.** A comparison of different formalized VHDL semantics

- **Shallow versus deep embedding**
  There exist two different approaches to formalize the semantics of a hardware description language (HDL) [10]:
  - *shallow embedding*
    A compiler translates a given HDL program into a logic formula. This method is easier to implement, but less powerful than the following.
  - *deep embedding*
    All constructs of the HDL are defined *within* a logic by definitions of formulas for the (usually abstract) syntax and formulas for the semantics of these constructs. Properties of the whole HDL can be proved because deep embedding allows quantification over program texts, which is not possible with a shallow embedding.

- **Elaboration and Execution**
  The IEEE standard semantics [1] for a VHDL program consists of two steps: In the first step, the program is *elaborated* to an unspecified *model*. One can interpret elaboration as compiling a program into some format fitting for the second step, where an operational semantics is given: the model is executed. One can interpret execution as a VHDL simulator.

- **Abstraction types**
  According to [17], there are four abstraction types in hardware: *behavior, data, time* and *structure*. VHDL contains all abstraction types. Especially considering time, hardware delay effects like inertial delay are modeled within VHDL by two discrete time dimensions. One dimension is the 'real' time, i.e. one unit of this dimension is the physical time passing. Between two time points of the physical time, there can be an arbitrary number of units of the other dimension, called *delta* unit, where a certain computation step is performed. Without the delta dimension, effects like inertial delay are missing.

The most critical point of formalizing VHDL is the complexity of the task. Thus most works have reduced the complexity by leaving out some security of the used method, leaving out difficult features (like two time dimensions), reducing or even leaving out abstraction types or leaving out the elaboration part.

Table 1 gives a comparison of this work with the most recent other works. Except arrays, this work covers all major aspects of a secure, formal VHDL semantics: it is deep embedded, formalizes elaboration and execution, contains both time dimensions, behavior, structure and scalar datatypes. The preceding work [13] was especially enhanced here by an extended method for deep embedding, by structure via hierarchical transition systems and by scalar datatypes. These enhancements are presented in the following.

## 2 Formalization method: generator for formal compilers

Deep embedding of formal systems $\Phi = (\Sigma, \Gamma, \Phi)$ consist of:

- A set of formulas $\Sigma$ is defined as the syntactic domain of $\Phi$ by using fitting type definitions.
- A set of formulas $\Gamma$ is defined as the semantic domain of $\Phi$ by using fitting type definitions.
- Semantics of syntactic objects of $\Phi$ is defined by a semantic function $\Psi$ : $\Sigma \to \Gamma$.

Having complex syntactic and semantic domains as for VHDL in mind, as much automation as possible is needed to realize deep embedding with justifiable effort. The basic idea and the first steps toward automation have been presented in [12]: the syntax as a context–free grammar is represented in HOL by derivation trees; the semantic function $\Psi$ is defined by a special formalization method: $\Psi$ is considered as a formal compiler. The definition of $\Psi$ is now simplified by using methods from compiler construction to automatically generate a definition of $\Psi$ from given attribution and translation rules.

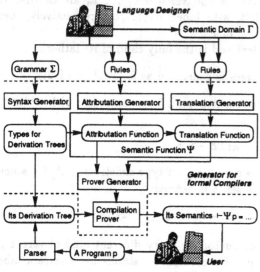

**Fig. 1.:** the *Generator*

Fig. 1 gives an overview of deep embedding with the generator for formal compilers. Instead of defining the complete semantics by hand, the language designer gives the context–free grammar, the attribution and translation rules based on the semantic domain. The definitions are then automatically generated. The semantics for a concrete program $p$ can then be formally derived by expanding the definitions. However, experimental results have shown that using standard rewriting of the HOL system leads to very inefficient proofs. Thus a specialized prover for deriving the semantics for a concrete program is automatically generated also.

The generation of derivation trees and of the translation function were presented in [12]. Thus only a short note on derivation trees is given here, which is needed to show the newly developed generation of an attribution function in more detail.

## 2.1   Derivation trees

A context-free grammar $G = (V, T, P, S)$ consists of a finite set of variables $V$, a finite set of terminals $T$, a finite set of productions $P$ and the start symbol $S \in V$. A symbol is a variable or a terminal. For variables we use $A, B, \ldots$, for terminals $a, b, \ldots$, for symbols $\alpha_i, \alpha_j, \ldots$, for words $\alpha \in (V \cup T)^*$ of length $n$ we use $\alpha_1 \alpha_2 \ldots \alpha_n$, for the empty word $\epsilon$. A production has the form $p : A \rightarrow \alpha_1 \alpha_2 \ldots \alpha_n$, where $p$ is a unique name for the production, $A$ the left side and $\alpha_1 \alpha_2 \ldots \alpha_n$ the right side of the production.

Derivation trees are defined as follows:

1. Every internal vertex has a label $A \in V$.
2. The label of the root is $S$.
3. Every leaf has a label $a \in T$ or $\epsilon$.
4. If an internal vertex $v$ has label $A$ and vertices $v_1, \ldots, v_n$ are children of vertex $v$, in order from the left, with labels $\alpha_1, \ldots, \alpha_n$, respectively, then $p : A \rightarrow \alpha_1 \ldots \alpha_n$ is a production of $P$.
5. If vertex $v$ has label $\epsilon$, $v$ is a leaf and is the only child of its father.

A formalization of derivation trees is generated as follows[2] (see [12]):

1. For every terminal $a \in T$, a type constructor a for a new type a with arbitrary attribute set $\mathfrak{A}$ is defined:

$$\text{a} : {}'\mathfrak{A} \rightarrow \text{a}$$

2. For every production $p_A^i : A \rightarrow \alpha_{i_1} \ldots \alpha_{i_{n_i}}$, a type constructor $p_A^i$ for a new type A with arbitrary attribute set $\mathfrak{A}$ is defined:

$$p_A^i : {}'\mathfrak{A} \rightarrow \alpha_{i_1} \rightarrow \ldots \rightarrow \alpha_{i_{n_i}} \rightarrow A$$

As a result, we get a type constructor for every different kind of vertex of the derivation tree: the result type of a constructor stands for a certain label of a vertex and the constructor name stands for the associated production or the terminal. For start symbol $S$, type S represents all arbitrarily attributed derivation trees of $G$.

---

[2] $'\mathfrak{A}$ denotes a type variable

## 2.2 Attributation

The key approach to get a fast and easy–to–define attributation is to narrow the scope of a transforming step of attributes to a very local scope, in fact to the attributes of a node, its parent and its children in the derivation tree. In that local scope, one distinguishes between sending the attribute of the parent *down* to the node and sending the attributes of the children *up* to the node.

This motivates an attribution system [18]. The definition of an attributation system is divided into two parts. The first part is the definition of locally restricted transformation rules $h^{\uparrow}_{p_A^i}$ for sending attributes up and $h^{\downarrow}_a$, $h^{\downarrow}_{p_A^i}$ for sending attributes down. The second part is a traversing strategy $k$, which traverses the derivation tree and, if possible, applies the previously defined transformation rules. $k$ is applied to a derivation tree again and again until a certain termination condition is met. The transformation rules are set up by the language designer. The traversing strategy $k$ is automatically generated.

## 2.3 Transformation rules

The representation of attributed derivation trees in HOL use *one* arbitrary attribute set $\mathfrak{A}$. In order to make transformation rules context–sensitive, $\mathfrak{A}$ is divided into a finite number of disjoint subsets by the language designer:

$$\mathfrak{A} \stackrel{def}{=} \mathfrak{A}_1 \cup \mathfrak{A}_2 \cup \ldots \cup \mathfrak{A}_l$$

For a representation in HOL, type constructors $s_u$ for every attribute subset $\mathfrak{A}_u$, $1 \le u \le l$ are automatically [8] defined:

$$s_u : \text{'}\mathfrak{A}_u \text{ -> } \text{'}\mathfrak{A}$$

Now, there are three sorts of possible transformation rules:

1. For a terminal $a$, a transformation rule $h^{\downarrow}_a$ can be partly specified, which is applicable if the attribute $a_u$ of the leaf labeled $a$ is a member of the attribute subset $\mathfrak{A}_u$ and the attribute $a_v$ of the parent vertex of the leaf labeled $a$ is a member of the attribute subset $\mathfrak{A}_v$. Then, a transformation $H$, whose result is an attribute of the attribute subset $\mathfrak{A}_w$, replaces the previous attribute $a_u$.

$$h^{\downarrow}_a (s_u\ a_u)\ (s_v\ a_v) \stackrel{def}{=} s_w\ (H\ a_u\ a_v)$$

Note that the type constructors $s_u$ and $s_v$ represent the context, to which the transformation rule $h^{\downarrow}_a$ is sensitive.

2. For a production $p^i_A : A \rightarrow \alpha_{i_1} \ldots \alpha_{i_{n_i}}$, a transformation rule $h^{\downarrow}_{p_A^i}$ can be partly specified, which is applicable if the attribute $a_u$ of the vertex labeled $A$ is a member of the attribute subset $\mathfrak{A}_u$ and the attribute $a_v$ of the parent vertex of the vertex labeled $A$ is a member of the attribute subset $\mathfrak{A}_v$. Then,

a transformation $H$, whose result is an attribute of the attribute subset $\mathfrak{A}_w$, replaces the previous attribute $a_u$.

$$h^{\downarrow}_{p^i_A} (s_u \, a_u) (s_v \, a_v) \stackrel{def}{=} s_w (H \, a_u \, a_v)$$

3. For a production $p^i_A : A \rightarrow \alpha_{i_1} \dots \alpha_{i_{n_i}}$, a transformation rule $h^{\uparrow}_{p^i_A}$ can be partly specified, which is applicable if the attribute $a_u$ of the vertex labeled $A$ is a member of the attribute subset $\mathfrak{A}_u$ and the attributes $a_{v_1} \dots a_{v_{n_i}}$ of the children labeled $\alpha_{i_1} \dots \alpha_{i_{n_i}}$ of the vertex labeled $A$ are members of the attribute subsets $\mathfrak{A}_{v_1} \dots \mathfrak{A}_{v_{n_i}}$, respectively. Then, a transformation $H$, whose result is an attribute of the attribute subset $\mathfrak{A}_w$, replaces the previous attribute $a_u$.

$$h^{\uparrow}_{p^i_A} (s_u \, a_u) (s_{v_1} \, a_{v_1}) \dots (s_{v_{n_i}} \, a_{v_{n_i}}) \stackrel{def}{=} s_w (H \, a_u \, a_{v_1} \dots a_{v_{n_i}})$$

Note that it is not demanded that transformation rules are defined for every possible context. So the definition of a transformation rule does not need to be complete, it might be partial (and usually is, because that is the what makes the definition of attributation easier). For simplicity, it is assumed that there is at least one transformation rule for every terminal and for every production.

## 2.4 Traversing strategy

One attribute subset $\mathfrak{A}_f$, $1 \leq f \leq l$ is defined as the final attribute subset, which is only allowed to be used as the resulting attribute subset of a transformation rule. Thus once an attribute of a node belongs to the final attribute subset $\mathfrak{A}_f$, it is not allowed to be changed anymore. For the ease of notation, predicates $\text{is\_}h^{\downarrow\uparrow}_x$ are defined, which are true if and only if $h^{\downarrow\uparrow}_x$ is applicable.

The task of attributation of a derivation tree is as follows: apply the applicable transformation rules again and again until all attributes in the derivation tree belong to the final attribute set $\mathfrak{A}_f$. A "Yo–Yo" strategy is used as the traversing strategy: the derivation tree is traversed first from the root down to the leaves applying transformation rules $h^{\downarrow}_a$, $h^{\downarrow}_{p^i_A}$ and then up from the leaves back to the root, applying transformation rules $h^{\uparrow}_{p^i_A}$. One down and up traversal is automatically defined as mutual recursive functions $k_\alpha$ as follows:

– For every terminal $a$, partly specify $k_a$ applied to a leaf labeled $a$ and attributed with $x_0$ ($x_0 = s_u \, a$ for some $u$ and some $a$), which gets an $x$ ($x = s_v \, a$ for some $v$ and some $a$) and returns the transformed attribute of the leaf labeled $a$ paired with the changed leaf:

$$k_a (a \, x_0) \stackrel{def}{=} \lambda x.\text{let } x^{\downarrow} = (\text{is\_}h^{\downarrow}_a \, x_0 \, x \Rightarrow h^{\downarrow}_a \, x_0 \, x \mid x_0) \text{ in } (x^{\downarrow}, a \, x^{\downarrow})$$

- For every production $p_A^i : A \to \alpha_{i_1} \ldots \alpha_{i_{n_i}}$, partly specify $\mathbf{k}_A$ applied to a vertex labeled $A$, attributed with $x_0$ ($x_0 = \mathbf{s}_u \, \mathfrak{a}$ for some $u$ and some $\mathfrak{a}$) and with subtrees $d_1 \ldots d_{n_i}$, which gets an $x$ ($x = \mathbf{s}_v \, \mathfrak{a}$ for some $v$ and some $\mathfrak{a}$) and returns the transformed attribute of the vertex labeled $A$ paired with the derivation tree that has the vertex labeled $A$ as its root. There are three cases:

1. only a transformation rule $\mathbf{h}_{p_A^i}^\downarrow$ was specified:

$$\mathbf{k}_A \, (\mathbf{p}_A^i \, x_0 \, d_1 \ldots d_{n_i}) \stackrel{def}{=} \lambda x. \mathtt{let} \, x^\downarrow = (\mathtt{is\_h}_{p_A^i}^\downarrow \, x_0 \, x \, \Rightarrow \, \mathbf{h}_{p_A^i}^\downarrow \, x_0 \, x \mid x_0) \, \mathtt{in}$$
$$(x^\downarrow, \mathbf{p}_A^i \, x^\downarrow \, (\mathtt{SND} \, (\mathbf{k}_{\alpha_1} \, d_1 \, x^\downarrow))) \ldots (\mathtt{SND} \, (\mathbf{k}_{\alpha_n} \, d_{n_i} \, x^\downarrow)))$$

2. only a transformation rule $\mathbf{h}_{p_A^i}^\uparrow$ was specified:

$$\mathbf{k}_A \, (\mathbf{p}_A^i \, x_0 \, d_1 \ldots d_{n_i}) \stackrel{def}{=} \lambda x.$$
$$\mathtt{let} \, x^\uparrow = (\mathtt{is\_h}_{p_A^i}^\uparrow \, x_0 \, (\mathtt{FST} \, (\mathbf{k}_{\alpha_1} \, d_1 \, x_0)) \ldots (\mathtt{FST} \, (\mathbf{k}_{\alpha_n} \, d_{n_i} \, x_0)) \, \Rightarrow$$
$$\mathbf{h}_{p_A^i}^\uparrow \, x_0 \, (\mathtt{FST} \, (\mathbf{k}_{\alpha_1} \, d_1 \, x_0)) \ldots (\mathtt{FST} \, (\mathbf{k}_{\alpha_{n_i}} \, d_{n_i} \, x_0)) \mid x_0) \, \mathtt{in}$$
$$(x^\uparrow, \mathbf{p}_A^i \, x^\uparrow \, (\mathtt{SND} \, (\mathbf{k}_{\alpha_1} \, d_1 \, x_0)) \ldots (\mathtt{SND} \, (\mathbf{k}_{\alpha_{n_i}} \, d_{n_i} \, x_0)))$$

3. transformation rules $\mathbf{h}_{p_A^i}^\downarrow$ and $\mathbf{h}_{p_A^i}^\uparrow$ were specified:

$$\mathbf{k}_A \, (\mathbf{p}_A^i \, x_0 \, d_1 \ldots d_{n_i}) \stackrel{def}{=} \lambda x. \mathtt{let} \, x^\downarrow = (\mathtt{is\_h}_{p_A^i}^\downarrow \, x_0 \, x \, \Rightarrow \, \mathbf{h}_{p_A^i}^\downarrow \, x_0 \, x \mid x_0) \, \mathtt{in}$$
$$\mathtt{let} \, x^\uparrow = (\mathtt{is\_h}_{p_A^i}^\uparrow \, x^\downarrow \, (\mathtt{FST} \, (\mathbf{k}_{\alpha_1} \, d_1 \, x^\downarrow)) \ldots (\mathtt{FST} \, (\mathbf{k}_{\alpha_n} \, d_{n_i} \, x^\downarrow)) \, \Rightarrow$$
$$\mathbf{h}_{p_A^i}^\uparrow \, x^\downarrow \, (\mathtt{FST} \, (\mathbf{k}_{\alpha_1} \, d_1 \, x^\downarrow)) \ldots (\mathtt{FST} \, (\mathbf{k}_{\alpha_{n_i}} \, d_{n_i} \, x^\downarrow)) \mid x^\downarrow) \, \mathtt{in}$$
$$(x^\uparrow, \mathbf{p}_A^i \, x^\uparrow \, (\mathtt{SND} \, (\mathbf{k}_{\alpha_1} \, d_1 \, x^\downarrow)) \ldots (\mathtt{SND} \, (\mathbf{k}_{\alpha_{n_i}} \, d_{n_i} \, x^\downarrow)))$$

the "repeat–until" behavior of an attributation is defined as the application of $f$ again and again, starting with $x$, until a condition *finished* is fulfilled:

$$(\mathtt{apply\_f\_cond\_n\_times} \, finished \, f \, x \, 0 \stackrel{def}{=} x) \land$$
$$(\mathtt{apply\_f\_cond\_n\_times} \, finished \, f \, x \, (\mathtt{SUC} \, n) \stackrel{def}{=}$$
$$(\mathtt{let} \, y = \mathtt{apply\_f\_cond\_n\_times} \, finished \, f \, x \, n \, \mathtt{in} \, finished \, y \, \Rightarrow \, y \mid f \, y)$$

There exist attributation systems which will not terminate. To select the result of terminating ones, Hilbert's choice operator $\epsilon$ is used.

$$\mathtt{apply\_f\_until\_finished} \, finished \, f \, x \stackrel{def}{=}$$
$$\epsilon y. finished \, y \land (\exists n. \mathtt{apply\_f\_cond\_n\_times} \, finished \, f \, x \, n = y)$$

An attribute subset $\mathfrak{A}_t$, $1 \leq t \leq l$ is defined as being the starting attribute subset. A derivation tree $d$ is attributed as follows: $map_1$ declares all attributes of $d$ as belonging to $\mathfrak{A}_t$ by applying $\mathbf{s}_t$ to all attributes. Then, $\mathbf{k}_S$ for start symbol $S$ with initial attribute $\mathbf{s}_t \, \mathfrak{a}_t$ is applied again and again until *end* indicates that all attributes belong to the final attribute subset $\mathfrak{A}_f$. Finally, $map_2$ deletes $\mathbf{s}_f$ from all attributes:

$$\mathbf{k} \, d = map_2 \, (\mathtt{apply\_f\_until\_finished} \, end \, (\lambda d. \mathtt{SND} \, (\mathbf{k}_S \, d \, (\mathbf{s}_t \, \mathfrak{a}_t))) \, (map_1 \, d))$$

## 2.5 Experimental results

Two examples were evaluated to demonstrate that time complexity for the traversing and translation scheme is small in comparison to the time needed to apply the transformation rules. Thus the time complexity is, informally spoken, mostly the complexity of the formalized language.

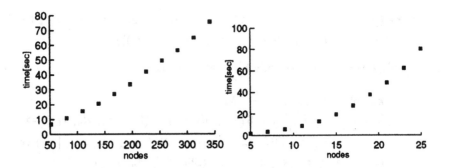

**Fig. 2.** proving time for examples postfix and net, respectively

Fig. 2 shows the experimental results for these two examples computed on a Sparc10: transformation of an arithmetic expression into postfix notation and transformation of the control flow of a list of sequential statements in a loop as a petri net. For both, the number of applications of the traversing strategy is constant $O(1)$. In the postfix example, the time complexity for the application of a transformation rule is $O(1)$, thus the time needed for the traversing and translation scheme predominates. In the second example, transformation rules with linear complexity $O(n)$ are used, thus the time needed for the transformation rules predominates, which can be clearly observed in fig. 2.

## 3 Semantic Domain for VHDL

### 3.1 Flowgraph model $fgm$

Describing state transition systems ($sts$) with arbitrary state spaces was simplified by the development of a flowgraph model $fgm$ [13] as a type ('VN, 'M)$fgm$. An external view of a $sts$ described by an $fgm$ consists of elements (called "nodes") storing data of arbitrary (data)type 'M, which are identified by an arbitrary name of type 'VN and are divided into input and output nodes. There is an internal state of some type not seen from the external view.

Fig. 3.: external view of a flowgraph

The complete state of an (`'VN`,`'M`)`fgm` is noted as `'M Q` for some `Q`. The state transition function `fgm_stf`:(`'VN`,`'M`)`fgm`->`'M Q`->(`'VN`->`'M`)->`'M Q` for some (`'VN`,`'M`)`fgm` takes a state `'M Q`, values for the input nodes `'VN`->`'M` and produces a new state `'M Q`. For some initial state, state sequences, safety and liveness properties are defined in the usual manner.

For *fgm*, 9 types and 133 definitions were made and 124 theorems were proven so far. For details, see [13].

## 3.2 Hierarchical state transition systems *hsts*

In order to get hierarchical flowgraphs, the model *hsts* for the hierarchical composition of arbitrary state transition systems `'STS` with the same external view as of (`'VN`,`'M`)`fgm` was developed as a type (`'STS`,`'VN`,`'M`,`'HN`)`hsts`.

**Structure** The representation type is based on mutual recursive types in order to get inductive composition theorems for verification. The abstraction predicate ensures consistency of identifiers and edges. For *hsts*, 7 types and 108 definitions have been made and 101 theorems have been proven so far.

An *hsts* is either a leaf component consisting of one *sts* or a node component. A leaf component consists of an sts `'STS` and input nodes `'VN` with *default values* `'M` and looks similar as in fig. 3.

A node component is shown in fig. 4. It contains a non–empty library of *hsts*, at least one subcomponent and directed edges connecting subcomponents. A subcomponent consists of a name of type `'HN`, which is unique within the node component, and a pointer to a hsts in the library. This division between subcomponent and its hsts reflects VHDL's component instantiation mechanism and simplifies formalization of scope and visibility of identifiers relevant in structural

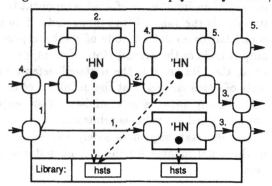

**Fig. 4.:** A node component of a *hsts*

VHDL. There are constructors for creating a node component with one named subcomponent with a new hsts; adding a subcomponent with a new hsts and adding a subcomponent with an hsts already existing in the library of the node component.

Three sorts of directed edges are allowed to connect the subcomponents: (1) an edge from an input node to an input node of a subcomponent, (2) an edge from an output node of a subcomponent to an input node of a subcomponent and (3) an edge from an output node of a subcomponent to an output node.

It is allowed to leave input (4.) and output nodes (5.) unconnected. Every input node should not have more than one edge pointing to it in order to ensure determinism for the semantics. See Fig. 4 for examples.

A node can be declared as being an input node with a default value.

Finally, the instantiation of the *hsts* with *fgm* may have a type as follows:
`(('VN,'M)fgm,'VN,'M,'HN)hsts`.

**Semantics** The *hsts* forms again a state transition system. Its state space has the same hierarchical structure as the *hsts*: it is either a state of a leaf component `leaf_hsts_state:'Q->'Q hsts_state` consisting of a state of its *sts* `'Q` or the states of the subcomponents of a node component `node_hsts_state :` `('Q hsts_state)list->'Q hsts_state`. Note that only leaf components have 'real' states. Node components do not have 'real' states but a composition of states of its subcomponents.

The definition of the initial state of an hsts ensures that for every leaf component of this hsts, there is a `leaf_hsts_state` and that for every node component, there is a `node_hsts_state` with states for every subcomponent. The definition of the state transition function ensures that for a given state of an hsts and a given input, the next state again meets the same hierarchical structure.

The next state of a leaf component is computed by applying the state transition function of an sts to the state of the sts 'stored' in the state of the leaf component. The next state of a node component is computed by computing the next states of the subcomponents, where the state transition function takes care that these computations get the input according to the edges of the node component: if an input node of a subcomponent is connected to an input node, then its value is the value of that input node; if an input node of a subcomponent is connected to an output node of a subcomponent, its value is the value of that output node in the current state; if an input node is not connected to any other node, then its value is the declared default value[3]. The choice of the allowed sorts of edges ensures that the definition of these computations of the inputs of the subcomponents is deterministic. If an output node of a node component is not connected, its value is arbitrary.

Finally, state sequences, safety and liveness properties for an hsts are defined in the usual manner for the output nodes of the top node component of that hsts.

## 3.3 VHDL scalar datatypes *scalar*

VHDL datatypes have usually been shallow embedded. E.g. [6] uses the HOL type for natural numbers to represent time and the HOL type for boolean values to represent VHDL's predefined enumeration type 'boolean'; [11] maps VHDL's datatypes into BDD's. As the semantics of the short–circuit operations[4] **and,**

---

[3] This was introduced because the same mechanism exists in structural VHDL.

[4] For a short–circuit operator, only the left operand is evaluated, if its value is sufficient to determine the result of the operator, even if the right operand might lead to an execution error. E.g. the semantics of expression 'FALSE and (x div 0 = y)' is 'FALSE', although the semantics of a division by zero is an execution error.

**nand, or, nor** depends on errors, these formalizations are not even complete for booleans.

Deep embedding needs a different approach here: VHDL types are not mapped to fitting HOL types, but a complete VHDL *type system* is defined *within* HOL.

Furthermore, value sets of all scalar datatypes in VHDL are isomorphic to a finite interval of integers. Thus they are finite. Even time is finite. However, a more abstract view is chosen: lower and upper bounds are skipped for VHDL type 'time' and 'integer'.

For *scalar*, 5 types and 95 definitions have been made and 254 theorems have been proven so far. A short overview is given in the following.

**Types** First, a representation of letters, digits and identifiers of VHDL in HOL is needed. Their original VHDL syntax was chosen as a representation and created by the method described in [12]. Except arbitrary physical and floating point types, all VHDL scalar types were represented as a HOL type **scalartype**, which includes:

- arbitrary *enumeration types*, which consist of an identifier for a type name and an unique list of enumeration literals. A unique list is a non–empty list, where each element appears in the list only once. Enumeration literals are either identifiers or character literals (which are either letters or digits).
- arbitrary *range constraints*, which consist of an identifier for a type name and a range, which consists of an integer for a lower bound, an integer for an upper bound and is either ascending or descending.
- physical type **TIME**, which is isomorph to integers[5].
- predefined range constraints **INTEGER, NATURAL**, which are isomorphic to the integers and natural numbers, respectively. As mentioned above, these types were defined without bounds here.
- anonymous range constraint *universal integer*, which is VHDL's anonymous type for decimal literals (see [1]).

**Values and Operations** A HOL type **scalar** was defined, which represents a typed value of VHDL including error cases and consists of either:

- an integer representing the value paired with a VHDL type represented by HOL type **scalartype**. The abstraction predicate ensures that the value belongs to the VHDL type.
- an abstract constructor **ELABORATION_ERROR:scalar**, which represents a type conflict as a result of an operation application.
- an abstract constructor **EXECUTION_ERROR:scalartype->scalar**, which represents that an operation application leads to an execution error (e.g. range conflicts, division by zero) and the result would have been of the given VHDL type. The last is needed for the definition of short–circuit operations (for an example, see below).

---

[5] Although arbitrary physical types might have been possible to formalize here, they were skipped here for simplification.

The definition of an operation is done in the following form:

*the types of the operands are not correct* $\Rightarrow$ `ELABORATION_ERROR`
| *application leads to range conflicts, etc.* $\Rightarrow$ `EXECUTION_ERROR` (*result type*)
| *resulting typed value*

This form including all subforms, where condition cases may be missing, is called *normal form* in the following. It was observed that constructors for literals and for only some 'basic' operations, namely `POS`, `VAL`, `SUCC`, `PRED` and the non–VHDL attributes `VALUEOF` (which returns the value of a typed value) and `TYPEOF` (which returns the VHDL type of a typed value and the VHDL type of an execution error) needed to be defined on the representation level and then 'lifted' to the abstraction level. All other operations were then defined in terms of these basic operations on the abstraction level. Additionally, VHDL's attributes concerning the datatypes here were defined, too.

As an example, consider the short–circuit operation **and**. Its operands must be of the same type, which may be either the predefined enumeration types **boolean** or **bit**. The right side is only evaluated if the left side is enumeration literal **true** or **'1'**. Note that the value of enumeration literals is their position number, which is `INT 0` for **false** and **'0'**.

S_and d1 d2 $\overset{def}{=}$
`IS_ELABORATION_ERROR d1` $\lor$ `IS_ELABORATION_ERROR d2` $\lor$
$\neg$((`TYPEOF d1` = `boolean`) $\lor$ (`TYPEOF d1` = `bit`)) $\lor$
$\neg$(`TYPEOF d1` = `TYPEOF d2`) $\Rightarrow$ `ELABORATION_ERROR`
| ($\neg$(`VALUEOF d1` = `INT 0`) $\land$ `IS_EXECUTION_ERROR d2`) $\lor$
`IS_EXECUTION_ERROR d1` $\Rightarrow$ `EXECUTION_ERROR (TYPEOF d1)`
| (`VALUEOF d1` = `INT 0`) $\Rightarrow$ d1 | d2

## 3.4 Verification

A conversion has been implemented, which proves that an expression consisting of literals and applied operations is equivalent to either following normalforms: `ELABORATION_ERROR` or to:

*a certain predicate over integers holds* $\Rightarrow$ `EXECUTION_ERROR` (*result type*)
| `VAL` (*result type*)
    (*constructor for making a decimal literal of out of an integer expression*)

This conversion makes use of the fact that as long as constant identifier are used for type names, the existence of type conflicts is decidable and that the operations are defined in normal form.

For proving with expressions over integers, one might use build–in proof methods of HOL. Additionally, a function for exporting expressions over integers to the interactive reasoning program RRL [3], which can be used for semi–automated proving of arbitrary expressions over integers, has been implemented.

# 4 VHDL semantics

A VHDL semantics along all abstraction types of hardware has been realized. A grammar for a subset with 46 terminals and 187 production rules has been used with 1–4 transformation rules per symbol. A detailed presentation is out of scope here. Instead, the basic ideas and basic properties of the formalized semantics are presented along these abstraction types and previously published results are only cited here. A new result of splitting the monolithic VHDL semantics of the IEEE standard [1] into an hierarchical structure is shown here in more detail.

## 4.1 Behavior

Behavioral descriptions in VHDL are *processes*, consisting of sequential statements executed in an infinite loop. A process reads signal *values* (here of HOL type *scalar*) and creates signal *drivers*, which are a list of transactions. A transaction is a future signal datatype value paired with a value of VHDL type TIME, at which the signal value is currently expected to become the value of that signal. A process has four possible states:

P1. executing its statements

P2. stopped because of an execution error

P3. waiting on certain signal value changes (*events*) and on certain predicates over signals

P4. waiting as in P3, but additionally waiting for at most a proposed *next simulation time*

In the semantics here, the semantics of processes are leaf components of an hsts with flowgraphs of *fgm* describing the state transition functions. For details, see [13].

## 4.2 Data

The basic idea to make data abstractions possible was to define the behavior of VHDL semantics to be independent from the used datatypes. Semantics definitions, which depended on certain parts of datatypes, were parametrised by these needed parts. Thus a 'signature' for a minimum subset of VHDL datatypes, which must be formalized, was found. It consists of types boolean and time, equivalence between datatypes, an order on values of type time, the adding operation on time, and special values for execution and elaboration errors.

The previously presented datatype *scalar* for almost all scalar datatypes was simply be plugged in. This formalization can easily be extended without changing the other semantics parts.

## 4.3 Time

Time in VHDL is represented by the *simulation cycle*. Unstructured 'sea of processes' are running in parallel. If at least one process is in state P2, then the whole simulation cycle stops. If all processes are in state P3 or P4, an *update* process determines the next simulation time by choosing the smallest time of the transactions and computes the next simulation times. Transactions might be deleted and signal values might change. Then, all processes will return to state P1, starting the next simulation cycle.

For details, see [13]. In those previous formalizations there are no structural descriptions since in [1] it is defined that all structure is flattended to the 'sea of processes' during elaboration. Thus they are not suited for structural abstraction and hierarchical verification. Therefore a new structuring of the simulation cycle is described in the following.

## 4.4 Structure

The smallest structural element in VHDL is a *design entity*, which consists of an interface description called *entity* and an implementation description called *architecture body*. An architecture body may consists of behavioral descriptions, i.e. processes, and instantiations of other design entities combined with connections between the processes and the interfaces of the design entities via signals.

The basic observation here is that if the interface of an entity contains only unidirectional, but not bidirectional signals, the 'scope' of the drivers needed to be dealt with by the update process is reduced to the architecture body, where they are created by processes of that architecture body. So concerning the drivers, the previously monolithic update process can be split into update processes for every design entity, preserving structural descriptions. An update process is formalized by a leaf component with a flowgraph of *fgm* describing the state transition function. A design entity is formalized by a node component, whose subcomponents are the leaf components for the processes, the node components of other instantiated design entities and the leaf component of the update process (fig. 5).

However, concerning the computation of the next simulation time, these update processes need to communicate. As an hsts represents a design entity, ususal parallel algorithms over trees are used to realize the communication between the update processes. A design entity has six possible states:

D1. waiting until all its processes are not in state P1 and all its instantiated design entities are in state D2 or D3.

D2. stopped if there is at least one of its processes in state P2 or at least one of its instantiated design entities in state D2.

D3. its update process computes its own proposal for the next simulation time out of the drivers and proposals of its processes and its instantiated design entities.

D4. waiting on the next simulation time which is sent from its parent design entity.

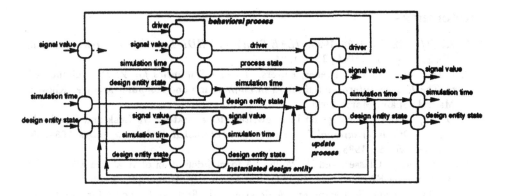

**Fig. 5.** A design entity formalized as a node component

D5. its update process updates its signal values and its drivers depending on the next simulation time.
D6. waiting until all its instantiated design entities have reached state, if there are any. Then, all its processes resume to state P1 and itself can resume to state D1.

Thus the next simulation time is computed bottom up from the leaves of the design hierarchy to the root and then propagated back down, successively starting the next simulation cycle.

## 5 Conclusion

A logical framework for deep embedding VHDL in HOL has been presented: the generation of formal compilers with the newly formalized attribution system, the flowgraph model, the newly formalized hierarchical transition systems and the newly formalized scalar datatypes. Within this framework, a significant subset of VHDL covering all abstraction types of hardware, namely behavior, data, time and structure was formalized, including the elaboration and the execution of VHDL programs.

Future works will concentrate on verification of VHDL programs. An approach to describe safety and liveness properties *within* VHDL itself has already been shown in [14]. The formalized different abstraction types of VHDL allow the use of specialized verification techniques for these abstraction types. A first verification method for the flowgraph model and thus for behavior has been presented in [14]. For the verification of datatypes, existing proving methods for integers as e.g. in RRL are used. Verification methods for structure and time have to be developed now.

# References

1. ANSI/IEEE Std 1076-1993. *IEEE Standard VHDL Language Reference Manual.* IEEE, New York, USA, June 1994.
2. C.D. Kloos and P.T. Breuer, editors. *Formal Semantics for VHDL*, volume 307 of *The Kluwer international series in engineering and computer science.* Kluwer, Madrid, Spain, March 1995.
3. D. Kapur and H. Zhang. RRL: a rewrite rule laboratory. In Lusk and Overbeek, editors, *9th International Conference on Automated Deduction*, pages 768-769. Springer Verlag, 1988.
4. E. Börger, U. Glässer, and W. Müller. A formal definition of an abstract VHDL'93 simulator by EA-machines. In C.D. Kloos and P.T. Breuer [2], chapter 4.
5. G. Dohmen and R. Herrmann. A deterministic finite-state model for VHDL. In C.D. Kloos and P.T. Breuer [2], chapter 6.
6. J.P. Van Tassel. A formalisation of the VHDL simulation cycle. In *International Workshop on Higher Order Logic Theorem Proving and its Applications*, pages 213-228. IFIP WG 10.2, September 1992.
7. M. Fuchs and M. Mendler. A functional semantics for delta-delay VHDL based on Focus. In C.D. Kloos and P.T. Breuer [2], chapter 1.
8. T.F. Melham. Automating recursive type definitions in higher order logic. Technical Report 146, University of Cambridge, Computer Laboratory, Cambridge CB2 3QG, England, September 1988.
9. P.T. Breuer, L.S. Fernandez, and C.D. Kloos. A functional semantics for unit-delay VHDL. In C.D. Kloos and P.T. Breuer [2], chapter 2.
10. R. Boulton, A. Gordon, M.J.C. Gordon, J. Herbert, J. Harrison, and J. van Tassel. Experience with embedding hardware description languages in HOL. In *Proc. of the International Conference on Theorem Provers in Circuit Design: Theory, Practice and Experience*, pages 129-156, Nijmegen, June 1992.
11. R. Herrmann and H. Pargmann. Computing Binary Decision Diagrams for VHDL Data Types. In *Proc. European Design Automation Conference (EURO-DAC94)*, pages 578-585, Grenoble, France, September 1994.
12. R. Reetz and T. Kropf. Simplifying Deep Embedding: A Formalised Code Generator. In T.F. Melham and J. Camilleri, editors, *International Workshop on Higher Order Logic Theorem Proving and its Applications*, pages 378-390, Malta, September 1994. Lecture Notes in Computer Science No. 859, Springer.
13. R. Reetz and T. Kropf. A flowgraph semantics of VHDL: a basis for hardware verification with VHDL. In C.D. Kloos and P.T. Breuer [2], chapter 7.
14. R. Reetz and Th. Kropf. A flowgraph semantics of VHDL: Toward a VHDL verification workbench in HOL. *Formal Methods in System Design*, 1995. (to appear).
15. D.M. Russinoff. Specification and verification of gate-level VHDL models of synchronous and asynchronous circuits. In E. Börger, editor, *Specification and Validation Methods.* Oxford University Press, Oxford, 1994.
16. S. Olcoz. A formal model of VHDL using coloured petri nets. In C.D. Kloos and P.T. Breuer [2], chapter 5.
17. T.F. Melham. Abstraction mechanisms for hardware verification. In G. Birtwistle and P.A. Subrahmanyam, editors, *VLSI Specification, Verification, and Synthesis*, pages 129-157. Kluwer Academic Publishers, 1988.
18. H. Zima. *Compilerbau I*, volume 36 of *Reihe Informatik.* B.I.-Wissenschaftsverlag, 1983.

# HOLCF: Higher Order Logic of Computable Functions

Franz Regensburger

regensbu@informatik.tu-muenchen.de

Technische Universität München

**Abstract.** This paper presents a survey of HOLCF, a higher order logic of computable functions. The logic HOLCF is based on HOLC, a variant of the well known higher order logic HOL, which offers the additional concept of type classes.

HOLCF extends HOLC with concepts of domain theory such as complete partial orders, continuous functions and a fixed point operator. With the help of type classes the extension can be formulated in a way such that the logic LCF constitutes a proper sublanguage of HOLCF. Therefore techniques from higher order logic and LCF can be combined in a fruitful manner avoiding drawbacks of both logics. The development of HOLCF was entirely conducted within the Isabelle system.

## 1 Introduction

This paper presents a survey of HOLCF, a higher order logic of computable functions. The logic HOLCF is based on HOLC, a variant of the well known higher order logic HOL [GM93], which offers the additional concept of type classes.

HOLCF extends HOLC with concepts of domain theory such as complete partial orders, continuous functions and a fixed point operator. With the help of type classes the extension can be formulated in a way such that the logic LCF [GMW79, Pau87] constitutes a proper sublanguage of HOLCF. Therefore techniques from higher order logic and LCF can be combined in a fruitful manner avoiding drawbacks of both logics.

The logic HOLC is implemented in the logical framework Isabelle [Pau94] and the development of HOLCF was conducted within the Isabelle system, too. The syntax, semantics and proof rules of HOLC together with the development of HOLCF are described in full detail in my thesis [Reg94].

In parallel with my development of HOLCF Sten Agerholm developed the HOL-CPO [Age94] system on the basis of Gordon's HOL System. Although the overall aim of the two theses is the same, namely the combination of HOL and LCF, the techniques used in the two approaches differ in many aspects. The availability of type classes had significant impact onto the development of HOLCF. Some problems Sten Agerholm had to deal with could be avoided[1] but

---

[1] See the discussion after the introduction of theory Cfun1 on page 12.

on the other hand new mechanisms for conservative (safe) theory extensions with respect to type classes had to be established first.

This paper is organized as follows. In section 2 I will give a brief survey of HOLC the higher order logic with type classes. The main focus of this survey is on the differences between Gordon's HOL [GM93] and HOLC which is implemented in the Isabelle system [Pau94]. Section 3 addresses the central issue of the paper, namely the development of the logic HOLCF. Finally section 4 draws a conclusion together with a survey of topics that have been formalized in HOLCF too but could not be presented in this paper due to space limitation. There is also a discussion of current and future work.

## 2  Higher order logic with type classes

The logic HOLC is a variant of Gordon's HOL [GM93]. It is formalized within the Isabelle system [Pau94] which is not only a logical framework but also a generic tactical theorem prover. In Isabelle terms HOLC is called an *object logic* which is formalized using Isabelle's *meta logic*, namely intuitionistic higher order logic. The meta logic is the formal language of the logical framework Isabelle [Pau89]. In Isabelle the logic HOLC is just called HOL but in this paper I use the name HOLC to avoid confusion with Gordon's HOL.

Besides some minor syntactic differences the main difference between Gordon's HOL and Isabelle's HOLC is the availability of *type classes* in HOLC. The concept of type classes is not specific to the object logic HOLC; it is derived from Isabelle's meta logic. In the beginning type classes were introduced in Isabelle by Nipkow [Nip91, NP93] as a purely syntactic device. They admit a fine grained use of polymorphism for the description of object logics. Since type classes are available in the meta logic they can be used in object logics, too. However, this is only sensible if the semantics of the object logic gives meaning to the concept of polymorphism with type classes.

### 2.1  What are type classes in HOLC?

This question is answered best by using some examples. As a basis for the following examples some knowledge of polymorphism in Gordon's HOL, which is Hindley/Milner polymorphism, is assumed. A detailed description of polymorphism in HOL, especially its semantics, can be found in [GM93].

In Gordon's HOL types are interpreted as inhabitants of a universe of sets which exhibits certain closure properties sufficient for the interpretation of types and type constructors. Polymorphic constants, such as the identity function $=::\alpha\Rightarrow\alpha\Rightarrow$bool or Hilbert's choice function $\varepsilon::(\alpha\Rightarrow$bool$)\Rightarrow\alpha$, are interpreted as families of interpretations (generalized cartesian products indexed by the sets of the universe). The syntax of HOL provides type variables, usually denoted by small Greek letters, e.g. in the type $\alpha\Rightarrow\alpha\Rightarrow$bool of the polymorphic equality $=$. Type terms are interpreted in an environment which maps every type variable to a member of the above mentioned universe of types.

The interpretation of types in HOLC is slightly more involved. In HOLC there may be different kinds of types or *classes* of types respectively in order to use Isabelle's terminologies. In the semantics of HOLC every type class is associated with its private universe of type interpretations. The most important type class in HOLC is the class **term** the semantics of which directly corresponds to the HOL universe of sets. Besides the type class **term**, which is mandatory, theories in HOLC may depend on additional type classes. The issue of additional type classes is discussed later on in this paper. For the moment, let us assume that there is just the class **term**. The semantics of a HOLC theory which respects this restriction corresponds to the semantics in Gordon's HOL.

Now I will present some examples which demonstrate the use of type classes. Suppose we want to formalize partially ordered sets (po's) so that we can address the ordering relation of the ordered sets via the polymorphic constant $\sqsubseteq::\alpha\Rightarrow\alpha\Rightarrow$bool like in LCF. A first attempt for a formalization, in which the power of type classes is not used, would be the following (the syntax is explained below):

```
Porder0_first = HOL +
default term
consts
        ⊑::α⇒α⇒bool
rules
refl_less          x ⊑ x
antisym_less       x ⊑ y ∧ y ⊑ x ⟶ x = y
trans_less         x ⊑ y ∧ y ⊑ z ⟶ x ⊑ z
end
```

The example shows a typical theory extension in Isabelle. The new theory is called **Porder0_first**. It extends the theory **HOL** with a new polymorphic constant $\sqsubseteq$ of type $\alpha\Rightarrow\alpha\Rightarrow$bool. The properties of the new constant are specified using the three well known axioms **refl_less**, **antisym_less** and **trans_less**. The phrase **default term** tells Isabelle's type inference mechanism that every type variable which occurs without an explicit qualification of the type class should be treated as a type variable of class **term**. In the above example we used this default mechanism to simply write $\sqsubseteq::\alpha\Rightarrow\alpha\Rightarrow$bool instead of the more verbose phrase $\sqsubseteq::\ \alpha::$**term** $\Rightarrow \alpha \Rightarrow$ bool.

In the three axioms of the example above it is not necessary to mention any types since the type inference mechanism can deduce all of the needed information. The technique of type inference for type class polymorphism is addressed in full detail in [NP93].

The theory **Porder0_first** is problematic for two reasons. First of all the theory constitutes an extension which is not *safe* in the sense of HOL.[2] We used three axioms to specify the notion of a partial order instead of using definitions which is prefered in HOL since definitions preserve models and therefore consistency. The second problem is that the above formalization is too strong. It means

---

[2] See page 5 for an explanation of *safe extension*.

that every type $\tau$ in the class **term** must be equipped with an ordering relation $\sqsubseteq::\tau\Rightarrow\tau\Rightarrow$bool. One could argue that there is always at least one trivial partial ordering for every type, namely the identity relation. However, this rather crude patch is ruled out immediately, once we add the additional constant $\bot::\alpha$ and a fourth axiom

    minimal          $\bot \sqsubseteq x$

With the help of type classes we can find an elegant way out of the second problem and with a little more effort we can make this way a safe one too, which also solves the first problem of conservativity. A discussion of solutions to this problem which stay in the framework of Gordon's HOL can be found in Agerholm's thesis [Age94]. We reformulate our first attempt in the following way:

PorderO_second = HOL +
default term
classes  po < term
consts
        $\sqsubseteq:: (\alpha::po)\Rightarrow\alpha\Rightarrow$bool
rules
refl_less          $x \sqsubseteq x$
antisym_less      $x \sqsubseteq y \wedge y \sqsubseteq x \longrightarrow x = y$
trans_less        $x \sqsubseteq y \wedge y \sqsubseteq z \longrightarrow x \sqsubseteq z$
end

The phrase **classes po < term** declares the new class identifier **po**. By convention a theory which mentions a class identifier in a context like **classes po < term** is interpreted as the definition of the class on the left hand side of the < symbol.[3] The properties of the new type class are entirely specified in the body of the class definition. First of all, the phrase **po < term** means that **po** is supposed to be a *subclass* of **term**.

In order to explain the semantics of the subclass relation <, we have to look at the constants and the axioms which are specified in the sections **consts** and **rules** respectively. The constants in the section **consts** of a class definition are called the *characteristic constants* of the new class. The axioms in the section **rules** are called the *characteristic axioms* of the new class. The semantics of the class **po** is now as follows.

It consists of a universe of mathematical structures (algebras if you like) such that each of these structures can be obtained by the following construction. First we take a structure out of the universe for the *superclass*. In the example the superclass is **term** and the structures in the universe for **term** solely consist of a carrier set. Then we add an interpretation for the characteristic constant $\sqsubseteq$, such that the characteristic axioms for this constant, here the axioms of a partial ordering, are fulfilled. In summary the interpretation of the class **po** is the

---

[3] This implies that there must be exactly one such occurrence per class identifier. The class **term** is the only exception to this rule.

universe of all partial orderings[4] that can be obtained by taking the carrier from the universe of the superclass term and adding an arbitrary ordering function.

In general the semantics for a class h with h < k is the universe of all structures that can be obtained from the structures in k by adding interpretations for the characteristic constants of h such that the characteristic axioms of h are fulfilled. This way, every structure in the universe carries along the particular interpretations for the characteristic constants of that class and all its superclasses.

This is sufficient since the characteristic constants and axioms[5] of a class must be polymorphic exactly in one type variable of the class.[6] As usual the polymorphic constants are interpreted as generalized cartesian products, but this time the products are indexed by the entire structures of the universe for the type class. This way it is easy to select always the right instance for the polymorphic constant, which is by construction the one carried along as part of the indexing structure. See [Reg94] for a formal treatment.

Now we come back to the first problem, namely the conservativity of the theory extension. In HOL an extension of a theory is called conservative (save, definitional) if and only if an arbitrary model of this theory can be extended in a *strongly persistent* [EGL89, GM93] way to a model of the extended theory. Therefore safe extensions also preserve consistency.

In our example we extended the theory HOL with a new class po together with a description of its characteristic constants and axioms. The only thing which can go wrong is that the new universe for the class po is empty. This means that there is no way to extend any structure of the universe term by an interpretation for $\sqsubseteq$ that fulfills the characteristic axioms. In order to prevent this failure we simply have to show in advance that there is at least one such extension.

In the example it is easy to find a witness. For the type of the witness we take bool and as ordering function we take the identity $=::bool \Rightarrow bool \Rightarrow bool$. Then we prove the following theorems in the theory HOL:

$$x \ (=::bool \Rightarrow bool \Rightarrow bool) \ x$$
$$x \ (=::bool \Rightarrow bool \Rightarrow bool) \ y \land y = x \longrightarrow x = y$$
$$x \ (=::bool \Rightarrow bool \Rightarrow bool) \ y \land y = z \longrightarrow x = z$$

Of course this is a trivial task but it shows that there will be at least one structure in the universe for the class po, namely the interpretation of the type bool together with the identity function on type bool as the ordering function.

---

[4] Carrier plus ordering function. In higher order logic we use functions $\tau \Rightarrow \tau \Rightarrow bool$ to model binary relations.

[5] These are precisely those constants and axioms which occur in the class definition.

[6] This restriction leads to a strictly weaker notion of class polymorphism than the one known e.g. from the functional programming languages HASKELL or GOFER. However, the restriction simplifies the semantics of class polymorphism since the interpretation of the instance of a characteristic constant is always non-polymorphic. It therefore can and is supposed to be an element of some appropriate carrier set in the universe of class term.

Now we are ready to give the final version of the theory Porder0 which is an example of an *extension by a new class*. It is as follows:

Porder0 = HOL +
default term
classes     po < term
arities     bool::po
consts
            $\sqsubseteq$:: $(\alpha$::po$)\Rightarrow\alpha\Rightarrow$bool
rules
refl_less           $x \sqsubseteq x$
antisym_less        $x \sqsubseteq y \wedge y \sqsubseteq x \longrightarrow x = y$
trans_less          $x \sqsubseteq y \wedge y \sqsubseteq z \longrightarrow x \sqsubseteq z$

inst_bool_po        $(\sqsubseteq$::bool$\Rightarrow$bool$\Rightarrow$bool$) = (=$::bool$\Rightarrow$bool$\Rightarrow$bool$)$
end

The only difference between the version Porder0_second and the final one is that we mentioned the witness type bool. In the section **arities** we specified that the type bool is a type in class po and the axiom inst_bool_po describes the *instance* of the characteristic constant $\sqsubseteq$ for the witness type bool. The new *arity* and the instance definition are validated by the theorems we proved before which also guarantee that the above theory extension is safe. A formal treatment of all these argumentations can be found in [Reg94].

In the examples above we saw how to introduce a new type class in a conservative way. Suppose now that we want to formalize the following. Given a type $\sigma$ in class term and a type $\tau$ in class po the type of functions $\sigma \Rightarrow \tau$ can be partially ordered too using the pointwise extension of the ordering in $\tau$. This time we make the theory extension safe from the beginning. First we define the pointwise ordering on the function space:[7]

Fun1 = Porder0 +
consts
            less_fun:: $(\alpha$::term $\Rightarrow \beta$::po$) \Rightarrow (\alpha \Rightarrow \beta) \Rightarrow$ bool
rules
less_fun_def        less_fun $= (\lambda$f1 f2.$\forall$x. f1$(x) \sqsubseteq$ f2$(x))$
end

In the theory Fun1 we just introduced the new constant less_fun. Since the only axiom less_fun_def is a definition the theory extension Fun1 is obviously conservative. The theory Fun1 is an example for an *extension by a new constant* which corresponds to the same notion in Gordon's HOL. Constants which are introduced in this way are different from characteristic constants of classes. Therefore they are not restricted in the degree of their polymorphism. Characteristic constants can only be introduced within a class definition. However, for

---

[7] In the example Fun1 is based on Porder0. In the full development of HOLCF the theory Fun1 is based on additional theories Porder and Pcpo. See figure 1.

the definition of a new constant characteristic constants of an already defined class can be used like in the example above. Next we prove the following three theorems in the theory Fun1:

$$\mathsf{less\_fun(x)(x)}$$
$$\mathsf{less\_fun(x)(y)} \land \mathsf{less\_fun(y)(x)} \longrightarrow \mathsf{x = y}$$
$$\mathsf{less\_fun(x)(y)} \land \mathsf{less\_fun(y)(z)} \longrightarrow \mathsf{less\_fun(x)(z)}$$

These theorems show that the function less_fun behaves like a partial ordering. Therefore we are allowed to formalize the following theory extension which is called an *extension by a new arity*.

```
Fun2 = Fun1 +
arities    ⇒ :: (term,po)po
rules
inst_fun_po        (⊑::(α::term ⇒ β::po) ⇒ (α ⇒ β) ⇒ bool) = less_fun
end
```

The phrase $\Rightarrow::\mathsf{(term,po)term}$ tells Isabelle's type inference mechanism that given a type $\sigma$ in class term and a type $\tau$ in class po the function space $\sigma \Rightarrow \tau$ is a type in class po. The axiom inst_fun_po fixes the instance of the characteristic constant $\sqsubseteq$ for the type $\sigma \Rightarrow \tau$. Due to the theorems we proved in advance the above extension is safe again.

This concludes the short survey on the logic HOLC. Besides the *extension by a new class* (theory Porder0), the *extension by a new arity* (theory Fun2) and the *extension by a new constant* (theory Fun1) there is the *extension by a new type* of class term. This last extension mechanism corresponds directly to the extension by type definition in Gordon's HOL and is therefore not discussed here. However, we will see an example for this extension mechanism in section 3.4.

# 3 Development of HOLCF

In this section I will present parts of the development of HOLCF using the higher order logic HOLC with type classes which was briefly described in the previous section. Figure 1 shows part of the hierarchy of theories which constitutes the logic HOLCF.

The theory Porder0 is known from section 2. In this theory the type class po of partial orders is introduced. In Porder the notions of upper bounds, least upper bounds and $\omega$-chains are introduced. In theory Pcpo the class pcpo of pointed complete partial orders is introduced as a subclass of po. The characteristic constant of this class is the symbol $\bot$ for the least element. The main parts of the theories Fun1 and Fun2 were already presented in section 2. Theory Fun3 contains just the arity and instance declarations for the function type constructor $\Rightarrow$ with respect to the type class pcpo.

In theory Cont the notions of monotone and continuous functions are defined as predicates on the full function space $\alpha \Rightarrow \beta$ over pcpo's $\alpha$ and $\beta$. Since

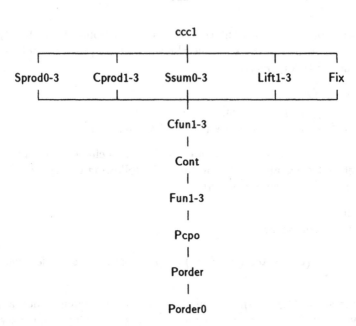

**Fig. 1.** The HOLCF theories

continuous functions play a central role in the logic LCF the theories Cfun1−3 introduce a special type constructor → for continuous functions. In order to avoid confusion elements of the type $\alpha\to\beta$ are called *operations*. The type $\alpha\to\beta$ is introduced via a type definition that fixes the interpretation of $\alpha\to\beta$ to be isomorphic to the subset of continuous functions of the type $\alpha\Rightarrow\beta$. We will discuss this further below.

The theories Sprod0−3, Cprod1−3, Ssum0−3 and Lift1−3 conservatively introduce the types of strict products, cartesian products, strict sums and lifting type over types in class **pcpo**. Since the extensions have to be safe there are always several steps needed for every type construction. In a first step the type construction itself is defined via an extension by a new type of the class **term**. In the next step, the ordering function is defined and it is proved that the function behaves like a partial ordering. The next step shows that there is a least element in the type under consideration and that there always exist least upper bounds for $\omega$-chains. This validates the last step which specifies that the type construction yields types in class **pcpo** provided the argument types of the type construction are in class **pcpo** too.

The theory Fix contains the fixed point theory of LCF. Central to this theory are the definitions of the fixed point operator and the definition of admissibility. Amongst others Kleene's fixed point theorem, Scott induction and various propagations of admissibility are proved.

The theory cccl is a union of the theories enumerated above and defines in addition the identity operation and the composition of operations. The name of the theory stems from the fact that the class pcpo together with operations as arrows forms a category. In addition this category can also be shown to be cartesian closed by taking the cartesian product as categorical product and the type of operations as exponential.[8] In the following we concentrate on the theories Porder, Pcpo, Cont, Cfun1−3 and Fix.

## 3.1 The theory Porder

In theory Porder the notions of upper bounds, least upper bounds (lub's) and $\omega$-chains are introduced. Due to the use of the type class po the polymorphism of the various constants can be restricted to the class of partially ordered types. This leads to a very natural formalization of the concepts above. Note how the characteristic constant $\sqsubseteq$ of the class po is used in the axioms. The default class is still term. Therefore the type variable $\alpha$ is explicitly qualified with the class po. However, one qualification per type term is sufficient. Note the difference between the infix predicate $\lhd\!\!\!\lhd$ and the function lub. The former is a *relation* which means that x is a least upper bound of set S whereas the latter is a *function* which yields *some* x that is a least upper bound of S provided there exists one. The type constructor set is the polymorphic powerset constructor. Applied to a type $\tau$ it constructs the powerset of $\tau$ that is isomorphic to the type $\tau\Rightarrow$bool.

Porder = Porder0 +
consts

| | | |
|---|---|---|
| $\lhd$ | :: $\alpha$ set $\Rightarrow$ $\alpha$::po $\Rightarrow$ bool | (infixl 55) |
| $\lhd\!\!\!\lhd$ | :: $\alpha$ set $\Rightarrow$ $\alpha$::po $\Rightarrow$ bool | (infixl 55) |
| lub | :: $\alpha$ set $\Rightarrow$ $\alpha$::po | |
| is_chain | :: (nat$\Rightarrow\alpha$::po) $\Rightarrow$ bool | |

rules

| | |
|---|---|
| is_ub | S $\lhd$ x = $\forall$y.y$\in$S $\longrightarrow$ y $\sqsubseteq$ x |
| is_lub | S $\lhd\!\!\!\lhd$ x = S $\lhd$ x $\wedge$ ($\forall$u. S $\lhd$ u $\longrightarrow$ x $\sqsubseteq$ u) |
| lub | lub(S) = ($\varepsilon$x. S $\lhd\!\!\!\lhd$ x) |

| | |
|---|---|
| is_chain | is_chain(Y) = ($\forall$i.Y(i) $\sqsubseteq$ Y(Suc(i))) |

end

It is convenient to have both of these notions to talk about least upper bounds. For technical reasons an $\omega$-chain is formalized as a function which enumerates the chain and not as the range of this enumeration.

## 3.2 The theory Pcpo

In this theory we introduce the new type class pcpo (*pointed complete partial orders*) as a subclass of po. The intention is that pcpo is inhabited by all types

---

[8] Together with suitable arrows.

which are not only partially ordered but in addition have a least element and are complete with respect to $\omega$-chains. This is the kind of types which is needed to formalize the logic LCF.

```
Pcpo = Porder +
classes   pcpo < po
arities   void :: pcpo
consts
          ⊥:: α::pcpo
rules
minimal            ⊥ ⊑ x
cpo                is_chain(Y) ⟶ ∃x. range(Y) ≪ x::(α::pcpo)

inst_void_pcpo     (⊥::void) = ⊥_void
end
```

The witness for the non-emptiness of the new class is the trivial type void which solely consists of one element ⊥_void. Clearly this type is partially ordered by the identity relation and has ⊥_void as least element. I did not provide the formalization of the type and its instance for the class po since it is trivial.

Note that due to the explicit qualification x::(α::pcpo) chain completeness is only required for types in class pcpo. Without this qualification the type inference mechanism would have computed α::po which would be too strong. The function range is the function which yields the range of its argument function. Here we get the range of the chain Y.

## 3.3   The theory Cont

We skip the theories Fun1, Fun2 and Fun3. In these theories it is shown that the full function space $\alpha \Rightarrow \beta$ over types $\alpha$::term and $\beta$::pcpo can be partially ordered by the pointwise ordering and has pcpo structure. We immediately skip to the theory Cont that introduces the notions of monotone and continuous functions.

```
Cont = Fun3 +
default pcpo
consts
          monofun :: (α::po ⇒ β::po) ⇒ bool
          contlub :: (α ⇒ β) ⇒ bool
          contX   :: (α ⇒ β) ⇒ bool
rules
monofun            monofun(f) = ∀x y. x ⊑ y ⟶ f(x) ⊑ f(y)

contlub            contlub(f) = ∀Y. is_chain(Y) ⟶
                           f(lub(range(Y))) = lub(range(λi.f(Y(i))))

contX              contX(f) = ∀Y. is_chain(Y) ⟶
                           range(λi.f(Y(i))) ≪ f(lub(range(Y)))
end
```

First of all note that we changed the default class to be pcpo. Therefore we need the explicit qualification po for the type of the predicate monotone. The first two axioms of the theory directly correspond to those which can be found in every text book.[9] This is due to the use of type classes which allows us to hide a lot of details behind the scenes. In a higher order logic without type classes the axioms would be cluttered with premises about the ordering relation which needs to be passed as an explicit argument to all of the predicates above. See [Age94] for more details.

Perhaps the definition of the third axiom contX is surprising. However, it can be proved, and indeed it was proved in this theory that the following holds.

$$\text{contX}(f) = (\text{monofun}(f) \wedge \text{contlub}(f))$$

## 3.4 Theories Cfun1 - Cfun3

The theory Cfun1 is central to the development of HOLCF. Here the expressive power of higher order logic with type classes is apparent in several places. In theory Cfun1 we introduce the type of operations such that its semantics is isomorphic to the subset of continuous functions. The theory is as follows:

```
Cfun1 = Cont +
types      → 2      (infixr 5)
arities    → ::     (pcpo,pcpo)term
consts
  Cfun   :: (α ⇒ β)set

  fapp   :: (α → β)⇒(α ⇒ β)         ( (_[_]) [1000,0] 1000)
  fabs   :: (α ⇒ β)⇒(α → β)         (binder Λ 10)

  less_cfun :: (α → β)⇒(α → β)⇒bool
rules
  Cfun_def                 Cfun = {f. contX(f)}

  Rep_Cfun                 fapp(g) ∈ Cfun
  Rep_Cfun_inverse         fabs(fapp(g)) = g
  Abs_Cfun_inverse         f ∈ Cfun ⟶ fapp(fabs(f))=f

  less_cfun_def            less_cfun(g1,g2) = ( fapp(g1) ⊑ fapp(g2) )
end
```

The theory Cfun1 is an example for a *conservative extension by a new type*. The constructor → is introduced as an infix type constructor. The three axioms Rep_Cfun, Rep_Cfun_inverse and Abs_Cfun_inverse state that the type $\alpha \to \beta$ is

---

[9] In a text book you probably will find $f(\bigsqcup i.Y(i)) = \bigsqcup i.f(Y(i))$ instead of $f(\text{lub}(\text{range}(Y))) = \text{lub}(\text{range}(\lambda i.f(Y(i))))$.

isomorphic to the set Cfun which is the set of all continuous functions of type $\alpha \Rightarrow \beta$. Of course it has been shown in advance that this subset is not empty.[10]

The interesting thing about this theory is that the new constructor is restricted to argument types which are in class pcpo. This restriction is vital since without a pcpo structure the 'subset of all continuous functions' is without any meaning. Due to the use of type classes the new constructor is 'total' on its argument classes. The same situation arises during the formalization of strict products and strict sums (theories Sprod0–3, Ssum0–3).

The technique used is similar to the one which can be found in languages with subtypes. There suitable subtypes are used to model partial functions. In a higher order logic without type classes there is no way to introduce a type constructor for continuous functions, strict products and strict sums since it would have to be partial. See [Age94] for a detailed discussion of the problem.

The mysterious phrases ( (_[_]) [1000,0] 1000) and (binder $\Lambda$ 10) introduce mixfix syntax for the new type. Instead of writing the less readable fapp(f)(x) for application and fabs($\lambda$x.t(x)) for abstraction of an operation, the user simply writes f[x] and $\Lambda$x.t(x). This syntactic sugaring yields a smooth embedding of LCF terms. A term is part of this LCF sublanguage if it is just built of variables, continuous constants, $\Lambda$-abstractions and _[_]-applications.

As a result of the above type definition $\beta$-reduction for operations is subject to a restriction which concerns the continuity of the abstraction. It can be shown that the following theorem about $\beta$-reduction of operations holds:

$$contX(t) \longrightarrow (\Lambda x.t(x))[u] = t(u)$$

This means that in order to do a $\beta$-reduction the continuity of the body has to be proved first. Fortunately this continuity proof can be done automatically if the body t(x) is a term in the LCF sublanguage.

The last axiom less_cfun_def defines the ordering relation for operations. Of course the ordering is inherited from the full function space. In the theories Cfun2 and Cfun3 it is shown that the ordering defined above really yields a pcpo-structure which finally is used to validate the instances

inst_cfun_po     $(\sqsubseteq::(\alpha \rightarrow \beta) \Rightarrow (\alpha \rightarrow \beta) \Rightarrow bool) = $ less_cfun
inst_cfun_pcpo    $\perp::\alpha \rightarrow \beta = \Lambda x.\perp$

and the arity definitions

arities $\rightarrow$      :: (pcpo,pcpo)po
arities $\rightarrow$      :: (pcpo,pcpo)pcpo

---

[10] The new Isabelle version provides a subtype package in the style of Gordon's HOL that produces this axiomatization behind the scenes. The package also checks whether the user supplied a theorem about the non-emptiness of the representing set.

### 3.5  The theory Fix

This theory introduces the fixed point theory of LCF. The main parts are shown below. The iterator iterate which iterates an operation n-times starting with value c is defined by primitive recursion. The parameter n in the third argument of primitive recursion is not really needed for the definition of the iterator. However, we have to supply it in order to confirm to the type of primitive recursion nat_rec.

Fix = Cfun3 +
consts
| | |
|---|---|
| iterate | :: $nat \Rightarrow (\alpha \rightarrow \alpha) \Rightarrow \alpha \Rightarrow \alpha$ |
| lfix | :: $(\alpha \rightarrow \alpha) \Rightarrow \alpha$ |
| fix | :: $(\alpha \rightarrow \alpha) \rightarrow \alpha$ |
| adm | :: $(\alpha \Rightarrow bool) \Rightarrow bool$ |

rules

| | |
|---|---|
| iterate_def | $iterate(n,F,c) = nat\_rec(n,c,\lambda n\ x.F[x])$ |
| lfix_def | $lfix(F) = lub(range(\lambda i.iterate(i,F,\perp)))$ |
| fix_def | $fix = (\Lambda f.\ lfix(f))$ |
| adm_def | $adm(P) = \forall Y.\ is\_chain(Y) \longrightarrow$ |
| | $\qquad (\forall i.P(Y(i))) \longrightarrow P(lub(range(Y)))$ |

end

The *function* lfix of type $(\alpha \rightarrow \alpha) \Rightarrow \alpha$ is just introduced as intermediate constant to ease the technical treatment. The interesting constant is the fixed point operator fix which has type $(\alpha \rightarrow \alpha) \rightarrow \alpha$ of an operation. In definition adm_def the notion of admissibility is defined. Some of the main theorems of the theory Fix are the following fixed point properties:

| | |
|---|---|
| fix_eq | $fix[F]=F[fix[F]]$ |
| fix_least | $F[x]=x \longrightarrow fix[F] \sqsubseteq x$ |
| fix_def2 | $fix[F] = lub(range(\lambda i.\ iterate(i,F,\perp)))$ |

The first two of them are well known from LCF. Note that the notation corresponds to the one used in LCF. Clearly the third theorem is beyond the expressive power of LCF. It is Kleene's constructive characterization of the least fixed point of a continuous function. It is already a theorem since only *functions* may be defined using an application context. Extensionality of functions guarantees the conservativity of such 'definitions'. In order to derive the theorems above, the continuity of the *function* lfix had to be proved first. The proof follows the argumentation that can be found in the literature about LCF. See [Win93] and [Gun92] for two different approaches.

Two other prominent theorems of domain theory are the principle of Scott-Induction and computational induction.

| | |
|---|---|
| fix_ind | $adm(P) \wedge P(\perp) \wedge (\forall x.\ P(x) \longrightarrow P(F[x])) \longrightarrow P(fix[F])$ |
| comp_ind | $adm(P) \wedge (\forall n.\ P(iterate(n,F,\perp))) \longrightarrow P(fix[F])$ |

They both immediately follow from the definition of admissibility. In addition various propagations of admissibility were derived. Some of these are listed below:

| | |
|---|---|
| adm_less | $contX(u) \wedge contX(v) \longrightarrow adm(\lambda x.u(x) \sqsubseteq v(x))$ |
| adm_subst | $contX(t) \wedge adm(P) \longrightarrow adm(\lambda x.P(t(x)))$ |
| adm_conj | $adm(P) \wedge adm(Q) \longrightarrow adm(\lambda x.P(x) \wedge Q(x))$ |
| adm_disj | $adm(P) \wedge adm(Q) \longrightarrow adm(\lambda x.P(x) \vee Q(x))$ |

In LCF these theorems are hard-wired as syntactic tests in the system since they cannot even be expressed inside the logic. See [Pau84] for a discussion of the drawbacks of this lack of expressive power. HOLCF is much more flexible since the admissibility of a predicate can often be derived by a special argumentation although the predicate does not directly fit into the syntactic schemes like the ones listed above. This is due to the fact that admissibility is definable in HOLCF.

# 4 Conclusion

In section 2 the central ideas of higher order logic with type classes were presented. In particular mechanisms for theory extensions with respect to type classes and their conservativity were illustrated using some simple examples. In section 3 the main steps of the development of HOLCF, a higher order version of LCF, were described.

Only a few theories were presented and almost no theorems. However, summing up it took more than 30 steps of conservative theory extensions and about 600 theorems to formalize and derive all the logical concepts that constitute LCF. The full formalization of HOLC, its syntax, semantics and proof rules together with a detailed description of the development of HOLCF and some applications can be found in my thesis [Reg94].

Due to the use of type classes and Isabelle's advanced syntactic capabilities the resulting formalization of LCF is smoothly integrated into HOLC. Higher order logic and logic of computable functions can freely be mixed which yields a higher order version of LCF, namely HOLCF. The advantage of this combination was briefly discussed during the presentation of fixed point theory. The concept of admissibility can be formalized inside the logic which remedies some drawbacks of LCF.

There are other advantages of the combination that could not be discussed due to a lack of space. In [Reg94] some recursive data types like strict lists or streams were formalized in HOLCF. For types with strict constructors (e.g. strict lists) structural induction principles can be derived that are not restricted by any admissibility considerations. In LCF only for strict types over chain-finite argument types can the admissibility proviso be eliminated [Pau87]. In addition, for all tree-like types a co-induction principle [Pit92] can be derived in HOLCF.[11]

---

[11] Usually this is only interesting for types with infinite elements, e.g. streams.

Currently HOLCF is tuned for use as the kernel language of a specification language for distributed systems in the style of [BDD+93]. A type definition package in the style of LCF that produces exclusively conservative axiomatizations is in preparation.

# 5 Acknowledgment

I am grateful for the constructive suggestions received from the referees. I would also like to thank Tobias Nipkow for his advice and many discussions about HOLCF.

# References

[Age94]   Sten Agerholm. *A HOL Basis for Reasoning about Functional Programs.* PhD thesis, University of Aarhus, BRICS Departement of Computer Science, 1994. BRICS Report Series RS-94-44.

[BDD+93]  Manfred Broy, Frank Dederichs, Claus Dendorfer, Max Fuchs, Thomas Gritzner, and Rainer Weber. The Design of Distributed Systems: An Introduction to FOCUS. Technical Report TUM-I9202-2, Institut für Informatik, Technische Universität München, 1993.

[EGL89]   H.D. Ehrich, M. Gogolla, and U.W. Lippeck. *Algebraische Spezifikation abstrakter Datentypen.* Teubner, 1989.

[GM93]    M.J.C. Gordon and T.F. Melham. *Introduction to HOL: A Theorem Proving Environment for Higher Order Logic.* Cambridge University Press, 1993.

[GMW79]   M. Gordon, R. Milner, and C. Wadsworth. *Edinburgh LCF: A Mechanised Logic of Computation,* volume 78 of *LNCS.* Springer, 1979.

[Gun92]   Carl A. Gunter. *Semantics of Programming Languages: Structures and Techniques.* The MIT Press, 1992.

[Nip91]   Tobias Nipkow. Order-Sorted Polymorphism in Isabelle. In G. Huet, G. Plotkin, and C. Jones, editors, *Proc. 2nd Workshop on Logical Frameworks,* pages 307–321, 1991.

[NP93]    Tobias Nipkow and Christian Prehofer. Type checking type classes. In *Proc. 20th ACM Symp. Principles of Programming Languages,* pages 409–418, 1993.

[Pau84]   L.C. Paulson. *Deriving Structural Induction in LCF,* volume 173 of *LNCS,* pages 197–214. Springer, 1984.

[Pau87]   L.C. Paulson. *Logic and Computation, Interactive Proof with Cambridge LCF,* volume 2 of *Cambridge Tracts in Theoretical Computer Science.* Cambridge University Press, 1987.

[Pau89]   L.C. Paulson. The foundation of a generic theorem prover. *Journal of Automated Reasoning,* 5(3):363–397, 1989.

[Pau94]   L.C. Paulson. *Isabelle: A Generic Theorem Prover,* volume 828 of *LNCS.* Springer, 1994.

[Pit92]   Andrew Pitts. A co-induction principle for recursively defined domains. Technical Report 252, University of Cambridge, Computer Laboratory, 1992.

[Reg94]   Franz Regensburger. HOLCF: Eine konservative Erweiterung von HOL um LCF, 1994. Dissertation, Technische Universität München.

[Win93]   G. Winskel. *The Formal Semantics of Programming Languages.* The MIT Press, 1993.

# A Mechanized Logic for Secure Key Escrow Protocol Verification

Tom Schubert and Sarah Mocas

Department of Computer Science
Portland State University

**Abstract.** Reasoning about key escrow protocols has increasingly become an important issue. The Escrowed Encryption Standard (EES) has been proposed as a US government standard for the encryption of unclassified telecommunications. One unique feature of this system is key escrow. The purpose of key escrow is to allow government access to session keys shared by EES devices. We develop a framework to formally specify and verify the correctness of key escrow protocols that we mechanize within the HOL theorem proving system. Our logic closely follows the logic, SVO , used for analyzing cryptographic protocols which was developed by Syverson and vanOorschot [13]. Using the HOL mechanization of SVO , we formally demonstrate the failure of the EES key escrow system by showing that it does not insure that the escrow agent receives correct information. This was previously shown experimentally [2]. Last, we offer an alternative escrow protocol and demonstrate its correctness.

## 1 Introduction

Several logics for analyzing cryptographic protocols and authentication schemes have been proposed (see for example [1, 3, 8, 13, 14]). The primary goal of this type of analysis is to verify the security of a protocol or authentication scheme thereby insuring that an unauthorized third party does not have access to secret information that is exchanged between the users of a protocol. In this paper, we develop a logic for analyzing key escrow systems. In considering a key escrow system, the goal is different in that we want to insure that a third party can decode encrypted information sent between two parties, hence insuring the "integrity" of the escrow agent.

Our logic for analyzing key escrow systems closely follows the logic used for analyzing protocols that was developed by Syverson and vanOorschot [13]. Syverson and vanOorschot's logic was chosen as it encompasses many of the desirable features from earlier developed logics in an integrated approach. Using our mechanization of SVO, we formally demonstrate the failure of the EES key escrow system to insure that the escrow agent receives correct information. This was previously shown experimentally [2]. We also present an alternative escrow protocol and demonstrate its correctness.

In this paper, we will first describe key escrow protocols and the EES key escrow mechanism. Then, we will present our mechanization of a belief logic based on the SVO logic and the extensions required to reason about key escrow protocols. After presenting examples of key escrow protocol verification, we will conclude with a brief description of future work.

## 1.1 Background

Reasoning about key escrow protocols has increasingly become an important issue. The Escrowed Encryption Standard (EES) defines a group of US Government cryptographic chips including both the Clipper and Capstone chips [7]. EES has been proposed as a US government standard for sensitive, but unclassified government and civilian telecommunications. One unusual and highly controversial feature of this system of encryption is key escrow. Key escrow is achieved as follows. Each EES device is assigned a unique identifier and secret key during the manufacturing process. This identifier and key are then stored in escrow. The initial communication for any encrypted session between two EES processors involves the transmission of a "Law Enforcement Access Field" (LEAF), which contains the unique identifier and an encrypted copy (encrypted with the escrowed device specific key) of a previously agreed upon session key. The purpose of this transmission is to allow legal government access to the session key generally via wiretapping. Many of the details of the Clipper chip and escrow scheme are classified, but much of the information that is available can be found in [2, 5].

Although the EES cryptographic algorithm "Skipjack," used by both the Clipper and Capstone chips, is classified, many of the details of the escrow scheme are public and open to scrutiny. Consequently flaws in this scheme have been detected. M. Blaze has outlined various techniques that allow secure communication between two users of EES devices without transmission of a valid LEAF. We reiterate some of his observations and subject them to more formal analysis. Most importantly, Blaze showed that there is a reasonable scenario under which a dishonest user can interoperate with an honest user without sending a correct LEAF.

Two models of cheating will be considered. In one model, two users of EES devices, A and B, cheat (intentionally preventing the escrow agent from receiving a correct session key). In the second model one user, A, is dishonest and user B is honest.

In the first model it is trivial for A and B to use EES devices without giving the escrow agent a valid key. For example, A can simply apply some other form of shared cryptography, such as exclusive-or with a random string (one-time pad) or encoding using a public-key system, to the LEAF when it is sent. Upon receiving the modified LEAF, B can reverse this process prior to passing the LEAF to its EES device. This may seem like an unreasonable action in that A and B could instead use their alternate, shared cryptosystem to communicate all of their messages but in fact their solution is very practical. Both Clipper and Capstone are designed for fast encryption/decryption of large quantities of information, whereas many alternative cryptosystems are slow or expensive.

The more interesting case to consider consists of one dishonest user, A, and an honest user, B. An ideal key escrow system would, in this case, insure that either the escrow agent has received a valid session key and chip identifier or that the honest user would not continue communication. Using our mechanization of the SVO logic, we can show that, given the available information on the data contained in the LEAF, this guarantee is not implicit in the government scheme. Further, we propose a modification to the LEAF that insures that, with high probability, either the escrow agent has received a valid session key and chip identifier or that the honest user will not continue communication. We verify that these properties hold for this modification using HOL.

## 1.2 Key Escrow

The key escrow mechanism used by EES devices is dependent on the transmission of the encoded session key and user identification. As stated above, both of these pieces of information are passed in one transmission in the LEAF. It is this transmission that we will focus on. The LEAF consists of 128 bits that are divided as follows:

UID       32 bit unique unit identifier (the serial number of the chip),
$\{SK\}_{UK}$    80 bit session key, $SK$, encoded using the device unit key, $UK$,
CS        16 bit LEAF checksum.

The information described above is then encoded using a fixed family key, $FK$, resulting in the LEAF, $\{UID, \{SK\}_{UK}, CS\}_{FK}$. Session keys are exchanged between two users prior to the transmission of the LEAF. The LEAF transmission is the direct result of initializing an EES device for a session by entering the session key.

An initialization vector, IV, is also transmitted with the LEAF and it appears that the calculation of the checksum depends on both the session key and the IV. M. Blaze observed that all bits in the LEAF change when the IV or session key changes [2]. The exact construction of the $CS$ and IV are not public.

During the manufacturing process each EES device is assigned a unique identifier $UID$ and unit key $UK$. Further, groups of devices are assigned a single family key, $FK$. The unit key is constructed from the exclusive-or of two other keys, $KU1$ and $KU2$ that are then each stored separately with the $UID$ by two designated government agencies, the escrow agents. For further detail see D. Denning's paper [5].

On intercepting a valid LEAF, a government agent may then decrypt the LEAF with the family key, use the $UID$ to retrieve the keys $KU1$ and $KU2$ from the escrow agents, and then exclusive-or these two keys to obtain the unit key $KU$. As each LEAF contains a session key encoded with $KU$ the government agent may now obtain the session key for two users of EES devices.

## 2 Formal Belief Logic

Belief logics are designed to consider what conclusions individual parties (principals) in a communication dialog can deduce based on messages received and a set of initial assumptions or beliefs. Logics devised to reason about cryptographic protocols generally consider only idealized protocols; there are no bit streams, but rather typed messages. Thus, all parties are presumed to recognize varying message formats, despite the lack of this information in the bit stream. Belief analysis attempts to show that only desired properties are guaranteed by the communication (data security, non-repudiation, no replayed transactions, etc). Note that proofs about idealized protocols are not a guarantee that the concrete protocols are correct. There are many implementation assumptions that if invalid, would cause a secure, idealized protocol to actually be insecure. For example, these logics all assume that the cryptoalgorithm is secure.

An interesting aspect to the SVO logic is its use of "abstracted protocols." SVO messages may also include propositions about held beliefs. For example, if a principal A were to pass a message to principal B that included a new encryption key that A believed to be a good secret key, the logic permits the message to include not just the key, but also A's (implicit) belief that it is a good key.

Using the HOL system type definition mechanisms, we define a number of application specific data types. HOL's recursive type definition facility [9] automates the process of defining new data types in terms of already existing types. Both new type constants and type constructors (operators) can be defined. Additional (recursive) functions can be defined to operate on the concrete data representation of the type. The properties of new types must be derived by formal proof. This guarantees that the type does not introduce inconsistency into the logic.

### 2.1 Types

To model protocols where possibly encrypted messages are passed among different principles, we construct new data types for principals, keys, and message items. Principals are easily defined as unique entities and will exhibit independent behavior and hold autonomous beliefs.

```
let PRINCIPAL = define_type 'principal' 'principal = PRINCIPAL num';;
```

There are two different types of cryptographic algorithms (symmetric key and public key) and several ways in which a cryptosystem and key may be used (data encryption, digital signature, key exchange agreement)[1]. To distinguish between these variations, types key and keyuse are created.

---

[1] A introduction to cryptographic algorithms may be found in [6] or [10].

```
new_type_abbrev( 'crypt', ":bool");;
define( "PUBLIC  = T");;
define( "PRIVATE = F");;

let KEY = define_type 'key' 'key = NONE | SECRET num | PUB crypt num';;
let KEYUSE = define_type 'keyuse' 'keyuse = DATA | SIG | XCHG';;
```

Messages consist of items that may be names of principals, keys, data, and nonces (an indicator of the timeliness of a message). Recall that the abstract protocol also allows assertions to be passed in messages. For example, the protocol permits a principal to indicate (**CLAIM**) its confidence in a public or shared secret key. Additionally, the abstract protocol allows messages to be signed without indicating how this is achieved. We have also added the possibility that a LEAF field may be sent (described in section 1.2). To simplify the LEAF specific inference rules (described in the following section), we add to the LEAF definition an indicator that shows which principal is the intended recipient. Since the exact construction of the *CS* field has not been publicly disclosed, we define this field as a :num.

```
let SIGNED = define_type 'signed' 'signed =  UNSIGNED
                                          | SIGNED principal';;

let KEYFACT = define_type 'keyFact' 'keyfact =
                          PK     principal keyuse key
                        | SHARE principal principal key';;

let ITEM = define_type 'item' 'item = EMPTY
                            | INFO   num
                            | NAME   principal
                            | KEY    key
                            | CLAIM  keyfact
                            | NONCE num
                            | LEAF   principal principal (key#key) num';;

new_type_abbrev('message', ":key# signed # (item)list");;
```

## 2.2   Propositions

The SVO logic defines a number of property predicates to describe the set of beliefs present in a system. This set may also include "facts" that may not be believed by all principals. Informally[2], a message can be *received*, but it may be encrypted. If a principal *has* the key, then the plaintext may be obtained. However, the principal has no assurance who the plaintext came from unless the message is signed. If a message is signed, then a principal can assume that the signing principal *said* the plaintext. But we don't know when the message was said, unless it is *fresh*, in which case, the principal can assume the signer effectively, *says* the message at the same time it is received (the message is timely,

---

[2] A discussion on the rational for these constructors is beyond the scope of this paper.

and thus not a replay). Still, the message may not come from an authority (e.g. "Simon says") unless the signer *controls* the message. If so, then whatever the message *asserts* is considered accurate.

```
let PROPOSITION = define_type 'prop'
    'prop =    ASSERT   keyfact
           | BELIEVES principal prop
           | CONTROLS principal keyfact
           | SEES     principal (item)list
           | SAYS     principal (item)list
           | RECEIVED principal message
           | SAID     principal (item)list
    .      | FRESH    (item)list
           | HAS      principal key';;
```

The SVO axiom schemata, with some small additions and omissions, is given next.

**Rules and Axioms:** The SVO logic has two primitive inference rules (modus ponens and necessitation) and a significant axiom schemata for reasoning about protocols. These rules are captured in HOL by an inductively defined relation [4], **INFER**. The necessitation rule states that anything derivable from axioms alone can be believed by a principal.

**Believing:** Principals may believe a proposition if it logically follows from already held beliefs.

$P$ *believes* $\varphi$ $\wedge$ $P$ *believes* $(\varphi \supset \psi)$ $\supset$ $P$ *believes* $\psi$

$P$ *believes* $\varphi \supset P$ *believes* $(P$ *believes* $\varphi)$

**Receiving:** If a principal has received an unencrypted message, then she has received the concatenates of the message. If a principal has received an encrypted message and has the decryption key, the principal has effectively received a unencrypted copy of the message.

$P$ *received* $(X_1, ..., X_n) \supset P\, received X_i$

$P$ *received* $\{X\}_k$ $\wedge$ $P$ *has* $\hat{K} \supset P$ *received* $X$

**Seeing:** If a principal receives an unencrypted message, then she sees all the concatenates and any function (encryption or decryption) that can be applied to the concatenates.

$P$ *received* $X \supset P$ *sees* $X$

$P$ *sees* $(X_1, ...X_n) \supset P$ *sees* $X_i$

$P$ *sees* $(X_1)$ $\wedge$ ... $\wedge$ $P$ *sees* $(X_n) \supset P$ *sees* $F(X_1, ...X_n)$

**Having:** If a principal sees a key, then the principal has the key and vice-versa (note only keys can be "had").

$P$ *has* $k \equiv P$ *sees* $k$

**Source Association:** Principals can deduce the identity of the sender of a message if the message is encrypted with the senders signature public key or if the message is signed using a shared secret key.

$$P \overset{k}{\leftrightarrow} Q \wedge R \text{ received } \{X^Q\}_k \supset Q \text{ said } X$$
$$PK_\sigma(Q, K) \wedge R \text{ received } \{X\}_{k-1} \wedge P \text{ received } \{X^Q\}_k \supset Q \text{ said } X$$

Saying: A principal sees anything that she said.

$$P \text{ said } (X_1, ... X_n) \supset (P \text{ said } X_i \rightarrow P \text{ sees } X_i)$$
$$P \text{ says } (X_1, ... X_n) \supset (P \text{ said } (X_1, ... X_n) \wedge P \text{ says } X_i)$$

Freshness: To insure that protocols are not susceptible to replay attacks, only recent (fresh) messages are considered valid. A message is fresh if any concatenate is fresh and encrypted fresh messages are also fresh.

$$\text{fresh}(X_i) \supset \text{fresh}(X_1, ... X_n)$$
$$\text{fresh}(X_1, ... X_n) \supset \text{fresh}(F(X_1, ... X_n))$$

Nonce-Verification: If a received message is fresh can be attributed to a particular principal, then we can infer that the principal "says" the message.

$$(\text{fresh}(X) \wedge P \text{ said } X) \supset P \text{ says } X$$

Jurisdiction: If a principal is a trusted authority on a property, then anything she says can be taken as true.

$$(P \text{ controls } \varphi \wedge P \text{ says } \varphi) \supset \varphi$$

Symmetric goodness of shared keys: If two principals share a secret key, then:

$$P \overset{k}{\leftrightarrow} Q \equiv Q \overset{k}{\leftrightarrow} P$$

EES LEAF Validation: If a leaf is received and the receiving principal has the secret keys, then the receiver may believe that the claimed assertion is true.

$$(E \text{ received } \{LEAF \, P \, Q \, \{sk\}_k \, n\}_{fk} \wedge E \text{ has } \hat{k} \wedge E \text{ has } \hat{f}k)$$
$$\supset E \text{ believes } P \overset{sk}{\leftrightarrow} Q$$

Note this rule is not in the original SVO logic and, in fact, we will show that it leads to undesirable conclusions. Using the SVO logic, we might state this rule as an assumption made by the escrow agent. A natural way to define principal specific inference rules might be to add an additional proposition type constructor that expected a list of propositions and a new proposition that could be drawn from the list of propositions. In our mechanization, we chose to defer the resulting recursive type problem to future work when we port our mechanization to HOL90.

SVO also defines rules for "Key Agreement" and "Comprehending," which we have chosen not to include.

$$PK_\sigma(P, K_P) \wedge PK_\sigma(Q, K_Q) \supset P \overset{K_{PQ}}{\leftrightarrow} Q$$
$$P \text{ believes } (P \text{ sees } F(X)) \supset P \text{ believes } (P \text{ sees } X)$$
$$P \text{ received } F(X) \wedge P \text{ believes } (P \text{ sees } X) \supset P \text{ believes } P \text{ received } F(X)$$

For our application, key agreement occurs external to the escrow protocol and thus, the rule is never used. The comprehending rules allow a principal to *see* a function of her inputs. This function can be any one-one function and in practice, is encryption or decryption. Since these functions can already be represented in the logic, it is unnecessary to provide a second form of representation.

## 2.3 Tactics and Lemmas

Performing proofs using only the above axioms is exceedingly tedious. We initially developed a collection of simple tactics to simplify proof manipulations. These tactics were used to prove a number of general lemmas that form the basis for derived inference rules. A more sophisticated tactic mechanism will be described in the next section.

In the box below, Lemma0 states that if a principal has a key and has received a message encrypted with that key, then the principal can infer he has received the message. Lemma1 states that if a key is part of a received, but encrypted message, the new key can be obtained if the receiving principal has the message decryption key. Lemma3 states that if a principal has the key for a received, encrypted message, then the principal can obtain any component of the plaintext.

```
Lemma0
 ⊢ ∀ P k s L 1. RECEIVED P (k,s,1) isIN L ∧
                HAS P (DKEY k) isIN L ⟹
                INFER L (RECEIVED P (NONE,s,1))
Lemma1
 ⊢ ∀ 1 A B fk sk s L. (KEY sk) subItem 1 ∧
     (RECEIVED B(fk,s,1)) isIN L ∧  (HAS B(DKEY fk)) isIN L ⟹
     INFER L(HAS B sk)

Lemma2
 ⊢ ∀ 1 x A B fk s L.
     x subItem 1 ∧  (RECEIVED B(fk,s,1)) isIN L ∧
     (HAS B(DKEY fk)) isIN L ⟹
     INFER L(RECEIVED B(NONE,s,[x]))
```

The lemmas below simplify the process of showing that a principal can believe a received claim when the claim is part of a source authenticated message from an appropriate authority.

```
Lemma3
 ⊢ ∀ 1 n k A B L cl.
     (NONCE n) subItem 1                ∧  (CLAIM cl) subItem 1      ∧
     (ASSERT(SHARE A B k)) isIN L        ∧  (CONTROLS A cl) isIN L    ∧
     (RECEIVED B(k,SIGNED A,1)) isIN L   ∧  (HAS B(DKEY k)) isIN L ⟹
     INFER L(ASSERT cl)

Lemma4
 ⊢ ∀ 1 n k s A B L cl.
     (NONCE n) subItem 1         ∧  (CLAIM cl) subItem 1 ∧
     (RECEIVED B(k,s,1)) isIN L  ∧  (ASSERT(PK A SIG k)) isIN L ∧
     (HAS B(DKEY k)) isIN L      ∧  (CONTROLS A cl) isIN L  ⟹
     INFER L(ASSERT cl)

Lemma5
 ⊢ ∀ 1 n k s A B L cl.
     NONCE n subItem 1         ∧  CLAIM cl subItem 1 ∧
     RECEIVED B (k,s,1) isIN L  ∧  ASSERT (PK A SIG k) isIN L  ∧
     HAS B (DKEY k) isIN L      ∧  CONTROLS A cl isIN L ⟹
     INFER L (ASSERT cl)
```

Using the mechanization developed above, it is straightforward to prove a number of simple inferences. For example, to show that an escrow agent believes it obtains the correct secret key that is passed between two principals A and B (where A and B are EES devices), requires a simple specialization of lemma4 above.

```
⊢ ∀ A B E fk sk n.
  let L =
      [HAS B(DKEY fk);BELIEVES E(HAS E(DKEY fk));ASSERT(SHARE A E fk);
       ASSERT(SHARE A B fk);CONTROLS A(SHARE A B sk);
       RECEIVED E(fk,SIGNED A,[NONCE n;CLAIM(SHARE A B sk)])]
  in
   INFER L(BELIEVES E(ASSERT(SHARE A B sk)))
```

## 2.4   General Support for Reasoning about Inductive Definitions

To better support the use of the mechanized SVO logic, we have developed a general infrastructure for creating automated and interactive procedures to prove goals about inductively defined relations. The infrastructure provides tools to build interactive functions and goal-directed support functions, procedures to eliminate existentially quantified variables from terms, and tactics to generalize existentially quantified rules. Many of the functions are generic, expecting as arguments the list of inference rules returned by **new_inductive_definition** and a list of the relevant combinators. In our case, the list of relevant combinators includes all of the type constructors from the newly defined types (**key, keyuse, keyfact, item,** and **proposition**).

We note that for many applications, validation of the inductively defined rules is, at least initially, of greater importance than automated proofs. For the automated proofs to be useful, their construction must be understood. Thus, it is valuable for the automated procedures to inform the user what rules are relevant and when rules are applied to a goal.

Interactive support is provided by a collection of parameterized functions that suggest an appropriate rule to apply based on the current goal or a term. (**suggested_rules()** and **rules_for_term()**, respectively). Using these mechanisms, users don't need to remember all the rules that may apply, but can request a list of potentially useful rules. Many protocol logics have a significant number of inference rules ([8] lists 44 rules). Even SVO logic, which was designed to significantly reduce the number of rules compared with other protocol logics, requires 23 rules.

We have also developed a tactic generating function, **make_suggest_tac** that returns a tactic specialized for a list of inference rules and a list of combinators. The tactic searches for and then applies, an appropriate rule to apply based on the current goal. In practice, this tactic appears to provide a significant improvement over tactics that try every rule. The tactic also outputs what rules where chosen to apply in a given situation. The displayed list of decisions can assist the user understand why a proof is correct and often, why it fails.

A second general tactic **EXISTS_ELIM_TAC** has also been developed to search for witnesses to replace existentially quantified variables. The tactic searches the assumption list for candidates that match the variable's usage in the goal. After replacement, goals can often be reduced by rewriting with the assumption list. For performance considerations, the tactic defers rewriting until all existentially quantified variables have been replaced.

Using the above two tactics, a general strategy was developed and programmed into a tactic (**IND_RULES_TAC**). The initial proofs of the lemmas required a fair number of steps, but with the use of the **IND_RULES_TAC**, the proofs are performed with only a few tactics. The resulting proofs also require less time with fewer intermediate theorems generated. At this point in development, the tactic does not backtrack and so occasionally, the user must select the correct rule.

The final general support mechanism developed supports generalizing some of the rules returned by **new_inductive_definition**. When creating a new inductive definition, there are often rules with existentially quantified variables in the antecedent that could instead be universally quantified. While this is not always beneficial, in many cases, proofs are simplified.. For example, the first Seeing rule states that a principal receives an unencrypted message, the principal also *sees* the message.

$$\vdash \forall L\ P\ X.(\exists s.\text{INFER } L(\text{RECEIVED } P(\text{NONE}, s, X))) \implies \text{INFER } L(\text{SEES } P\ X)$$

Note the variable **s** is a placeholder for the signed message field and can be converted to a universally quantified variable. The general (forward proof) function can convert this rule to:

$$\vdash \forall L\ P\ X\ s.\text{INFER } L(\text{RECEIVED } P(\text{NONE}, s, X)) \implies \text{INFER } L(\text{SEES } P\ X)$$

## 3 Key Escrow Protocol Verification

This section will describe how the logic developed in the previous section can be used to reason about the EES key escrow protocols. In modeling the proposed government protocol, the principles used in this section are:

| | |
|---|---|
| A , B | two users, |
| E | a government agent (loosely referred to as the escrow agent), |
| $C_A$ , $C_B$ | the EES chips used by A and B, respectively. |

The principals $C_A$ and $C_B$ have been included so that it is clear that users, not encryption devices, have the potential for dishonest actions. Under certain scenarios, we rely on the "honesty" of the EES devices to insure the integrity of the messages received through the protocol. We have intentionally not included $C_E$, the escrow agent's EES technology. The key escrow protocol is strictly for the benefit of the escrow agent and so we can assume that E is honest.

Protocol analysis typically occurs at an abstract level, making a fair number of assumptions about the properties of the cryptographic system underlying the protocols. In practice, cryptographic systems do not necessarily guarantee the assumptions that are made by protocol analysis tools[12]. Below we state our set of assumptions. Ideally, we would like to represent the concrete cryptographic system details in HOL to validate these assumptions.

1. $C_A$ and $C_B$ are tamper resistant and not faulty,
2. A and B have previously used a secure method to exchange the session key $SK$,
3. The cryptographic algorithms are computationally secure,
4. The transmission between an EES device, $C_X$, and the owner of the device, X, is secure.

Initially we will use a simpler model consisting of the principals:

M      a potentially dishonest agent,
E      a government agent (loosely referred to as the escrow agent),
$C_A$      an EES chip,
$C_B$      an EES chip.

In this case we will add the following assumptions about the initialization of the EES devices:

1. On behalf of $C_A$ and $C_B$ a secure method to exchange the session key $SK$ has been executed,
2. $C_A$ and $C_B$ have both been given $SK$.

In all examples, the message that is passed is referred to as the LEAF. In the government key escrow system, the initialization vector, IV, is passed with the LEAF. We have omitted explicit reference to the IV as it is passed in the clear and any cryptographic function that it serves is not publicly known.

## 3.1 Government Key Escrow

As previously stated, ideally we want to insure that the escrow agent receives a valid chip identifier, $UID$, and encoded session key, $\{SK\}_{UK}$, from which $SK$ can be retrieved. With respect to our logic, this translates into showing that the escrow agent, E, believes that she has a valid $UID$ and encoded session key. That is, E *believes* (E *sees* $UID$) and E *believes* (E *sees* $\{SK\}_{UK}$) where $UID$ and $\{SK\}_{UK}$ are from a valid LEAF and $SK$ is the session key that is shared by the EES devices.

Using our simplified model, where A and B are omitted, the basic escrow protocol consists of the following message:

Message 1    $C_A \rightarrow C_B, E : LEAF$

indicating that $C_A$ sends $C_B$ and E a LEAF. The EES protocol expects the following relationships to exist prior to transfer of the LEAF.

– The sharing of the unit keys is established when the EES devices are manufactured. We also assume E recognizes the device identifiers $UID_A, UID_B$.

$$E \stackrel{UK_A}{\leftrightarrow} C_A \text{ and } E \stackrel{UK_B}{\leftrightarrow} C_B$$

$$E \text{ believes } (E \stackrel{UK_A}{\leftrightarrow} C_A) \text{ and } C_A \text{ believes } (E \stackrel{UK_A}{\leftrightarrow} C_A)$$

$$E \text{ believes } (E \stackrel{UK_B}{\leftrightarrow} C_B) \text{ and } C_B \text{ believes } (E \stackrel{UK_B}{\leftrightarrow} C_B)$$

– $C_A$ and $C_B$ have a secure session key.

$$C_A \stackrel{SK}{\leftrightarrow} C_B$$

$$C_A \text{ believes } (C_A \stackrel{SK}{\leftrightarrow} C_B) \text{ and } C_B \text{ believes } (C_A \stackrel{SK}{\leftrightarrow} C_B)$$

– EES devices control the messages that they generate.

$$C_A \text{ controls LEAF and } C_B \text{ controls LEAF}$$

– We assume that each principal that *shares* a key, *has* the shared key.

The LEAF is constructed by $C_A$ (which by (P *controls* $\varphi$ $\wedge$ P *says* $\varphi$) $\supset \varphi$ gives the LEAF) and the transfer of the LEAF takes place as follows:

$$C_A \text{ said LEAF}$$
$$C_B \text{ received LEAF}$$
$$E \text{ received LEAF}$$

Note the idealization process required by SVO and other logics results in a loss of information about the order in which messages are sent and received. To avoid this step and retain sequence information, a more sophisticated representation could be achieved by using a process algebra notation.

Consider the conclusions that can be drawn by the escrow agent E. Any principal who receives a LEAF and has the LEAF (symmetric) family key can also see the contents of the LEAF. Given $C_B$ and E have the family key and applying the second **Receiving** rule, both principals then also receive $(UID_{C_A}, \{SK\}_{UK-C_A}, CS)$, the list of components of the LEAF. By the first **Seeing** rule, $C_B$ and E also *see* the list. Further applying the second **Seeing** rule, $C_B$ and E see the components $UID_{C_A}, \{SK\}_{UK-C_A}$ and $CS$.

Since E sees both $UID_{C_A}$ and $\{SK\}_{UK-C_A}$ and has $(UID_{C_A}, UK - C_A)$ in escrow, then by the second **Receiving** rule, we can conclude that E sees the session key $SK$ and that E believes that the session key that he has is the same session key that was sent by $C_A$ and shared between $C_A$ and $C_B$. Using our mechanization in HOL, a proof of this behavior can be easily constructed.

```
⊢ ∀ A B E fk sk ka n s.
  let L =
        [HAS E(DKEY ka);HAS E(DKEY fk);ASSERT(SHARE A E ka);
         ASSERT(SHARE A E fk);ASSERT(SHARE A B sk);
         RECEIVED E(fk,s,[LEAF A B(ka,sk)n])]
  in
    INFER L(BELIEVES E(ASSERT(SHARE A B sk)))
```

Unfortunately, as M. Blaze has shown, it may not be the same session key that $C_A$ and $C_B$ will eventually use. We will use the word *correct* to refer to the LEAF that was said by $C_A$.

Consider the following modification of our simplified protocol where a malicious principal, $M$, intercepts and modifies the LEAF.

Message 1   $C_A \rightarrow M$ :LEAF
Message 2   $M \rightarrow C_B, E$ :LEAF'

Here the message LEAF', is not the message sent by $C_A$, but is sufficiently similar so that $C_B$ accepts LEAF' and continues to interoperate with $C_A$. This is represented in SVO as:

$C_A$ *said* LEAF
M *received* LEAF
M *said* LEAF'
$C_B$ *received* LEAF'
E *received* LEAF'

Using the same series of steps given above, we can conclude that E sees a session key $SK'$ for user $UID'$ (retrieved from LEAF'). Consequently, if E believes that LEAF is a message that was sent from $C_A$, then E believes that LEAF' is also a message that was sent from $C_A$. The substitution of LEAF' for LEAF is justified by the experimental work done by Blaze.

The government key escrow protocol extends the number of principals to include A and B, the owners of the EES devices.

Message 1   $C_A \rightarrow A$ : LEAF
Message 2   $A \rightarrow B, E$ : LEAF
Message 3   $B \rightarrow C_B$ : LEAF

Letting A play the role of M, our HOL mechanization allows us to prove that the escrow agent E fails to obtain the shared key. Thus, the government protocol fails for one honest user and one dishonest user.

```
⊢ ∀ A B E fk sk sk' ka n s.
  let L =
        [HAS E(DKEY ka);ASSERT(SHARE A E ka);HAS E(DKEY fk);
         ASSERT(SHARE A E fk);BELIEVES B(ASSERT(SHARE A B sk));
         BELIEVES A(ASSERT(SHARE A B sk));
         RECEIVED E(fk,s,[LEAF A B(ka,sk')n])]
  in
    (INFER L(BELIEVES E(ASSERT(SHARE A B sk')))  ∧
     INFER L(BELIEVES B(ASSERT(SHARE A B sk)))  ∧
     INFER L(BELIEVES A(ASSERT(SHARE A B sk))))
```

M. Blaze showed that any randomly generated 128 bit string will have a $1/2^{16}$ chance of appearing valid for the current session key and $IV$. An attack that replaces a correct LEAF with a bogus LEAF appears to be feasible in practice [2]. We will therefore assume that an EES device does not currently use the checksum to validate the contents of the LEAF to an extent that would allow E to believe that the correct LEAF had been transmitted.

We consider two additional assumptions each of which can be used to convince E that she had received the correct information.

1.    E *believes* (B *believes* (B *received* CORRECT_LEAF)).
2.    E *believes* ($C_B$ *believes* ($C_B$ *received* CORRECT_LEAF)).

The first assumption assumes that B has the computational resources to either decode a message encoded with the family key and then determine that the message is correct or to determine that it is correct without using the family key. In either case, these are extremely strong assumptions. The second assumption does offer a solution that will be outlined in the next section. The rule E *believes* (A *said* CORRECT_LEAF) should not convince E of the correctness of the LEAF since there is no guarantee that the LEAF that E received is the LEAF that A sent. It appears that E must use $C_B$ to verify the protocol.

We note though, given the government protocol, not only can we not conclude that $C_B$ believes that $C_A$'s $UID$ and encoded session key were transmitted we can not even show that $C_B$ believes that the LEAF that he received was sent by $C_A$ only that it was sent by some EES device. This is not a surprising result as M. Blaze has already shown experimentally that a dishonest user can send an honest user an invalid LEAF that will be accepted as correct.

### 3.2   Modified Government Key Escrow

We propose that the LEAF be modified to include an additional field that contains an encrypted version of the $UID$ and encrypted session key. The LEAF is now:

$$\{UID, \{SK\}_{UK}, \{\{SK\}_{UK}, UID\}_{SK}, CS\}_{FK}.$$

We also propose that any so modified EES device that receives a LEAF check that:

$$\text{if } (\{SK\}_{UK}', UID') = D_{SK}(\{SK\}_{UK}, UID)$$

$$\text{then } \{SK\}_{UK}' = \{SK\}_{UK} \text{ and } UID' = UID.$$

Further, if these equalities do not hold, then we require that the device terminate the session. Given these modifications we argue that E believes that she has a valid $UID$ and encoded session key. In fact, E's belief is solely based on the belief that an honest user, in this case $C_B$, believes that a valid LEAF has been transmitted and received. This is the second assumption listed above, namely E *believes* ($C_B$ *believes* ($C_B$ *received* CORRECT_LEAF)).

In this case, $C_B$ *believes* principal($C_B$ *received* CORRECT_LEAF) since only $C_A$ and $C_B$ share $SK$ and $\{SK\}_{UK}' = \{SK\}_{UK}$ and $UID' = UID$ where $(\{SK\}_{UK}', UID') = D_{SK}(\{\{SK\}_{UK}, UID\}_{SK})$ The modification adds a level of authentication for $C_A$ that appears (given known information) to be missing from the government EES protocol. This modification is enough to verify, with high probability, that the LEAF came from $C_A$ and has not been modified.

The mechanization of this protocol and the authentication of $C_A$ in HOL can be accomplished by modifying the type definition for items so that the LEAF type constructor also includes the new authentication field to represent the encryption of the $\{\{SK\}_{UK}, UID\}_{SK}$. The LEAF authentication proof rule must also then be modified to check that the authentication field is correct.

It is fairly obvious that a message using a small hash function to validate the message content is weak (collisions are easy to find). Our modification is stronger than simply using a larger hash function since additional authentication is provided for the sender through the use of a shared key. Device $C_B$ can now verify that the LEAF is from $C_A$, since only $C_B$ and $C_A$ share the "hash function" $E_{sk}$. It should be noted that our modification does increase the length of the LEAF by 112 bits. Even though key exchange occurs only once during a session, this may complicate a hardware implementation. We are unsure as to whether or not the checksum field can be removed. Though our modification replaces the known use of this field, its interaction with the IV field is not public.

# 4  Conclusions

In this paper we have described the embedding of the SVO logic within the HOL theorem proving system. The logic has been extended to support reasoning about key escrow protocols and we have demonstrated its use with the original EES key escrow protocol and an improved protocol. As a byproduct of this development, we have also created an infrastructure to support proofs using inductively defined relations.

As observed by Syverson and vanOorschot[13], the SVO logic is still a beginning step in the development of a logic that reasons about cryptographic protocols. Likewise, our adaptation of SVO as a tool to reason about escrow protocols and specifically our mechanization of this logic in HOL are only initial steps. Our current mechanization only allows propositions about keys to be passed as part of the abstract protocol. We would like to define propositions as functions of items passed in messages and also permit messages to include propositions. We expect that when porting to HOL90, the mutually recursive data types can be more easily defined.

We also plan on further developing semi-automated decision procedures. Many of the tactics developed greatly assisted in simplifying the proof development process, but it is clear that the tactics can be more fully automated.

Finally, we intend to extend the mechanization so that the behavior of principals are defined as within a process algebra. This work will be based on earlier work by one of the authors[11]. Analysis using SVO and other belief logics constructs the set of messages sent and beliefs by hand as part of the *protocol idealization* step. However, the behavior of a system can be modeled as a value passing process algebra where *agents* can perform *actions* send, receive, and generate. Put in this framework, the process of creating the belief set from principal definitions could be automated. This will allow us to model networks where messages are broadcast, private networks, and malicious principals that replace part of the network (*man-in-the-middle attacks*).

# References

1. M. Abadi and M. Tuttle. A semantics for a logic of authentication. In *Proceedings of the Tenth ACM Symposium on Principles of Distributed Computing*, pages 201–216. ACM Press, 1991.

2. Matt Blaze. Protocol failure in the escrow encryption standard. In Lance J. Hoffman, editor, *Building in Big Brother*. Springer-Verlag, 1995.

3. Michael Burrows, Martin Abadi, and Roger Needham. A logic of authentication. *ACM Transactions on Computer Systems*, 8(1), Feb. 1990.

4. Juanito Camilleri and Tom Melham. Reasoning with inductively defined relations in the HOL theorem prover. Technical Report 265, University of Cambridge Computer Laboratory, 1992.

5. D. Denning. The U.S. key escrow encrytion technology. *Computer Communications*, to appear.

6. Dorothy Denning. *Cryptography and Data Security*. Addision-Wesley, 1982.

7. National Institute for Standards and Technology. Escrow encrytion standard. *Federal Information Processing Standards Pubulication 185*, 1994.

8. L. Gong, R. Needham, and R. Yahalom. Reasoning about belief in cryptographic protocols. In *Proceedings of the 1990 IEEE Computer Society Symposium on Research in Security and Privacy*. IEEE Computer Society Press, 1990.

9. Tom Melham. Automating recursive type definitions in higher order logic. In G. Birtwhistle and P.A Subrahmanyam, editors, *Current Trends in Hardware Verification and Automated Theorem Proving*, pages 341–386. Springer-Verlag, 1989.

10. Bruce Schneier. *Applied Cryptography*. John Wiley & Sons, Inc, 1994.

11. E. Thomas Schubert. A hybrid model for reasoning about composed hardware systems. *Conference on Computer-Aided Verification*, June 1994.

12. Gustavus J. Simmons. Cryptanalysis and protocol failures. *Communications of the ACM*, 37(11), November 1994.

13. Paul F. Syverson and Paul C. van Oorschot. On unifying some cryptographic protocol logics. In *Proceedings of the 1994 IEEE Symposium on Research in Security and Privacy*, May 1994.

14. P. vanOorschot. Extending cryptographic logics of cryptographic logics of belief to key agreement protocols. In *Proceedings of the First ACM Conference on Computers and Communications Security*, pages 232–243, 1993.

# A New Interface for HOL - Ideas, Issues and Implementation

Donald Syme

The Computer Laboratory
University of Cambridge
E-Mail: Donald.Syme@cl.cam.ac.uk

**Abstract.** TkHolWorkbench is a new set of interface tools for HOL implemented using the Tk toolkit. It aims to be robust, extensible, lightweight and user-friendly. The tools are designed to augment the existing HOL interface. The project applies rapid prototyping and the use of an interpreted toolkit to the field of theorem proving interfaces. The topics considered in this paper are: the motivations for a new interface for HOL; the design objectives and usability targets for TkHolWorkbench; a description of the TkHolWorkbench tools as they now stand; and the extensible design architecture used in the implementation.

## 1 Introduction

This paper describes a new interface for the HOL theorem proving system called TkHolWorkbench. This interface has been under development at the University of Cambridge for the last 6 months, and the author hopes that this interface, or some derivative of it, will eventually become the interface of the HOL2000 project.

The aim of this paper is to give an overview of the functionality of TkHol-Workbench, and to explain some of the motivations and ideas that have contributed to its development. It also seeks to establish a set of *design objectives* and *usability targets* for TkHolWorkbench. A discussion of the extensible design architecture used in the implementation is also included.

TkHolWorkbench is not the first new interface proposed or developed for the HOL system. The most significant efforts have been CHOL by Laurent Thery [11], an unnamed interface by Sara Kalvala [4] and xhol by Thomas Schubert [9]. There are, as one would expect, many similarities in motivation and design between these systems and TkHolWorkbench. However, the reader familiar with these systems will see significant differences too.

TkHolWorkbench takes a *tool-oriented* approach to interface design decomposition. The idea is to augment the command line HOL interface with *helper* tools. The tools facilitate various user activities, such as defining a new recursive type or proving a theorem using backward proof. Four tools are described in this paper:

- TkTheoryViewer for browsing theories.
- TkTRS for searching the theory hierarchy for particular theorems.

- TkDefineType for making recursive type definitions.
- TkGoalProof for backward goalstack proof.

Other tools to be included in the upcoming release of TkHolWorkbench are:

- TkNewDef for making new, non-recursive definitions.
- TkRecDef for making primitive recursive definitions.
- TkIndDef for defining inductive relations.

Section 2 of this paper describes some of the motivations for the TkHolWork-bench project. These are made more concrete in a series of design objectives and usability targets in section 3. Section 4 gives an overview of TkHolWorkbench itself and describes the tools mentioned above. Section 5 describes the design architecture of the system, and section 6 looks at how how user extensions can be added to the interface.

# 2 Informal Motivations

## 2.1 Real theorem provers deserve real interfaces

The fundamental motivation for building a new interface for HOL is simple: HOL is an excellent theorem proving system, but has a crude, command-line interface. HOL deserves a good interface, or so the reasoning goes, just as it deserved a good implementation in ho190 [10]. This was summarized succinctly by Laurent Thery [11]: "Real theorem provers deserve real interfaces".

This seems almost self-evident, but is worth being more specific about the expected benefits of producing a graphical user interface (GUI) for HOL. Good GUIs do not come without cost in time and effort, and, after all, it is always possible that theorem proving will turn out to be one of the application domains where GUIs have little impact. However, the following list of possible benefits indicates why this author believes that HOL users can benefit from a GUI. It draws from several sources [11, 4, 9, 7].

- The current interface to HOL is very weak, particularly in the area of visu-alisation. *There must be better ways of visualising theory development* than are currently used.
- Theory development by less skilled people will become more common if the goals of formal methods research are realised. HOL *needs to be more accessible* if it is to meet these needs.
- HOL will be *easier to learn* with a good GUI interface. The present text interface provides little help in reducing the steepness of the learning curve.
- Increasingly, *users expect high quality data presentation* to accompany the packages they use. HOL does not currently provide this.
- Also, *users expect a high degree of interoperability between packages*. HOL does not currently provide this.

The following sections will consider in more detail three further motivations for the construction of TkHolWorkbench:

- A GUI will make more functionality readily available to HOL users.
- A good interface can generate new ideas about HOL itself.
- "The time is right", since user interfaces are finally becoming easier to build.

## 2.2   Making more functionality accessible

It is often claimed that GUIs restrict the amount of functionality accessible to users, in comparison to command line interfaces. However, it is the author's belief that this need not be the case, and that a GUI can make *more* HOL functionality available. We shall consider two examples where this should be the case: the TRS library and the Window Proof library.

The TRS (theorem retrieval system) library was developed several years ago by Richard Boulton [1]. Though well designed and implemented, this library has rarely been used, since it is clumsy to construct the necessary search patterns using a command line interface. TkHolWorkbench includes a simple tool which gives access to this functionality. Adding a GUI interface to this library has made it more accessible, and indeed it is now easy to do dynamic searches through the theory hierarchy.

Alternative proof packages such as the Window Proof library [3] do not seem to get used as often as they deserve. This is probably because of the steepness of the initial learning curve involved. GUI interfaces to such packages should make them easier to learn, and thus more accessible for both new and old HOL users. Window inference seems particularly well suited for use with a GUI. [1]

## 2.3   Interfaces can generate new ideas

In a research setting, it seems to be the case that a new interface to a tool has the potential to generate new ideas about the research area itself. Already with TkHolWorkbench, users have suggested changes which really require changes to HOL. Typically these relate to higher level actions that users want to perform, yet are not directly supported by HOL. Some examples are the need to support the alteration of definitions, the unloading and reloading of theories, and the management of theory scripts. These represent needs brought to light by the different mental model a new interface provides.

## 2.4   GUIs are now easier to build

The last motivations we shall mention are pragmatic and personal ones. GUI interface building is simply becoming much more feasible, due to the power brought to developers by tools such as John Ousterhout's Tk/Tcl package [8] and Microsoft's Visual Basic. The use of Tk/Tcl to develop TkHolWorkbench

---

[1] It is hoped that some kind of window inference tool will be available for use with a future release of TkHolWorkbench, possibly based on work in progress by Thomas Långbacka et al. at Åbo Akademi University[12].

has been instrumental in enabling both rapid prototyping and a robust final implementation.

Lastly, the author has spent much of the last five years developing interfaces for a wide range of systems. A personal desire to apply some of the things I have learnt to my favourite system, HOL, was certainly a significant motivation in this work.

# 3 Design Objectives and Usability Targets

In this section we follow the methodology of interactive system design described by Newman and Lamming [7] to develop a set of *design objectives* for TkHol-Workbench. These objectives seek to clarify the problem domain with which TkHolWorkbench is concerned. As part of these objectives we establish a set of *usability targets*, which describe the levels of support to be offered by the system.

We will state the design objectives for TkHolWorkbench in terms of the following areas:

- The *human activity* it will support.
- The *prospective users* who will perform the activity.
- The *levels of support* that it will provide, also known as the *usability targets* of the system.

## 3.1 The Human Activity

In a nutshell, TkHolWorkbench aims to support "theory development". Contrast this with other definitions we might consider using:

1. TkHolWorkbench will support "backward proof"
2. TkHolWorkbench will support "alternative proof techniques"
3. TkHolWorkbench will support "ML program development"

"Theory development" is a broader concept than (1) and (2) but narrower than (3). There is a deliberate choice being made here. TkHolWorkbench aims to do more than support proof techniques, but does *not* yet aim to be a full ML programming environment, except in the sense that it continues to support the command line HOL interface. Theory development tasks such as backward proof and type/constant definitions are often small and well defined, and seem to be more amenable to improvement using a GUI than arbitrary ML programming tasks. Furthermore, theory development is what HOL is fundamentally concerned with, and so must be the major focus of a user interface.

## 3.2 The Prospective Users

The prospective users of TkHolWorkbench are, broadly speaking, the current "end-users" of HOL. These presently include people performing hardware and software verification using HOL. Newcomers to the HOL system are implicitly amongst the target user group, because the first tasks a new user performs is invariably some kind of theory development. The user base is not initially expected to include those whose main work is programming new proof tools for HOL.

## 3.3 Usability Targets

The usability of an interface can be measured in several different ways. One of the most useful bottom line metrics is simple, yet often overlooked by developers: does the interface help users to accomplish their tasks more quickly? Other usability metrics are also important. Those cited by Newman and Lamming [7] are:

- The *incidence of errors* while performing the activity.
- The user's *ability to recover from errors* that occur.
- The magnitude of the user's task in *learning to use the system*.
- The user's *retention of learned skills*.
- The user's *satisfaction* with the system.
- The user's ability to *customize* the system.
- The *mental burden* placed on the user when using the system.

For these usability metrics it is easy to develop a set of usability targets for TkHolWorkbench. In particular:

- TkHolWorkbench must make theory development faster. For instance, theorems should be quicker to formulate and prove. Activities performed with the graphical tools should not take longer than with the command line interface.
- Errors must not be more common than with the command line interface, and must it be easier to recover from errors when they are made.
- TkHolWorkbench must be comparatively easy to learn. It is unrealistic to expect people who know the HOL logic to need more than about 60 minutes learning time, spread over the first days of use.
- Users must be able to customize TkHolWorkbench to their needs.
- TkHolWorkbench should reduce the mental burden of using HOL.

These targets provide a mechanism for measuring the success of an implementation, and also contribute to the formation of design guidelines that help the developer during the design and implementation process.

This paper does not evaluate TkHolWorkbench against these targets, but rather states them as goals to be achieved in current and forthcoming development. The field of Human Computer Interaction provides methodologies for performing usability evaluations such as Cognitive Walkthroughs and Keystroke-Level Analysis. Such methods will be used to evaluate TkHolWorkbench rigorously in the future.

# 4 An Overview of TkHolWorkbench

In this section, we move on to describing TkHolWorkbench itself. We shall consider some of the design principles which have featured in its development so far, and then some particular features of TkHolWorkbench.

From the user's viewpoint, TkHolWorkbench can be seen as set of *independent yet cooperating tools*. This means:

- The interface is a series of optional modules, exhibiting a low degree of interdependence.
- Tools do have to be used if the user does not find them helpful.
- Tools support standard X-Windows interaction techniques.

More will be said in section 5 on the internal mechanisms used to support this. TkHolWorkbench is also designed using the principle of *non-intrusiveness*. This means:

- The interface supports existing modes of interaction. The user is not forced to adopt the new modes offered.
- No end product of the interface relies on the interface itself. For example, proof scripts do not rely on tactics which are peculiar to the interface. Scripts can be rerun without the interface present.
- Running TkHolWorkbench does not have any side effects except those exhibited by running a normal HOL session.

Lastly, TkHolWorkbench is implemented using the principle that *lightweight* systems are more likely to meet their usability targets. This is with regard to installation time, tool startup time and runtime memory usage. The idea here is to balance resource usage against the functionality provided.

## 4.1 The Command Line

An important feature of TkHolWorkbench is that *the standard HOL interface is still available to the user*. In terms of implementation, a TkHolWorkbench program has a Unix stdin/stdout which behaves identically to a normal HOL process. A screen picture illustrating this is shown in figure 1. This feature has important ramifications:

- Users can start TkHolWorkbench as a substitute for their regular HOL session. The only difference is the appearance of a window in addition to the HOL prompt.
- Existing users are not forced to change their work patterns.
- TkHolWorkbench can be used from within editors like Emacs in the same fashion as the normal HOL interface.
- The interface developer is relieved of a great burden, as the interface no longer needs to provide access to *all* of the HOL system. The developer can concentrate on developing tools aimed at specific needs.

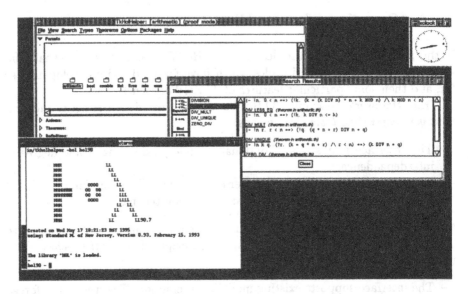

**Fig. 1.** TkHolWorkbench supports the existing HOL interface

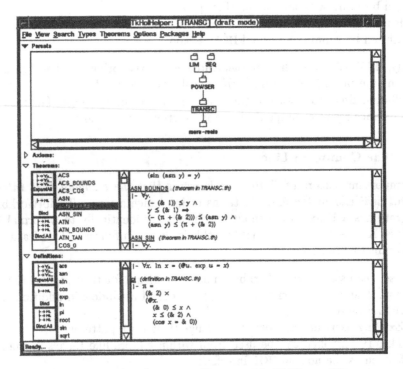

**Fig. 2.** Browsing the reals library with TkTheoryViewer

## 4.2   TkTheoryViewer

We now introduce some of the tools in the current release of TkHolWorkbench. TkTheoryViewer and TkTRS are both *visualisation* tools. This means they are primarily concerned with representing data to the user, rather than allowing state transitions. The primary role of TkTheoryViewer is to visualise two things:

- The theory graph of the theories loaded into the HOL session.
- The theorems, definitions and axioms of a selected theory.

More than one TkTheoryViewer may be created, if the user wants to view multiple theories simultaneously.

TkTheoryViewer provides an ideal mechanism to browse unfamiliar theories. This is a common task for both new and old HOL users. An example of this is shown in figure 2, where TkTheoryViewer is being used to browse through the reals library. The tool also makes an excellent proof assistant tool, as it is an easy process to look through a theory for a theorem which may be needed for a proof.

The theory graph of the theories loaded into the HOL session is visualised as a tree. This behaves in a manner similar to a file browser directory hierarchy:

- Nodes of the tree may be expanded and collapsed.
- A set of theories may be selected for operating on. In particular, the selected set of theories may be used as the basis of a TkTRS search.

Given that the theory structure in HOL is a graph it may seem odd that a tree is used to represent the information. Initially this was done for ease of implementation, but as it happens it seems to be an adequate and efficient representation, even for large theory structures.

The theory graph is also somewhat modified, allowing the tree to be "flattened" at certain points. It is particularly useful to flatten the tree at collector theories such as HOL, since the details of the theory structure above this point is not normally important. This feature can be inhibited if the user desires.

The theory graph automatically updates its contents when a new theory or library is loaded into the system. This happens even if the load command was issued from the command line interface. The mechanism used to implement this functionality is described in section 5.5.

TkTheoryViewer displays the contents of the currently selected theory in the windows below the theory graph:

- Theorems may be displayed in *Rich Text*. This uses different fonts and colors for various components of terms.
- Each of the windows for axioms, definitions and theorems are collapsible, and when no objects of a particular kind occur in the theory, the window is automatically collapsed. This minimizes the amount of screen real-estate used by the viewer.

The theory contents update automatically when theorems are saved or new definitions introduced.

There are several other actions available from TkTheoryViewer, as it is the normal main window of a TkHolWorkbench session. In future releases it will be possible to locate these actions on any top level window, as most of them are not directly related to the visualisation of theories. These actions include:

- Access to theory operations: loading a theory; adding a new parent; creating a new theory; and switching between proof mode and draft mode.
- Loading a HOL library. The list of all libraries available to be loaded is shown to the user.
- Loading an ML file.
- Saving the image of the HOL session.
- Access to other packages such as TkTRS and TkGoalProof via configurable menus.
- Access to a database of configuration options for TkHolWorkbench.
- Access to the values of global variables and flags in the HOL session.
- Help menus which use a WWW browser to access local or remote help files.

## 4.3  TkTRS

The TkTRS tool provides a simple interface to Richard Boulton's Theorem Retrieval System (`trs`) library [1]. The `trs` library provides the following functionality to the user:

- The ability to search loaded theories for theorems and definitions.
- Search patterns may be based on names or theorem structure.
- Arbitrary lists of theorems may be searched as well.

The TkTRS tool provides fairly primitive access to a subset of this functionality. Figure 3 shows an example of using TkTRS to find all the theorems concerning division in all the loaded descendants of the `arithmetic` library.

In the top windows of the search window, the set of source theories to search is shown in the right hand list, and all other theories are shown on the left. Theories may be transferred to and from these lists.

The search pattern is specified in the lower boxes. It may be any of the following:

- Find theorems which contain a particular term structure.
- Find theorems whose name matches a particular wildcard pattern.
- Find theorems which match an arbitrary `trs` pattern.

The search results are presented in a separate window, with similar properties to the windows used to display theorems in the TkTheoryViewer.

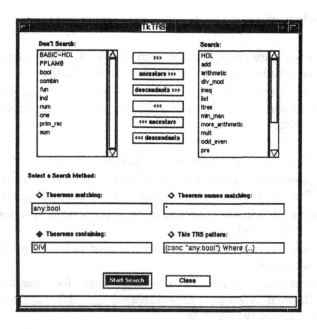

**Fig. 3.** Searching descendants of `arithmetic` for theorems about division

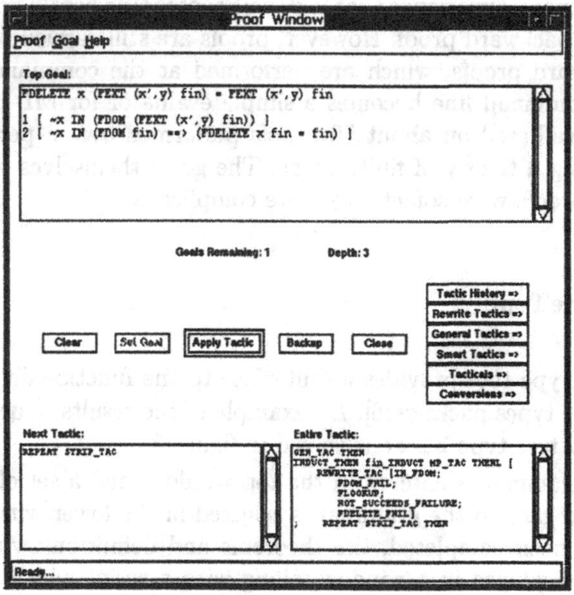

**Fig. 4.** Using TkGoalProof

## 4.4 TkGoalProof

The next two tools are examples of constructive tools. They support the user activities of backward goal proof and concrete type definition.

TkGoalProof is essentially an interface to a HOL goal stack. The operations conducted in the window will be familiar to anyone who has performed a backward proof in HOL:

1. The user enters the goal in the top window, and selects "Set Goal".
2. The user enters tactics into the "Next Tactic" window and selects "Apply Tactic". The goal is redisplayed after each application of the tactic.
3. As the user applies tactics, a complete tactic is constructed automatically in the "Entire Tactic" window.
4. The user may backup through the application of tactics. The entire tactic is adjusted accordingly.
5. After the theorem is proved, the user is prompted for a name by which to save the theorem. A script is then displayed by which the theorem may be reproven in a later HOL session.

As with all TkHolWorkbench tools, more than one TkGoalProof window may be created. Each window accesses different goal stacks, and hence extra windows can be used to prove lemmas needed for the current proof.

It is the author's experience that TkGoalProof is sufficiently useful to replace command line backward proof. However, proofs are still carried out in conjunction with forward proofs, which are performed at the command line. In this scenario the command line becomes a simple evaluator for ML phrases. These observations are based on about 50 proofs performed over a period of 3 days, while developing a theory of finite maps. The goals themselves were not large, although the proofs were sometimes quite complicated.

## 4.5 TkDefineType

The TkDefineType tool provides an interface to the functionality of Tom Melham's recursive types package [6]. An example of the results of using the tool to define a binary tree type btree is shown in figure 5.

The type grammar is entered in the top window, and a set of operations to perform with regard to the grammar is selected in the lower window. After the operation has been completed, the theorems and definitions which have been produced are displayed in a window, along with a script which will reproduce the operation in a later HOL session.

It is hoped that similar tools will soon be available to allow users to access other type definition packages. The extensibility of TkHolWorkbench means that these packages could come with their own interface tools.

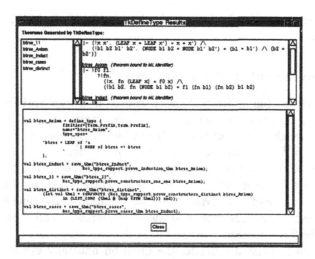

**Fig. 5.** The results window after defining btree using TkDefineType

# 5 Implementation Techniques

This section discusses some of the implementation techniques and the design architecture used in the current release of TkHolWorkbench. The two most important features of the architecture are its *modularity* and its *extensibility*. An understanding of the techniques used will benefit anyone wanting to extend TkHolWorkbench with a new tool.

## 5.1 Expect, Tk and HOL

TkHolWorkbench is implemented using Tk/Tcl [8]. The architecture of TkHol-Workbench is not dependent on the implementation language. However, it is facilitated by it, and hence the reader unfamiliar with Tk and Tcl may find the following facts helpful.

Tcl (Tool Command Language) is a small, interpreted language developed by John Ousterhout at Berkeley [8]. It takes a "nihilistic" approach to language design, allowing arbitrary additional command constructs and using string values only. It is sufficiently efficient and flexible for many purposes, though much of its value lies in the many extensions which have been built on top of it.

Tk is a C extension to Tcl which implements a widget set for the X window system. The standard Tk interpreter is the program wish.

TkHolWorkbench is implemented by the well known technique of driving the HOL session as a slave process of the Tk/Tcl process. The two process are run in tight synchronisation, and cross-language calls from Tcl to ML can be made. Thus, for most practical purposes, the two processes may be considered one. The implementation of this mechanism uses "Expect" [5], a process control extension to Tcl. Expect is specifically designed for the control of command line processes such as HOL.

## 5.2 Separating out the User Interface

The most basic notion of design separation used in TkHolWorkbench is that of dividing the program in two halves - the "computation engine" and the "user interface". Though often touted as a "good idea", this ideal is often ignored in practice. TkHolWorkbench tries to adhere to it strictly.

The basic relation between the components is client-server. The computation engine acts as a "data server", and the user interface is a client of the services it provides. In the case of TkHolWorkbench the computation engine roughly corresponds to the HOL system itself, and the user interface to all those components written in the Tcl language.

When such a division is made, the question naturally arises as to exactly what goes where. An obvious test of separation is whether it is feasible for the computation engine to be run in an environment where the windowing system is not present. Another good rule is that all non-interface related functionality used by interface components should lie in the HOL system, preferably in HOL libraries. These libraries should be available for all to use, including other interfaces.

A common problem is that the computation engine ends up being "incomplete" without the interface being present. The approach used in TkHolWorkbench is to treat the user interface as little more than a fancy method of performing transitions in the underlying system. Naturally, the interaction techniques used may themselves be complex, but the end result of an action needs to be series of well defined transition in the underlying computation engine. The end result of a TkHolWorkbench action must be ML code which can be executed without the interface being present.

## 5.3 Modules in the User Interface

Modularity can be difficult to achieve in user interfaces. The basic tension is often between modularity and interoperability.

In the GUI setting, interface components will usually be widgets. Some of the basic principles of designing widgets are:

- Each widget must be given a well defined, unambiguous purpose.
- Widgets which visualise data that changes should not store their own data. Instead they should "reflect" data that comes from an external data source. Often this data will be stored in the computation engine.
- Widgets should be configurable. The Tk toolkit is a good set of examples of configurable widgets.
- Widgets should interoperate by standard techniques where possible. These include clipboards, selections, drag-and-drop and modality. This will reduce the need for dependencies between widgets.

By applying these rules carefully, the implementation of TkHolWorkbench has turned out to be highly modular. This has resulted in a set of reusable, configurable widgets from which to build other tools.

## 5.4 Modules and Dynamic Loading

The architecture of TkHolWorkbench allows modules containing new tools and widget types to be *dynamically loaded*. This is achieved primarily by the autoloading facilities available in Tcl. The benefits of dynamic loading are enormous. It makes rapid development far easier. Also, it removes a considerable obstacle to achieving extensibility - the need to recompile or reconfigure the system to include a new module. Extensibility is covered further in section 6.

## 5.5 Anonymous Notification of State Change

One of the principles of widget construction listed in section 5.3 was that widgets should rarely store their own data but should instead "reflect" data from an external source. This, however, begs the following question: how do widgets know when the data they are representing has changed? An example of this problem is the case of the theory tree display in the TkTheoryViewer tool. An interface should update this display when the theory tree changes. Note that a change in the data may be induced by some external source, for instance from an interaction at the command line prompt.

The solution to this problem lies in expanding the range of services provided by the computation engine, i.e. HOL, to include *notification*. In particular:

- The computation engine must include facilities to notify clients, i.e. interface components, when changes occur in the state of the engine. Where possible, *all* items of state which may be dynamically changed should be traceable in this fashion.
- This notification must be anonymous, in the sense that the computation engine must allow runtime, rather than compile-time, binding of clients. This is necessary if the integrity of the computation engine is to be maintained.
- More than one client must be able to register its interest in each data item.

Thus, to solve the problem of maintaining an up-to-date representation of the theory graph, TkHolWorkbench uses modified versions of several side-effecting functions in the HOL system. These functions support the registration of notification clients. Ideally this kind of functionality should be built into the core of HOL.

## 6 Extending TkHolWorkbench

### 6.1 Packages and User Extensions to TkHolWorkbench

The last section described how TkHolWorkbench can be thought of as a set of essentially independent modules (or objects). In TkHolWorkbench jargon, the code implementing a set of object classes is called a "package". Packages usually contain several files of Tcl code. Packages are very similar to HOL libraries, and indeed it is conceivable that HOL libraries can act as packages. Often a package will implement a new tool, just as a HOL library will contain a theory.

In the implementation of TkHolWorkbench, each component is simply one such package. The program tkhol is just an invocation of a small Tcl program which runs several packages together.

Extension is a simple process. Additional packages may be specified from the command line:

```
> tkholhelper -package ~/mytool -win MyTool
```

The -package argument specifies a directory which contains the package, and the -win argument specifies that a toplevel window of the given class should be created at startup. In Tcl, this means a procedure "MyTool" which creates a toplevel instance of the tool should be defined somewhere in the package. All the functionality of MyTool will have been implemented somewhere in the package mytool.

It is also possible to load packages dynamically from a menu within the system, or from the HOL command line. In future versions it is hoped that the interface will be able to detect and automatically load packages provided in HOL libraries.

The author does *all* development of new tools as package extensions. Like HOL libraries, packages can come with documentation and online help and so can provide facilities upon which other tools can be built.

## 6.2 Adding a Virtual Theory Controller

To test this functionality, a small extension was added to TkHolWorkbench by the author and Paul Curzon working together. The idea was to write a small interface to some "virtual theory" functionality being developed by Curzon [2]. TkHolWorkbench comes with an example "skeleton" package, which was used as a starting point. Within two hours, an interface had been developed from existing widgets which controlled the current "virtual theory" being accessed. The ease with which this package was created and integrated illustrates the potential that an extensible architecture holds.

## 7 Summary

This paper has presented the motivations, objectives and current status of TkHol-Workbench. A rigorous evaluation of TkHolWorkbench has not yet been performed. However it is the author's belief that it is well on the way to meeting the design objectives and usability targets outlined in section 3. It is hoped that given time it will become an accepted part of a standard HOL environment under X Windows.[2] Already it is a useful tool, and future development should only

---

[2] Note that Tk/Tcl is also supported on MS Windows platforms, which should soon be able to support HOL.

improve the situation. Importantly, the architecture has been both practical and extensible, and facilitates rather than hinders development.

It would seem that the user activity within HOL is surprisingly complicated, incorporating, as it does, arbitrary programmability. TkHolWorkbench does not yet aim to support all HOL user activities. This may become a significant issue in future development. The HOL community is increasingly paying more attention to the larger issues of theory development, and interface issues are very relevant to this area.

Future work on TkHolWorkbench will concentrate largely on using the extensibility of the product to develop interface support for language embeddings in HOL. A further test of the TkHolWorkbench architecture will come with some cooperative work on a Window Inference package being planned with Thomas Långbacka et al. at Åbo Akademi University [12].

# References

1. Richard Boulton. *The HOL trs library - A Theorem Retrieval System*. From the HOL88 system distribution.
2. Paul Curzon. Virtual theories. Submitted to the 8th International HOL Workshop on Higher Order Logic and Its Applications.
3. Jim Grundy. *The HOL window library*. From the HOL88 system distribution.
4. Sara Kalvala. Developing an interface for HOL. *Proceedings of the 1991 HOL Workshop*, 1991.
5. Don Libes. *Exploring Expect*. O'Reilly & Associates, January 1995.
6. Tom Melham. Automating recursive type definitions in higher-order logic. In G. Birtwhistle and P.A. Subrahmanyam, editors, *Current Trends in Hardware Verification and Automated Theorem Proving*, pages 341–386. Springer-Verlag, 1989.
7. William Newman and Mik Lamming. *Interactive Systems Design*. Addison-Wesley, December 1995.
8. John Ousterhout. *Tcl and the Tk Toolkit*. Addison-Wesley, April 1994.
9. Tom Schubert and John Biggs. A tree-based, graphical interface for large proof development. *Supplementary Proceedings of the 1994 HOL Workshop*, 1994.
10. Konrad Slind. An implementation of higher order logic. *Master's Thesis*, January 1991. Research Report No. 91/419/03, Department of Computer Science, University of Calgary.
11. Laurent Thery. Real theorem provers deserve real interfaces. In *Software Engineering Notes*, volume 17. ACM Press, 1992.
12. Rimvydas Rukšėnas Thomas Långbacka and Joakim von Wright. TkWinHOL - a tool for doing window inference in HOL. Submitted to the 8th International HOL Workshop on Higher Order Logic and Its Applications.

# Very Efficient Conversions

Morten Welinder*

Carnegie Mellon University
School of Computer Science
5000 Forbes Avenue, Pittsburgh, PA-15213, USA
Email: Morten.Welinder@cs.cmu.edu

**Abstract.** Using program transformation techniques from the field of partial evaluation an automatic tool for generating very efficient conversions from equality-stating theorems has been implemented.

In the situation where a HOL user would normally employ the built-in function `GEN_REWRITE_CONV`, a function that directly produces a conversion of the desired functionality, this article demonstrates how producing the conversion in the form of a program text instead of as a closure can lead to significant speed-ups.

The HOL system uses a set of 31 simplifying equations on a very large number of intermediate terms derived, e.g., during backwards proofs. For this set the conversion generated by the two-step method is about twice as fast as the method currently used. When installing the new conversion, tests show that the *overall* running times of HOL proofs are reduced by about 10%. Apart from the speed-up this is completely invisible to the user. With cooperation from the user further speed-up is possible.

## 1   Introduction

A *conversion* in HOL [GM93] is a function that takes as its argument a term, called the *object term,* and produces a theorem, presumably stating that the object term equals some other term. Conversions span from the trivial `REFL` simply expressing that any term equals itself to functions that repeatedly traverse a term doing rewriting on subterms.

In many applications of HOL, for instance during backwards proofs, it seems natural to use conversions to simplify terms because these simplifications can be very tedious to do by hand. By using simplifying conversions the user can focus on the problematic parts of the proofs.

Given one conversion, say beta reduction, the HOL system provides the user with a number of conversion augmenting functions. For example there is `REPEATC` which from beta conversion will produce a conversion that will do repeated beta conversion at a term's top level, and there is `REDEPTH_CONV` that will produce a conversion which will perform exhaustive beta conversion in a term and its

---

* On leave from Department of Computer Science, University of Copenhagen, Denmark. Partially supported by The Danish Research Academy.

subterms. Since these augmenting functions, thoroughly described by Paulson in [Pau83], are both elegantly and quite efficiently implemented in the current HOL system this article will focus on conversions that do whatever they do once at the top of a term. The implementation relies on ideas by Roger Fleming and Richard Boulton, see [Bou94, Sli95].

Several versions of HOL exist and the methods presented here should work with any of them, and even with some other theorem provers. The presentation and the implementation, however, are based on a single version namely HOL-90.7. Since the efficiency of HOL-functions varies from implementation to implementation, arguments about the relative efficiency of methods should be understood in the HOL-90.7 setting. They are, however, believed to hold for all HOL implementations.

## 1.1 Overview

We will start with describing the particular class of conversions that we will deal with in this paper. This is done in Section 2 and is followed by a short description of the method that HOL currently uses for generating such conversions.

In Section 3 we will then describe how a variant of the HOL system's current one-step method can be turned into a two-step method. Pros and cons of doing so will be discussed.

Then, in Section 3.3, we will describe how the HOL system can benefit from conversions generated by the two-step method. We will also show how a cooperating user can gain even better results.

Finally we will draw our conclusions in Section 4 and sketch directions for future work.

## 2 Conversions from Equalities

An important subset of all conversions is the set of *rewriting conversions*. Such a conversion is based on one or more equality-stating theorems, possibly universally quantified, and works by matching the object term against the left-hand sides of the theorems until a match is found. When a match is found, the free (or universally bound) variables of the theorem are instantiated according to the match yielding the desired result. If none of the given theorems can be matched an exception is raised. As we are looking for instantiations (i.e., substitutions of terms for free variables) of the theorems' left-hand sides, we will call these left-hand sides *patterns*.

It is important to realize that the result of applying a rewriting conversion to a term is *not* just the simplified term, but a theorem evincing the validity of the rewriting.

Consider, for example, the following two theorems describing addition on Peano numbers:

$$\vdash \forall n.0 + n = n \tag{1}$$

$$\vdash \forall m.\forall n.\ \text{SUC}\ m + n = \text{SUC}(m + n) \tag{2}$$

(As theorems often come in groups we will from now on sometimes present them as conjunctions. This should be considered syntactic sugar and a pre-processor will eliminate it. We will still count each equality as a theorem.) The rewriting conversion based of the two addition theorems will exhibit the following behaviour:

$$0 + \text{SUC}(\text{SUC}(0 + 0)) \Rightarrow \vdash 0 + \text{SUC}(\text{SUC}(0 + 0)) = \text{SUC}(\text{SUC}(0 + 0)) \tag{3}$$

$$\text{SUC}(\text{SUC}\,0) + \text{SUC}\,0 \Rightarrow \vdash \text{SUC}(\text{SUC}\,0) + \text{SUC}\,0 = \text{SUC}(\text{SUC}\,0 + \text{SUC}\,0) \tag{4}$$

$$\text{SUC}(\text{SUC}(0 + 0)) \Rightarrow \text{HOL\_ERR}\ (\text{whatever}) \tag{5}$$

The number of theorems used to create conversions in this manner can be quite large. For example, the HOL system employs a basic set of 31 theorems for general simplification. These are listed below to give the reader an idea of what kind of theorems to expect.

$$\forall x.(x = x) = T \tag{6}$$

$$\forall t.((T = t) = t) \land ((t = T) = t) \land ((F = t) = \neg t) \land ((t = F) = \neg t) \tag{7}$$

$$(\forall t.\neg\neg t = t) \land (\neg T = F) \land (\neg F = T) \tag{8}$$

$$\forall t.(T \land t = t) \land (t \land T = t) \land (F \land t = F) \land (t \land F = F) \land (t \land t = t) \tag{9}$$

$$\forall t.(T \lor t = T) \land (t \lor T = T) \land (F \lor t = t) \land (t \lor F = t) \land (t \lor t = t) \tag{10}$$

$$\forall t.(T \Rightarrow t = t) \land (t \Rightarrow T = T) \land (F \Rightarrow t = T) \land (t \Rightarrow t = T) \land (t \Rightarrow F = \neg t) \tag{11}$$

$$\forall t_1.\forall t_2.((T \to t_1 \mid t_2) = t_1) \land ((F \to t_1 \mid t_2) = t_2) \tag{12}$$

$$\forall t.(\forall x.t) = t \tag{13}$$

$$\forall t.(\exists x.t) = t \tag{14}$$

$$\forall t_2.\forall t_1.(\lambda x.t_1)t_2 = t_1 \tag{15}$$

$$\forall x.(\text{FST}\,x, \text{SND}\,x) = x \tag{16}$$

$$\forall x.\forall y.\,\text{FST}(x, y) = x \tag{17}$$

$$\forall x.\forall y.\,\text{SND}(x, y) = y \tag{18}$$

Note that more than one rule may apply: the term $T = F$ may be rewritten into either $F$ or $\neg T$ by the first and fourth conjunct of (7). The "correct" behaviour for a rewriting conversion is unspecified in this situation and proofs shouldn't depend on which choice is made. As we will see later, however, they occasionally do.

The list suggests that theorems with lambda-abstractions are rare; in fact only (15) contains one. This is supported by the fact that the most common additions to the above list are definitions of functions like (1) and (2) above. Actually, the last three theorems above represent such an addition that is done during boot-strapping. We will assume that abstractions are indeed rare and consider ourselves lucky because matching abstractions with their freedom in naming could be expensive. Thus relatively little effort will go into optimizing matching of abstractions.

## 2.1 Direct Conversion Generation

The naïve way to turn a list of theorems into a rewriting conversion would be to simply try each theorem in turn:

```
fun naive_make_conv thl = FIRST_CONV (map REWR_CONV thl)
```

Recall that REWR_CONV is a primitive that will turn one equality theorem into a conversion, and that FIRST_CONV composes a list of conversions by trying them in turn until one succeeds.

Obviously conversions generated by the naïve method get slow as the number of theorems grows. Handling 31 theorems would be painfully slow. Therefore HOL uses a reduction method based on what we will call the *face* of a term:

**Definition 1.** By the face of a term we mean: (i) for a constant, the pair of symbol CONST and the constant's name; (ii) for a variable, the symbol VAR; (iii) for a combination, the symbol COMB; (iv) for an abstraction, the symbol ABS.

The face of a given term expresses what we can learn about the term from just looking at its top node. Computing a term's face is inexpensive. When a face is used for pattern matching we can use the HOL primitive dest_term for computing the face. It returns a little more than the face, but as we will see that is an advantage.

Observe now, that a term can only match a given pattern (the left-hand side of a theorem's conclusion) when they have the same face or when the face of the pattern is VAR; if the faces are different and if the pattern is not a variable then no instantiation of the pattern will equal the object term. This important property is what the term net module in HOL uses to make efficient conversions. Term nets are a variation of the discrimination net technique used in artificial intelligence, see [CRMM87, Chapters 8 and 11]. The following is a simplified description of how the matching in the term net module goes on.

In order to produce an efficient conversion from a list of theorems we start out by dividing the theorems into groups with respect to the patterns' faces. As variables match everything we will then add the theorems with VAR-face to the other groups. The conversion we are looking for will then examine the face of the object term and pick the right group. In case of the COMB and ABS groups matching can then recursively take place on subterms, but with the reduced list of theorems. At the leaves of this matching tree we are left with a (hopefully) small set of possible theorems with which we use the naïve method above.

Note that just because a theorem is present at some leaf there is no guarantee that it actually does match; the above method is only a quick-and-dirty method for reducing a problem size. There are several things that might go wrong with the actual matching: (1) If a pattern is not linear (i.e., some free variable occurs more than once) the assumption that a variable at any point matches anything does not hold. (2) The types may fail to match. (3) If the pattern contains an abstraction there is no keeping track of bound variables.

All the grouping of the theorems might at a first glance look very expensive, but since it is completely independent of the object term it can be done once and for all. This leaves a quite efficient conversion that quickly selects a small number of possible theorems — typically none or just one — then tries those in turn.

## 3 Staging the Matching Process

Given the observations that (1) a lot of the work in the matching process only depends on the list of theorems, and (2) the list of theorems is known well in advance, and (3) the conversion will be used a lot of times, it seems natural to ask whether we can produce an optimized version for one particular list of theorems. More specifically, can we from the program (text, not closure),

$$\text{\underline{makeconv}} : \underline{\text{thmlist}} \to \underline{\text{term}} \to \underline{\text{thm}},$$

(We use the underlining to stress that we are talking about the program text of a function that when considered an SML-program has the non-underlined type.) and the value for its first parameter,

$$\text{thl} : \text{thmlist},$$

(the actual value or a suitable encoding of it) produce a new program (text again),

$$\underline{\text{conv}} : \underline{\text{term}} \to \underline{\text{thm}},$$

where the theorems are "frozen" in the new program? The answer is "Yes," automatic program transformators of this kind do exist and are called partial evaluators [JGS93, BW93, BW94]. Unfortunately, for various technical reasons to be elaborated on in a separate paper, the resulting programs from using currently implemented partial evaluators are not satisfactorily efficient. See also [Wel94].

Having failed the automatic approach we turn to a related technique: handwriting a program that given a list of theorems will produce the wanted conversion as a program text:

$$\text{convgen} : \text{thmlist} \to \underline{\text{term}} \to \underline{\text{thm}}.$$

Note that the types of makeconv and convgen differ only in the underlining. Using terminology from the field of partial evaluation, convgen is called a generating extension of makeconv. We say that evaluation using convgen is *two-stage* because we have to invoke the SML-interpreter/compiler twice: once on convgen itself and once on the generated conversion.

Writing convgen directly is not nearly as elegant as the automatic approach, if for no other reason then because we could produce something functionally equivalent to convgen automatically from makeconv by evaluating the expression (mix mix makeconv) where mix is the partial evaluator. (That was indeed the application of mix to its own program text.) This variant would unfortunately produce conversions suffering from the same lack of efficiency as observed above.

## 3.1 Staging Versus Non-Staging

Since the two-stage method is somewhat more complicated than the one-step there must be some benefits to make it worth while. And in fact there are:

**Speed.** The two-stage conversions are about twice as fast, see the timings below. Several factors contribute to the increase in speed: (1) The actual compiled code in the one-stage conversions is something that uses the object term to work through a term net, i.e., it sort-of interprets the term net; in contrast the same thing is accomplished by control flow in the two-stage version. (2) When the matching is over, the one-stage conversion always uses REWR_CONV which redoes the matching while the two-stage version usually uses SPEC which is [implemented by] a relatively cheap beta conversion. (3) Evaluation involves considerably more closures in the one-stage version.

**Memory.** Conversions produced by the two-stage method ought to produce less garbage on the heap. "Ought" here means that I have been unable to verify that experimentally, possibly because I have been unable to get sufficiently stable readings of garbage collection times.

There are unfortunately also some drawbacks:

**Time to construct.** Since the compiler has to be re-invoked for the two-stage method, construction takes longer. Obviously this means that the two-stage method should only be used when the conversions are expected to be used often.

**Extensibility.** Conversions produced by the one-step method (actually the term nets behind) are easily and efficiently extended with extra theorems. This is what takes place when, e.g., REWRITE_TAC is used. This is not directly possible with the two-step approach.

**Code size.** The code size of a two-step conversion is approximately five lines per theorem plus a minor overhead. I fail to see how this could become a problem in real-life use, even if hundreds of theorems were used.

Note, that all these advantages and disadvantages are special cases of the general problematics between using interpreters and compilers.

## 3.2 Generating Conversions

This section describes how to construct a program that generates the conversions we are looking for, i.e., how to construct a conversion generator. The method described here has actually been implemented in the form of The MW Conversion Generator, currently version 0,21. It is available from the author.

The job is, given a list of patterns, to write a function that matches a term against the patterns. Since we will be dealing with term combinations which have more than one (namely two) subterm each which must all match, we generalize the situation to matching a list of terms against a list of pattern lists. The terms here cannot be real terms (i.e., of type Term) since the terms are not available at the time we are generating the conversion; instead they are names of SML variables, e.g. `"tm122"`, of variables that in the resulting conversion will hold the terms in question.

At a given point in the matching we therefore have a list of options each consisting of a list of termvar-pattern pairs, the theorem we are matching against, and some extra information to be discussed later. For ease of presentation only the termvar-pattern pairs are considered in the following and the list of options therefore has SML-type `(string * term) list list`. The code to be generated for a given list of options can now be calculated as in the following where `typewriter` style is used for the generated code and plain roman style is used for calculations performed when generating.

$\text{code}([\,]) = $ **raise bad**
$\text{code}([\,] :: \_) = $ *(* Match found, see below. *)*
$\text{code}(\text{opts}) = $
  let
    $(\text{var}, \text{tm}) = \text{hd}\,(\text{hd}\,\text{opts})$
    $(\text{var}_1, \text{var}_2) = (\text{var} \wedge \text{``1''}, \text{var} \wedge \text{``2''})$
    $\text{rest}\,f = \text{map}\,\text{tl}\,(\text{filter}\,(\text{hasface}\,f)\,\text{opts})$
  in
    `(case dest_term var of`
     `CONST` $c_1$ `=>` $\text{code}\,(\text{rest}\,(\text{CONST}\,c_1))$
     `| ...`
     `| CONST` $c_n$ `=>` $\text{code}\,(\text{rest}\,(\text{CONST}\,c_n))$
     `| ABS {Body=`$\text{var}_1$`,...} =>` $\text{code}\,((\text{var}_1, \text{body}\,\text{tm}) :: \text{rest}\,\text{ABS})$
     `| COMB {Rator=`$\text{var}_1$`,Rand=`$\text{var}_2$`} =>`
       $\text{code}\,((\text{var}_1, \text{rator}\,\text{tm}) :: (\text{var}_2, \text{rand}\,\text{tm}) :: \text{rest}\,\text{COMB})$
     `| _ =>` **raise bad**
    `) handle _ =>` $\text{code}\,(\text{rest}\,\text{VAR})$
  end

The appearance of an expression like "var" in "`dest_term` var" is to be understood as that the result of evaluating that expression while generating the conversion — the result will be a string — is to be inserted in the generated program at the place where the expression occurs.

In the branches of the generated case expression only those options with matching face need to be considered. If there are no matching options, the corresponding branch can be eliminated. The exception handling is for variables which at this point are assumed to match anything that the other patterns did not match. If there are no patterns with variable face, then the exception handling construct can be eliminated.

Since the patterns are finite the above function will produce a finite tree of nested case expressions. At the leaves we are left with a typically small number of theorems and we can simply return `(REWR_CONV thm`$_1$`) ORELSEC ... ORELSEC (REWR_CONV thm`$_n$`)` applied to the object term. This can be seen as an unfolded version of the naïve method.

There are a few optimizations apart from the obvious above that are needed in order to get really good performance. First of all, using REWR_CONV means that the matching is redone, so we will want to avoid that. This definition tells of when that is possible:

**Definition 2.** A theorem is said to be SPEC-safe when (1) it has no free variables, i.e., all variables have been universally quantified over; (2) the pattern is linear, i.e., all variables occur exactly once; (3) the pattern contains no abstractions; and (4) the variables are all monomorphic.[2]

When a theorem is SPEC-safe it means that all the assumptions we have made, i.e., that variables match everything and that the names of $\lambda$-bound variables need not be checked, are indeed valid. We can therefore *guarantee* that the object term matches the pattern. Furthermore, since we have already done the matching of variables against terms, we can use (SPEC $tm_a$ (... (SPEC $tm_z$ thm)...) to get the resulting theorem. For this to work, the conversion generator must keep track of what variables are matched against.

The second optimization worth making is that when all theorems handled by the case expression are SPEC-safe then it is often possible to replace the underscore branch by the code for the variable case. This eliminates the expensive exception handling.

One final optimization worth making is to install a variant of dest_term in the term module. The current version is quite expensive for abstractions because of the deBruijn representation used internally. A version that doesn't do variable renaming is just as good for us since that does not change the faces of subterms. The term net module uses exactly the same trick.

As an example, consider again the theorems (1) and (2) for addition on Peano numbers. The complete conversion produced for these two theorems is shown in appendix A. The essential matching part is this piece of code

```
val conv:conv = fn tm =>
 (case dest_term tm of
   COMB {Rator=tm1,Rand=tm2} =>
   (case dest_term tm1 of
    COMB {Rator=tm11,Rand=tm12} =>
    (case dest_term tm11 of
      CONST {Name="+",...} =>
      (case dest_term tm12 of
       CONST {Name="0",...} => SPEC tm2 thm_1
       | COMB {Rator=tm121,Rand=tm122} =>
        (case dest_term tm121 of
          CONST {Name="SUC",...} => SPEC tm122 (SPEC tm2 thm_2_1)
          | _ => raise bad)
       | _ => raise bad)
      | _ => raise bad)
    | _ => raise bad)
   | _ => raise bad)
```

---

[2] This can be relaxed to either monomorphic or just a type variable with the restriction that type variables must then be different. This change would require that INST_TYPE be used to instantiate the types of the theorem. This is easy, but has not been implemented.

This is a very efficient combined matching and destruction of the object term. Furthermore, when matching succeeds, the beta-conversion-like SPEC is used to produce the wanted equality theorem. Except for the use of dest_term instead of the older term interface is_... this code is what a user would write by hand.

## 3.3 Integration with Hol

To test the conversions in practice we wish to replace the standard 31-theorem rewriting conversion by one produced by the two-step method. It turns out that it is quite easy to integrate such a conversion with HOL: we have to produce two conversions, one without and one with the pair theorems (16), (17), and (18). The former is only needed during boot-strapping until pairs are defined. We need to add 40 lines and change another 20 lines in the Rewrite functor and signature, change 5 lines in the boot-strapping scripts, and optionally add 10 lines in the Term functor and signature (to install a faster dest_term function). It seems reasonable to characterize this as minor changes.

There is one tricky point to the integration. As mentioned above the one-step conversions are extensible and the two-step conversions are not. We can solve this by noting that conversions can be composed sequentially with ORELSEC and that we can use the old machinery to handle the extra *ad hoc theorems*. In the presence of $n$ ad hoc theorems, the following method was used for tests. $n = 0$: just the two-stage conversion; $n = 1, \ldots, 4$: two-stage followed by naïve conversion; $n = 5, \ldots, 10$: two-stage followed by term net conversion; $n > 10$: revert completely to term net conversion. Exactly when to use which method is the result of hunches and a little testing, not deep theoretical insight or elaborate benchmarking.

Testing shows that some of the library proofs depend on the conversions to choose a certain one of several possible theorems to rewrite by. Consider the object term $(x_1 + x_2) * (x_3 + x_4)$ with the theorems $\vdash x * (y + z) = x * y + x * z$ and $\vdash (x + y) * z = x * z + y * z$ for integer or real numbers. The term net based conversions will always rewrite the term with respect to the latter theorem regardless of whether it is mentioned before or after the former theorem in the list of theorems. The proof of, e.g., theorem POW_PLUS1 from the real number library depends on this. I consider such a proof just as broken as one that depends on the order of assumptions. Therefore it is not a problem that the two-step method chooses a different theorem. Fixing the proof is trivial.

## 3.4 Some Timing Results

For reference, the timings below have been run on a Sun 4/75 (also know as a Sparc-Station 2) with 64MB Ram running little but the SunOS kernel and HOL-90. In particular this means that no swapping took place during the tests. The timings were done with SML–NJ version 0.93's built-in timers.

There seems to be no standard reference proof used for timings, so for the purpose of this article a part of the HOL library will be used for timing. More specifically, the code in `library/real/theories/src/real.sml`, a part of HOL-90.7's real number library, was used. This is approximately a two minute proof.

To test the stand-alone speed of a generated conversion the 31 theorems from the basic HOL conversion were used. From the test proof, the first 10000 terms to which the basic conversion is applied were used. Of these only 284 can actually be rewritten. These rewritable terms were duplicated to form another 10000 terms.

| Terms | One-stage | Two-stage | Reduction |
|---|---|---|---|
| 10000 first | 5.6 | 2.5 | 55% |
| 10000 rewritable | 23.4 | 12.9 | 45% |

Since obviously the proof process does other things apart from rewriting, the entire proof (i.e., `real.sml`) was also timed. This gives an estimate of the overall reduction in the running times for proofs.

| Proof | One-stage | Two-stage | Reduction |
|---|---|---|---|
| `real.sml` | 134.8 | 120.0 | 11% |

We consider a speed-up of over ten percept to be satisfactorily since the only user intervention needed was fixing the proof of `POW_PLUS1` as discussed previously.

## 3.5 Cooperation

When the internal rewriting done by the HOL system is changed to use a two-stage conversion, the ad hoc theorems used for rewriting, i.e., those passed to for example `REWRITE_TAC` become relatively expensive because they have to be handled by the old method.

A cooperating user will therefore have to reduce the number of ad hoc theorems used for rewriting in order to get maximal benefit. Since rewritings are a very convenient tool during proof this is somewhat awkward. However, there is one class of rewriting that comes up repeatedly, namely rewriting with respect to some function's definition. The theorems defining addition on Peano numbers, (1) and (2), serve again as an example. In a large theory it would be beneficial to create a two-step conversion for such a definition and to use `CONV_TAC` instead of using `REWRITE_TAC` with the defining equations as ad hoc theorems.

This method is not feasible for existing theories since it would require major changes in the proofs. It might, however, be very worth while for new theories.

## 4 Conclusions and Future Work

It has been shown in this article that evaluation of a function in some situations can be staged with significant speed-up as result. This observation has led to the construction of a fully automatic tool that generates conversions from equality-theorems. Since such conversions are used by the HOL core a general speed-up of HOL of about ten percent was possible.

The principle of staging can be used in other situations, for example the library **generator** seems promising with its interpretive style and extended running times. (This library helps with the embedding of programming languages' syntax and semantics in HOL, see [RK94].) It would be interesting to see this hypothesis proven in practice, if true.

The present work could be improved by a closer study of the garbage collection aspects involved. This is pending, awaiting the availability of a reliable method of measurement.

The work described in this paper has stayed away from combining the top-level rewriting with depth conversions like TOP_DEPTH_CONV. Doing so would mean giving up some of the modularity, but the potential gain is to avoid many fruitless attempted rewritings since a term resulting from one rewriting have a well-known structure.

# Acknowledgements

The author wishes to thank Thomas F. Melham for suggesting the conversion generation as a target for partial evaluation techniques and for help with an earlier version of the conversion generator. Further thanks goes to Jesper Jørgensen for valuable discussions of some of the partial evaluation aspects involved.

# References

[Bou94]     Richard J. Boulton. *Efficiency in a Fully-Expansive Theorem Prover*. PhD thesis, University of Cambridge Computer Laboratory, New Museums Site, Pembroke Street, Cambridge CB2 3QG, United Kingdom, May 1994.

[BW93]      Lars Birkedal and Morten Welinder. Partial evaluation of Standard ML. Technical Report 93/22, DIKU, Department of Computer Science, University of Copenhagen, October 1993.

[BW94]      Lars Birkedal and Morten Welinder. Handwriting program generator generators. In Manuel Hermenegildo and Jaan Penjam, editors, *Programming Language Implementation and Logic Programming. 6th International Symposium, PLILP '94, Madrid, Spain*, volume 844 of *Lecture Notes in Computer Science*, pages 198–214. Springer Verlag, September 1994.

[CRMM87] Eugene Charniak, Christopher K. Riebeck, Drew V. McDermott, and James R. Meehan. *Artificial intelligence programming*. Lawrence Erlbaum Associates, 2nd edition, 1987.

[GM93]      Michael J. C. Gordon and Thomas F. Melham. *Introduction to HOL*. Cambridge University Press, 1993.

[JGS93]     Neil D. Jones, Carsten K. Gomard, and Peter Sestoft. *Partial Evaluation and Automatic Program Generation*. Prentice-Hall, 1993.

[Pau83]     Lawrence Paulson. A higher-order implementation of rewriting. *Science of Computer Programming*, 3:119–149, 1983.

[RK94]      Ralf Reetz and Th. Kropf. Simplifying deep embedding: A formalised code generator. In Thomas F. Melham and Juanito Camilleri, editors, *Proc. of the 7th International Workshop on Higher Order Logic Theorem Proving and its Applications*, volume 859 of *Lecture Notes in Computer Science*, pages 378–390. Springer Verlag, September 1994.

[Sli95]    Konrad Slind. Hol-90.7, 1995. An implementation of the Higher-Order Logic Theorem Prover in Srandard ML.

[Wel94]    Morten Welinder. Towards efficient conversions by use of partial evaluation. In Thomas F. Melham and Juanito Camilleri, editors, *Supplementary Proceedings of the 7th International Workshop on Higher Order Logic Theorem Proving and its Applications*, September 1994.

# A   Peano Addition Conversion

```
(* ---------------------------------------------------------------- *)
(* The following code is machine generated. For this reason it is not *)
(* wise to edit it by hand. The main function is a conversion that  *)
(* rewrites left-to-right with respect to one of the theorems below. *)
(*                                                                   *)
(* The code was generated by MW Conversion Generator version 0,21.  *)

(* Make sure the conversion is compiled properly: *)
val save_System_Control_interp = !System.Control.interp;
val _ = System.Control.interp := false;

local
 open Term Thm Conv Drule Dsyntax

 (* This or another HOL_ERR exception will be raised if none of
    the theorems matches: *)
 val bad = HOL_ERR {message = "No matching theorem",
                     origin_function = "conv",
                     origin_structure = "top level"};

 (* Utility functions: *)
 (* We check that the theorems are the right ones: *)
 fun check thm tm =
   if concl thm=tm then
     thm
   else
     raise HOL_ERR {message = "Environment changed",
                    origin_function = "",
                    origin_structure = "top level"};
 val thm_1 = check (ADD1) (--'!n. 0 + n = n'--)
 val thm_2 = check (ADD2) (--'!m n. SUC m + n = SUC (m + n)'--)
in
 val conv:conv = fn tm =>
  (case dest_term' tm of
     COMB {Rator=tm1,Rand=tm2} =>
     (case dest_term' tm1 of
        COMB {Rator=tm11,Rand=tm12} =>
        (case dest_term' tm11 of
           CONST {Name="+",...} =>
           (case dest_term' tm12 of
              CONST {Name="0",...} => SPEC tm2 thm_1
            | COMB {Rator=tm121,Rand=tm122} =>
              (case dest_term' tm121 of
                 CONST {Name="SUC",...} => SPEC tm2 (SPEC tm122 thm_2)
               | _ => raise bad)
            | _ => raise bad)
         | _ => raise bad)
      | _ => raise bad)
   | _ => raise bad)
end;
val _ = System.Control.interp := save_System_Control_interp;

(* Machine generated code ends here.                              *)
(* ---------------------------------------------------------------- *)
```

# Recording and Checking HOL Proofs

*Wai Wong**

Department of Computing Studies
Hong Kong Baptist University
Kowloon Tong, Hong Kong

### Abstract

Formal proofs generated by mechanised theorem proving systems may consist of a large number of inferences. As these theorem proving systems are usually very complex, it is extremely difficult if not impossible to formally verify them. This calls for an independent means of ensuring the consistency of mechanically generated proofs. This paper describes a method of recording HOL proofs in terms of a sequence of applications of inference rules. The recorded proofs can then be checked by an independent proof checker. Also described in this paper is an efficient proof checker which is able to check a practical proof consisting of thousands of inference steps.

## 1 Introduction

Formal methods have been used in the development of many safety-critical systems in the form of formal specification and formal proof of correctness. Formal proofs are usually carried out using *theorem provers* or *proof assistants*. These systems are based on well-founded formal logic, and provide a programming environment for the user to discover, construct and perform proofs. The result of this process is usually a set of theorems which can be stored in a disk file and used in subsequent proofs. HOL is one of the most popular theorem proving environments. The users interact with the system by writing and evaluating ML programs. They instruct the system how to perform proofs. A proof is a sequence of inferences. In the HOL system, it is transient in the sense that there is no object that exists as a proof once a theorem has been derived.

In some safety-critical applications, computer systems are used to implement some of the highest risk category functions. The design of such a system is often formally verified. The verification usually produces a large proof consisting of tens of thousands, even up to several millions, of inferences. [Won93a] describes a proof of correctness of an ALU consisting of a quarter of a million inference steps. In such situations, it is desirable to check the consistency of the sequence

---

*The work described in this paper was carried out by the author while he was in the University of Cambridge Computer Laboratory supported by a grant from SERC (No. GR/G223654)

of inferences with an independent checker. The reasons for requiring independent checking are:

- the mechanically generated formal proofs are usually very long;

- the theorem proving systems are usually very complex so that it is extremely difficult (if it is not impossible) to verify their correctness;

- the programs that a user develops while doing the proof are very often too complicated and do not have a simple mapping to the sequence of inferences performed by the system.

An independent proof checker can be much simpler than the theorem prover so that it is possible to be verified formally. The U. K. Defence standard 00 – 55 calls for such an independent proof checker when the 'highest degree of assurance in the design' is required [oD91].

The necessary condition for a HOL proof to be checked by an independent checker is to have the proof expressed as a sequence of inferences. To achieve this, a method of recording HOL proofs has been developed and implemented. This comprises a proof file format, a small modification to the HOL core system and a library of user functions for managing the proof recorder. The approach of adding the proof recording feature to the HOL system is discussed in Section 3. This is followed by a section describing briefly the proof file format and a section on the proof recording library. Section 6 discusses some issues of implementing a proof checker. An efficient proof checker has been implemented and will be described in Section 7 and 8.

## 2   Proofs in HOL

A detailed description of the HOL logic and its proof theory, together with several tutorial examples of using the HOL system can be found in [GM93]. For the benefit of the readers who are not familiar with HOL, an overview of the HOL deductive system and the theorem-proving infrastructure is given in this section.

A *proof* is a finite sequence of *inferences* $\Delta$ in a deductive system. Each inference in $\Delta$ is a pair $(L, (\Gamma, t))$ where $L$ is a (possibly empty) list of sequents $(\Gamma_1, t_1)\ldots(\Gamma_n, t_n)$ and $(\Gamma, t)$ is known as a *sequent*. The first part $\Gamma$ of a sequent is a (possibly empty) set of terms known as the **assumptions**. The second part $t$ is a single term known as the **conclusion**. A particular deductive system is usually specified by a set of schematic *rules of inference* (also known as *primitive inference rules*) written in the following form

$$\frac{\Gamma_1 \vdash t_1 \quad \ldots \quad \Gamma_n \vdash t_n}{\Gamma \vdash t} \tag{1}$$

The sequents above the line are called the *hypotheses* of the rule and the sequent below the line is called its *conclusion*. Each inference step in the sequence of inferences forming a proof must satisfy one of the inference rules of the deductive system. There are eight primitive inference rules in HOL. They are described in detail in Section 16.3.1 of [GM93]. In HOL, rules of inference are implemented by ML functions.

More complex inference can be created by combining the primitive inference rules. For example, the rule of *symmetry of equality*([SYM]) can be specified as

$$\frac{\Gamma \vdash t_1 = t_2}{\Gamma \vdash t_2 = t_1}. \tag{2}$$

This can be derived using the primitive rules as follows:

1.  $\Gamma \vdash t_1 = t_2$                                      [Hypothesis]
2.  $\vdash t_1 = t_1$                                       [Reflexivity]
3.  $\Gamma \vdash t_2 = t_1$                        [Substitution of 1 into 2]

This style of presenting a proof is known as *Hilbert* style. Each line is a single step in the sequence of inferences. The first column is the line number. The middle column is the theorem(s) derived in this step. The right-hand column is known as the *justification* which tells which rule of inference is applied in each step.

Derived rules are also represented by ML functions. They are implemented in terms of the primitive rules. A theorem prover in which all proofs are fully expanded into primitive inferences is known as *fully-expansive*[Bou92]. The advantage of this type of theorem prover is that the soundness of the proof is guaranteed since every primitive inference step is actually performed. However, this is very expensive in terms of both time and space for any sizable proof. To improve the efficiency of HOL, some of the simple and frequently used derived rules, such as SYM, are not fully expanded, but are implemented directly in ML.

These rules, including the primitive rules and derived rules that are implemented directly in ML, will be referred to as *basic inference rules* or simply *basic rules* below. When recording a proof, all inference steps in which a basic inference rule is applied should be included so that any error resulting from bugs in the implementation of the inference rules can be caught.

Simple proofs can be carried out in HOL by calling the inference rules in sequence. However, these inference steps are far too small for any sizable proof. Another more powerful way of carrying out proof, known as *goal-directed* or *tactical* proof, is often used. In this proof style, a term in the same form as the required theorem is set up as a goal, tactics are used to reduce the goal to simpler subgoals recursively until all the subgoals are resolved. In such a proof, the user does not call the inference rules directly. However, a correct sequence of inferences is calculated and performed by the system behind the scenes automatically to derive the theorem.

A proof in HOL as described above is carried out within an environment which consists of a type structure $\Omega$ and a signature under the type structure $\Sigma_\Omega$. The type structure $\Omega$ is a set of type constants, each of which is a pair $(\nu, n)$ where $\nu$ is the name and $n$ is known as the arity. Type constants include both the atomic types and the type operators. For example, the name of the atomic type : *bool* is the string `bool` and its arity is 0, and the name of the type operator *list* is `list` and its arity is 1. The signature $\Sigma_\Omega$ is a set of constants, each of which is a pair $(c, \sigma)$ where c is the name and $\sigma$ is its type and all the type constants that occur in the $\sigma$s must be in $\Omega$. This provides a context against which the well-typedness of terms can be checked.

A formal theory of the HOL proof system has been developed by J. von Wright [vW94]. The notion of types, terms, inferences and proofs are captured in his theory. This provides a formal base for developing a proof checker.

# 3  Recording HOL Proofs

In the HOL systems, there exists no object as a HOL proof once a theorem is derived whatever the proof style used in the derivation. In order to check the consistency of a HOL proof by an independent system, one needs to preserve the proof.

Since the HOL system is a fully-expansive theorem prover, it is possible to record the sequence of inference rules together with the hypotheses and the conclusion in the derivation of a theorem. The recording can be done at the time the system performs each inference. Thus, the proof can be preserved as a sequence of inferences and saved into a disk file. A proof file format has been defined for this purposes. While the complete definition of the proof file format can be found in [Won93b], it is described briefly in Section 4 below.

The approach suggested above requires that the HOL system be modified so as to enable each inference rule to be recorded at the time it is performed. The principle of implementing the recording feature is to make as little change to the core system as possible. Furthermore, the modification to the HOL system should have as little penalty on the system performance as possible especially when the recording feature is not enabled.

The actual modifications to the core system were

- to define a new ML data type to represent the recorded inference rules;

- to modify all basic inference rules to save their names and arguments to an internal list;

- to add a small number of functions to enable/disable the recording and to access the list of saved inferences.

These modifications have been implemented in HOL88 version 2.02 as a small set of low level functions in the core system. The details are described in [Won93b].

In order to use the proof recording feature for practical proofs, a flexible and convenient user interface should also be provided. Such an interface was implemented as a HOL library in HOL88. It is described briefly in Section 5 while the details can be found in [Won94].

A benchmark of the proof recorder was carried out on the correctness proof of a simple multiplier described in [Gor83] which is often used as a benchmark for the HOL system. The results can be found in Section 9.

# 4 Proof File Format

A recorded proof is saved in a disk file in a format known as **prf** format. Proof files in this format are intended primarily for automatic checkers. They follow the Hilbert style of proofs as described in Section 2. It is a linear model which simplifies both the generation and the checking of proofs.

The proof file format **prf** has two levels: the *core* level allows only primitive inference rules in a proof, and the *extended* level allows all basic inference rules.

The syntax of the proof file format **prf** is similar to LISP S-expressions. Objects, such as proof lines, theorems, terms and so on, are enclosed in a pair of matching parentheses. The first atom in an object is a tag indicating what kind of object it is. A file in this format begins with a format expression which identifies the name, version and level of the format it conforms to. This is followed by an environment expression. The environment consists of all the types and constants known in the current theory. The remainder of the file consists of one or more proofs. Each proof expression begins with the **PROOF** tag identifying the expression as a proof. This is followed by the name of the proof and a list of theorems. These theorems are the goals of the proof. A checker checking the proof may stop processing the remaining proof lines after it has found all the theorems matching the goals in the theorem fields. The last part of a proof expression is a sequence of proof lines.

Although proof files in the **prf** format are text files, they are primarily for use by programs such as proof checkers. They are not for showing proofs to a human reader.

# 5 The record_proof Library

The **record_proof** Library serves as an interface to the proof recording feature in the core system. It is organised into two levels: the upper level is the user interface intended for users to record proofs and manage proof files; the lower level is for developers who may develop other utilities using the proof recording feature.

## 5.1 The User Interface

To a user, recording proof is a feature which can be enabled or disabled. Whatever the state the system is in, it performs proofs in the same way except that the extra step of recording the proofs in a file is carried out only if the feature is enabled. The typical use of this feature is

1. the user carries out a proof in the usual manner;

2. when he/she is satisfied with the proof, the proof recording feature is enabled by loading the library **record_proof**. Then, the proof is re-done once more in batch mode, and a proof file is generated.

While one is developing the proof, one will not require the system to record and save the proof in a disk file. To disable the proof recording feature, the library part **disable** can be loaded instead of the whole library. This is done by the command:

```
load_library'record_proof:disable';;
```

Usually, the proof script is saved in a script file. It can then be loaded into the system to perform the proof in a batch processing fashion. By loading different parts of the library as required, the same script file can be used to perform normal proofs and to generate proof files without any modification.

## 5.2   The Developer's Interface

In addition to the user interface, the **record_proof** library also provides a lower level interface to the proof recorder. This interface consists of a small number of ML functions to allow finer control of the proof recorder. They are useful for developing alternative user interfaces or applications other than proof checking.

The process of recording proofs and generating proof files can be divided into three stages:

1. recording inference steps;

2. generating a proof;

3. outputting to a text file.

In Stage 1, once the proof recorder was enabled by calling an ML function, every application of a basic inference rule is recorded in an internal buffer. Each inference is represented by an ML object of type **step**. The recording can be temporarily suspended and resumed later. The current state of the recorder and the internal buffer can be accessed by calling ML functions. The ML functions available to the developer for managing the proof recorder are documented in [Won94].

# 6   Checking HOL Proofs

Having modified the HOL system to incorporate the proof recorder and developed the **record_proof** library, HOL proofs can now be saved in proof files. The next phase is to develop an efficient proof checker to check the proofs.

The dominant requirement of this checker is to be able to check large proofs generated from real applications which consist of thousands or tens of thousands of inference steps. This means that the implementation should be fast and efficient, and should be able to perform reasonably well with limited resources, i.e., limited amount of physical memory and disk space. With the eventual formal verification in mind, the checker follows fairly closely von Wright's formal theory, especially the critical part, i.e., the checking of the inferences.

A checker accepting the core level proof file will be relatively simple, so it may possibly be verified formally. A checker for the extended level proofs could be implemented in two different ways. The first approach is to write a program to expand the inference steps involving derived rules into a sequence of primitive steps before being sent to the core checker. This approach has the advantage of utilising the core checker which may be formally verified, therefore, achieving higher confidence in the consistency of the proof. However, this approach can increase the number of inference steps considerably so the amount of time required to check the proof will be much longer.[1] The second approach checks all basic inference rules directly. This approach can result in a more efficient checker since the basic derived rules are relatively simple to check.

No matter which approach is used to implement a checker, its memory requirement is very large for large proofs because all theorems derived in the sequence have to be kept in memory. This is because a theorem derived in an earlier step may be referred to by the very last step. Logically, many modern systems are able to address many gigabytes, even up to terabytes of virtual memory, but physical memory is still limited. When large numbers of theorems are kept in memory, thrashing occurs, thus slowing down the process. This problem has been solved in the efficient checker by processing the proof file in two passes (see Section 8).

In fact, two versions of a checker have been implemented in Standard ML of New Jersey. One of them, the more formal version, accepts the core level proofs only. Its critical part was implemented by translating von Wright's HOL proof theory directly into SML functions. The other version, the more efficient version, can check proofs in the extended level using the direct approach mentioned in a previous paragraph. It is also based on the formal theory but with optimised use of memory. The major differences between these two versions are in the internal representation of types and terms, and the handling of theorem reference. The critical part — the checking of each inference — in both versions follows very closely to the formal theory. The non-critical parts of the two versions, for instance the proof file parser, the I/O handling and so on are identical, and use the same SML source files.

# 7   Using the Proof Checker

To a user, the checker is a program which reads a proof file, checks the proofs in it and reports back with either a success which means the proofs are correct or a failure which means the opposite. It creates a log file containing information of what hypotheses and stored theorems have been used and the resulting theorems of the proofs. The log file is in a format similar to the proof file.

## 7.1   Loading the Checker

Currently, the checker program modules have to be loaded into SML by evaluating the expression

---

[1] By examining the derivations of the derived rules, one can see that each derived rule may be expanded into five to twenty primitive rules.

```
use "join1.sml";
```

This will compile and link the modules to form the checker. After loading the modules, a top-level function **check_proof** is defined as the entry point to the checker.

When the program becomes stable, it will be possible to save an executable image. Then, the checker will be invoked as a shell command.

## 7.2 Invoking the Checker

The checker is invoked in SML by evaluating the function **check_proof** which takes a string as its sole argument. The string is the proof file name which, by convention, has the suffix **.prf** but the checker accepts any name. If the filename has a suffix **.gz**, the checker will assume it is a compressed file. It will run a decompresser automatically, and the log file will also be stored in a compressed form. The default compression/decompression utilities are the GNU **gzip/gunzip** programs. Below is a sample session of using the checker to check a compressed proof file named **MUL_FUN_CURRY** in the directory **proofs** parallel to the current directory.

```
- check_proof "../proofs/MULT_FUN_CURRY.prf.gz";          1

Current environment: MULT_FUN_CURRY

Proof: MULT_FUN_CURRY
Proof MULT_FUN_CURRY has been checked

Proof: MULT_FUN_CURRY_THM
Proof MULT_FUN_CURRY_THM has been checked


Using the following hypotheses:
<-8>    |- T :bool

   {... theorems deleted}

Proof: MULT_FUN
Proof MULT_FUN has been checked

Proof: MULT_FUN_DEF
Proof MULT_FUN_DEF has been checked

Using the following hypotheses:

   {... theorems deleted}

val it = () : unit
-
```

# 8    Implementation of the Checker

This section describes briefly the implementation of the checker. The full details including the complete source code can be found in [Won95].

The checker is structured into a number of modules as shown in Fig. 1. The modules can be divided into two groups: the core group and the auxiliary group. Modules in the core group are shown in the figure with thick border, whereas other modules are shown with thin border.

## 8.1    The Core Modules

The core modules implement the internal representation of HOL types, terms, theorems and proofs, and the checking of the inferences. In the formal version, these modules are translated directly from von Wright's formal theory of HOL proofs. In the efficient version, the internal representation is slightly different. De Bruijin's name-free representation is used for terms.

The **Check** module contains all functions for checking the consistency of inference rules. In the formal version, this module contains only eight checking functions for the eight primitive rules. These functions are translated directly from the formal theory. As an example, the function for checking the primitive rule **ASSUME** and its formal definition in the HOL proof theory is shown in Fig. 2. The SML function and the HOL definition are very close.

In the efficient version, the **Check** module contains functions for checking all basic inference rules. The functions for the primitive rules are the same as the formal version except very minor changes to take care of the slightly different representation of HOL types and terms. The functions for checking other basic rules are derived from specification of these rules found in [GM93]. Fig. 3 shows the basic inference rule **SYM** and its checking function.

One major difference between the two versions of the checker is that the formal version processes the proof file in one pass while the efficient version in two passes. Since a theorem derived in an inference step may be referred to by any subsequent steps, the checker has to cumulate all theorems in main memory while processing a proof. A practical proof may consist of thousands of inference steps. The storage for theorems will be huge. To overcome this problem, the efficient version processes the proof twice.

In the first pass, it builds a theorem reference table. This table consists of two dynamic arrays whose elements are integers as shown in Fig. 4a. Each element represent a proof line. The indices to the elements are the proof line numbers. Since the proof lines are numbered with both positive and negative numbers, but only non-negative numbers are allowed in indexing the array, two arrays are used. The **TabHyp** array is for the hypothesis lines whose line numbers are negative, and the **TabLine** array is for proof lines whose numbers are positive. These arrays are created using the **DynamicArray** module in the SML/NJ library. Using dynamic arrays instead of static ones releases the upper limit of the number of lines in the proof.

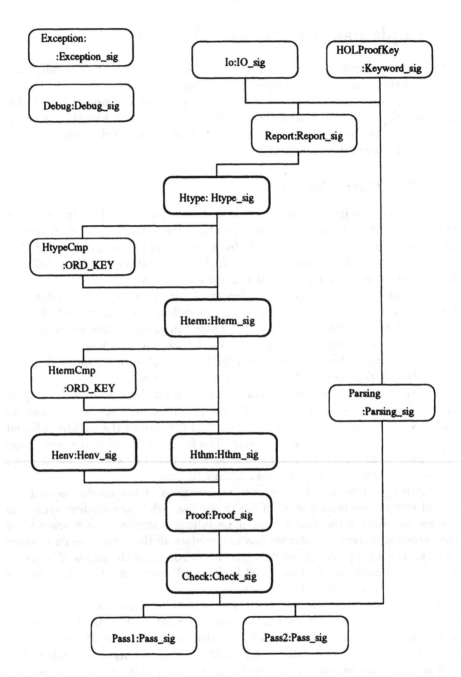

**Fig. 1.** Checker module structure

```
|- !Typl Conl as t tm. PASSUME Typl Conl (Pseq as t) tm =
   Pwell_typed Typl Conl tm /\
   Pboolean tm /\ (t = tm) /\ (as = {tm})
```
a) Formal definition in HOL

```
fun  PASSUME Typl Conl pseq tm =
        Pwell_typed Typl Conl tm andalso
        Pboolean tm andalso
        (Pseq_concl pseq = tm) andalso
        (Pseq_assum pseq = (SetOf [tm]))
; (* END FUN_DEC *)
```
b) Checking function in the formal version

**Fig. 2.** Checking function and formal definition of the primitive rule ASSUME

In the first pass, the checker looks at the justification part of the proof lines. When it encounters a reference to a theorem in a previous proof line, it enters the current line number into the element corresponding to the referred line in the table. For example, when the checker is at Line 3, it finds that this line refers to the theorem in hypothesis Line 1. It enters 3 into the first element of TabHyp. To speed up Pass 1 process, the checker can skip over other parts of the proof line quickly. This is done by scanning the input and looking for matching parentheses only. At the end of this pass, each element of the theorem reference table will contain the highest line number which is the latest line referring to the theorem. In the table shown in Fig. 4a, Line 5 is the last line referring to the theorem derived in Line 2 and Line 4.

In the second pass, the checker stores theorems referred to by other proof lines in a theorem table. This table is implemented by a dictionary in the Dict module of the SML/NJ library. The key of each entry is the line number. Since the dictionary is represented by balanced splay tree, searching for a theorem is fast. After checking a proof line, the checker examines the theorem reference table, if the value of the current element is greater than the current line number, i.e., it will be referred to later, the theorem is saved in the theorem table. Fig. 4b illustrates the situation in which the checker has just stored the theorem derived in Line 2. when the checker retrieves theorem, it also examines the theorem reference table. If the current line is the last one to refer to the theorem, i.e.,

$$\frac{\Gamma \vdash t_1 = t_2}{\Gamma \vdash t_2 = t_1}$$

```
fun chk_Sym(line, n, thm) =
let val thm1 = get_thm(line, n)
    val (left,right) = dest_eq (concl thm1)
in
    ((right,left) = dest_eq (concl thm)) andalso
    (HtermSet.equal((hyp thm), (hyp thm1)))
end
```

**Fig. 3.** Basic rule SYM and its checking function

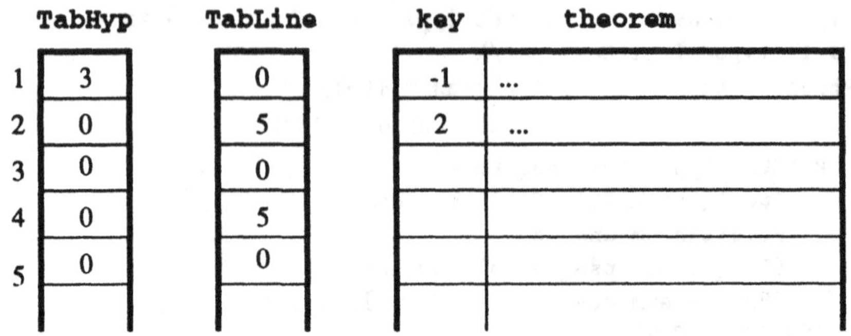

| TabHyp | | TabLine | | key | theorem | |
|:---:|:---:|:---:|:---:|:---:|:---:|:---:|
| 1 | 3 | | 0 | -1 | ... | |
| 2 | 0 | | 5 | 2 | ... | |
| 3 | 0 | | 0 | | | |
| 4 | 0 | | 5 | | | |
| 5 | 0 | | 0 | | | |

a) theorem reference table            b) theorem Table

**Fig. 4.** Data structures for theorem references

the current line number is equal to the value in the table, the theorem is removed from the dictionary. Continuing the scenario in Fig. 4b, the next line is Line 3, it refers to hypotheses Line 1. Since this is the last line referring to the theorem, the checker removes it from the table. This arrangement minimises the number of theorems stored in the table, thus reduces the memory requirement.

## 8.2 Auxiliary Modules

The **HOLProofKey** module defines the concrete syntax, i.e., the tags, of the proof files. The **Parsing** module consists of several higher order parsing functions. The parser proper is in the modules **Pass1** and **Pass2**. It is a recursive descend parser.

The **Exception** and **Debug** modules are responsible for handling errors. The **Debug** module maintains a debug flag for each module. The values of these flags are non-negative integers. Higher the value, more the information will be display while checking a proof. The **Report** module is for formatting the output to the log file.

The **Io** module handles all file input and output. When the checker is invoked, it creates a decompression process running as a filter in the background. The communication between this process and the checker is via a UNIX domain socket. In the case the file is uncompressed, no decompression is needed, but a dummy **cat** process is created, and the communication is still via a socket. This arrangement simplifies the checker as its input routine always reads from the input socket. Similarly, an output socket is created with a compression process to compress the output to the log file on the fly. This arrangement is illustrated in Fig. 5.

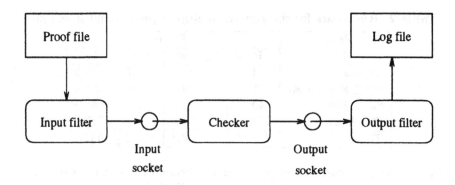

**Fig. 5.** Checker input/output arrangement

# 9 Benchmarking

A proof of correctness of a simple multiplier described in [Gor83] is often used as a HOL benchmark. This is a small to medium size proof which generates 14500 intermediate theorems. This proof has been used to test the proof recorder and the checker.

The multiplier proof consists of four ML files. A proof file is generated for each ML file. It contains all the sub-proofs in the corresponding ML file. Table 1 lists the time taken to record this proof and the proof file size. Two tests was carried out: the first with the proof recorder disabled; and the second with it enabled. The run time and the garbage collection time (GC) reported by the HOL system are listed under the columns headed **DISABLED** and **ENABLED**, respectively. The tests ran on a SUN Sparc 10 Server.

As the figures in the table show, the time (2412.9 seconds ≃ 40 minutes) required to record the proof and generate the proof files is considerable longer than to perform the proof only, but it is not excessive. Most of the extra time is spent in converting the internal presentation to the textual format and actually writing the disk files. This extra time is acceptable since the proof files will only be generated after the proof is completed satisfactorily (probably once) and be ran in batch mode.

The sizes of the proof files are also listed in Table 1. They are very large (43

**Table 1.** Benchmark of recording the multiplier proof (Time in seconds and size in bytes)

| FILE | No. of THMS | DISABLED RUN | GC | ENABLED RUN | GC | SIZE Raw | Compr'd |
|---|---|---|---|---|---|---|---|
| mk_NEXT | 2972 | – | – | 116.0 | 12.8 | 2693853 | 62202 |
| MULT_FUN _CURRY | 670 | – | – | 83.3 | 7.1 | 1553642 | 29103 |
| MULT_FUN | 6943 | – | – | 488.1 | 120.0 | 8101358 | 188201 |
| HOL_MULT | 3946 | – | – | 1675.5 | 250.1 | 31200001 | 447036 |
| **TOTAL** | 14531 | 65.3 | 11.8 | 2412.9 | 390.0 | 43627841 | 726542 |

Table 2. Benchmark for checking the multiplier proof (Time in seconds)

| Proof | Time | | | |
|---|---|---|---|---|
| File | Run | System | GC | Real |
| mk_NEXT | 139.3 | 15.3 | 3.7 | 170.0 |
| MULT_FUN_CURRY | 77.3 | 9.6 | 2.6 | 100.4 |
| MULT_FUN | 406.3 | 44.6 | 15.4 | 488.8 |
| HOL_MULT | 1472.1 | 152.0 | 98.5 | 1783.3 |
| Total | 2095.0 | 138.8 | 120.2 | 2542.5 |

Mbytes in total) because every intermediate theorem has to be saved. The size per theorem is comparable to the theory files in HOL88. However, the proof files are intended for automatic tools not for human readers, and they can be stored in compressed form. The size of the compressed files is much smaller. It amounts to less than 2% of the raw size. As the compression is done automatically, this does not pose too much burden to the user.

The multiplier proof files were successfully checked by the checker. No error was found from the proof files. Table 2 lists the time taken to check the proof files. This test ran on a SUN SparcStation 20.[2] The time is in the same order of magnitude as the recording. One important observation is that the process size is relatively small when performing the checking. The process size of the checker when it is just loaded is 14 Mbytes. The maximum size when performing the checking is only 16 Mbtyes. This shows that the implementation does keep the memory usage very small.

## 10 Conclusions

The research described in this paper shows a method of independent checking to ensure the consistency of mechanically generated proofs in LCF-like systems. A benchmark has shown that the proof checker is able to check a practical proof consisting of several thousands of inference steps. The application of this method is mainly in the formal verification of safety-critical and high-integrity systems. After a formal proof has been generated by a contractor, there is a need for an independent means of assessing the consistency of the proof.

There has been little work in the area of verified proof checking. The notable exception is the work of Boyer and Dowek on proof checking for Nqthm proofs [BD93]. Other theorem prover, such as nurpl[ea96], which makes use of proof objects in runtime to implement transformation tactics, may utilise similar approach to do proof checking.

If the proof checker itself can be formally verified, it will greatly boost the confidence in the consistency of checked proofs. Since the checker has been implemented using von Wright's formal theory of the HOL proof system as its

---

[2]The recording tests was carried out much earlier than the checking test. In fact, it was done before the checker was developed. The author was not able to access the same type of machine to do the checking test.

specification, it is possible to formally verify the checker provided that a formal semantics of Standard ML is developed. Attempts have been made to establish a formal semantics of Standard ML in HOL [Sym93] [VG93]. One approach to formally verify the proof checker is to reason based on the formal semantics directly, but there still more work needs to be done before one can attempt a formal verification of a practical program such as the checker. Another approach of using refinement has been suggested by von Wright [vW95].

In addition to being a format to communicate proofs between the HOL system and the proof checker, the proof file format may be used to communicate between different theorem prover systems with similar logic, such as between HOL88 and HOL90. It can also make the proofs themselves become deliverables. Having a proof saved as a sequence of inference steps allows new tools to be developed to analyse the proof, for instance, a tool generating a dependency graph to show the use of theorems in deriving a new theorem.

Although the recorder and checker is able to handle medium size proofs, there are several improvements could be made. Firstly, the proof file format could be changed to eliminate some redundancy and to reduce the file size. Secondly, the proof recorder could be improved so that the proof could be written into a file while it is being generated. This improvement will likely be done in HOL90. The proof checker now checks individual proofs in a flat environment. It is important to develop a system to manage the proofs and theorems in a more structured and hierarchical way if the checker is to be used in real applications which may contain many subproofs and their dependency is very complex.

# Acknowledgement

The idea of recording inference steps and generating proof lines has been suggested by many people including Malcolm Newey and Keith Hanna. Mike Gordon implemented a prototype of the recording functions in HOL88. The proof recorder described in this paper improved and enhanced this prototype. The translation of the formal HOL proof theory into ML functions used in the formal version was carried out by John Herbert. It was a great pleasure to work with Mike, Paul, Brian and others in the Hardware Verification Group in the Computer Laboratory in Cambridge. The work described here would not have been completed without their invaluable advice, help and company.

# References

[BD93]   Robert S. Boyer and Gilles Dowek. Towards checking proof checkers. In *Workshop on types for proofs and programs (Type '93)*. 1993.

[Bou92]  R. J. Boulton. On efficiency in theorem provers which fully expand proofs into primitive inferences. Technical Report 248, University of Cambridge Computer Laboratory, 1992.

[ea96]     Constable et al. *Implementing Mathematics with the Nuprl proof development system.* Prentice-Hall, 1996.

[GM93]     M. J. C. Gordon and T. F. Melham, editors. *Introduction to HOL— a theorem proving environment for higher order logic.* Cambridge University Press, 1993.

[Gor83]    M. J. C. Gordon. LCF_LSM, A system for specifying and verifying hardware. Technical Report 41, University of Cambridge Computer Laborartory, 1983.

[oD91]     Ministry of Defence. *Requirements for the procurement of safety-critical software in defence equipment.* Interim Standard 00-55, April 1991.

[Sym93]    D. Syme. Reasoning with the formal definition of standard ML in HOL. In *Higher Order Logic Theorem Proving and Its Applications,* Lecture Notes in Computer Science No. 780, pages 43–58. Springer-Verlag, 1993.

[VG93]     M. VanInwegen and E. Gunter. HOL-ML. In *Higher Order Logic Theorem Proving and Its Applications,* Lecture Notes in Computer Science No. 780, pages 59–72. Springer-Verlag, 1993.

[vW94]     J. von Wright. Representing higher order logic proofs in HOL. In Thomas F. Melham and Juanito Camilleri, editors, *Higher Order Logic Theorem Proving and Its Applications: 7th International Workshop,* volume 859 of *Lecture Notes in Computer Science,* pages 456–470. Springer-Verlag, September 1994.

[vW95]     J. von Wright. Program refinement by theorem prover. In *Proceedings of the 6th Refinement workshop,* Lecture Notes in Computer Science. Springer-Verlag, 1995.

[Won93a]   W. Wong. Formal verification of VIPER's ALU. Technical Report 300, University of Cambridge Computer Laboratory, New Museums Site, Pembroke Street, Cambridge CB2 3QG, ENGLAND, May 1993.

[Won93b]   W. Wong. Recording HOL proofs. Technical Report 306, University of Cambridge Computer Laboratory, New Museums Site, Pembroke Street, Cambridge CB2 3QG, ENGLAND, July 1993.

[Won94]    W. Wong. *The HOL record_proof Library.* Computer Laboratory, University of Cambridge, 1994.

[Won95]    W. Wong. A proof checker for HOL proofs. Technical report, University of Cambridge Computer Laboratory, New Museums Site, Pembroke Street, Cambridge CB2 3QG, ENGLAND, 1995. to be published as technical report.

# Formalization of Planar Graphs

Mitsuharu Yamamoto[1], Shin-ya Nishizaki[2], Masami Hagiya[1], and Yozo Toda[3]

[1] Department of Information Science, The University of Tokyo,
Hongo 7–3–1, Bunkyo-ku, Tokyo 113, Japan
E-mail: {mituharu,hagiya}@is.s.u-tokyo.ac.jp
[2] Department of Information Technology, Okayama University,
Tsushima-Naka 3–1–1, Okayama 700, Japan
E-mail: sin@momo.it.okayama-u.ac.jp
[3] Information Processing Center, Chiba University,
Yayoicho 1–33, Inageku, Chiba, Japan
E-mail: yozo@yuri.ipc.chiba-u.ac.jp

**Abstract.** Among many fields of mathematics and computer science, discrete mathematics is one of the most difficult fields to formalize because we prove theorems using intuitive inferences that have not been rigorously formalized yet. This paper focuses on graph theory from discrete mathematics and formalizes planar graphs. Although planar graphs are usually defined by embeddings into the two-dimensional real space, this definition can hardly be used for actually developing a formal theory of planar graphs. In this paper, we take another approach; we inductively define planar graphs and prove their properties based on the inductive definition. Before the definition of planar graphs, the theory of cycles is also introduced and used as a foundation of planar graphs. As an application of the theory of planar graphs, Euler's formula is proved.

## 1 Introduction

Needless to say, graph theory has many applications in mathematics and computer science. In particular, planar graphs, i.e., graphs that can be embedded into the two-dimensional plane are of great importance. For example, electrical circuits that are implemented on a single layer comprise planar graphs. Various diagrams that are used in mathematical reasoning such as Venn's diagrams are also planar graphs.

Besides these applications, planar graphs have deep mathematical structures that are observed with respect to properties such as colorability and duality. There are a number of theorems on planar graphs that have long and difficult proofs. The famous four-color theorem is an example of such theorems.

In this paper, we give a new formalization of planarity of graphs that is suitable for machine-assisted reasoning. Our goal is to give machine-checkable formal proofs of the above mentioned theorems on planar graphs. We started the present work based on the observation that there is no axiomatization of planarity of graphs that is actually used in reasoning about planar graphs.

Planar graphs are usually defined as graphs that can be embedded into the two-dimensional plane. However, this definition only serves for giving naive intuition and is never actually used in mathematical proofs. In addition, it seems extraordinarily difficult to formally prove properties of planar graphs based on this definition because it refers to many advanced notions concerned with the two-dimensional plane. For example, we must have formalized real numbers to define the two-dimensional plane. We must also have defined the usual topology on the two-dimensional plane before we have an embedding of a graph into the plane, which is defined in terms of continuous functions. In general, axiomatization of any theory should be at an appropriate level of abstraction. It is obvious that formalization by the two-dimensional real space is not at an appropriate level for working with planar graphs.

Among many areas of mathematics and computer science, some are easy and some are difficult to formalize. For example, it is relatively easy to give an axiomatization to an algebraic theory and give formal machine-checkable proofs according to the axiomatization. Some notions in the areas of geometry, on the other hand, are usually understood in a naive manner and have never been axiomatized. Giving formal definitions to those notions that can be actually used in machine-checkable formal proofs is considered as an important research direction in this field of verification technology.

Besides the difficulty of formalizing planarity, graph theory is a typical field of discrete mathematics in which proofs of meaningful theorems tend to be long and difficult, full of complex case analyses. Therefore machine assistance may be practical and useful once we have established good axiomatizations.

In this paper, as a result of the above considerations, we define planar graphs by using inductive definition and not by embeddings into the two-dimensional plane. We first define the base case that a cycle is a planar graph. We then define the operation of extending a given planar graph by adding an outer region to the graph. A graph is then defined to be planar if and only if it can be constructed from a cycle by successive applications of the operation. Note that our definition does not use any advanced notion about the two-dimensional plane but is only based on the theory of finite sets.

Since our definition of planar graphs is only concerned with finitely constructed objects, proof-checkers such as Nqthm that are based on first-order logic on constructive objects may seem sufficient. However, we found that graphs and their components are much more flexibly constructed than trees or lists in the sense that there is usually no standard way of constructing a given graph. Therefore, it is important to be able to compare graphs in an extensional way. This is one of the reasons why we use set theory and why we chose HOL as a logic to formalize planar graphs.

This paper is organized as follows. After giving preliminaries and briefly discussing how to define planarity, we introduce a theory of cycles. Since cycles are main components for constructing graphs and since they are more fundamental than lists for our purpose, we have developed the theory of cycles on HOL.

After the theory of cycles, we give our inductive definition of planar graphs, and prove some basic properties of planar graphs. We then show Euler's theorem based on the inductive definition.

We began this project with the ultimate goal of formally proving Kuratowski's theorem on planar graphs. It says that a graph is planar if and only if it contains as a subdivision neither $K_{3,3}$ nor $K_5$. We have already carefully examined the proof of Kuratowski's theorem and are sure that our definition is sufficient to carry out the whole proof of Kuratowski's theorem. At the end of the paper, we also give some prospects on this enterprise.

# 2   Higher-Order Logic

We introduce two notions in this paper, the notion of cycles and that of planar graphs. They are formalized in the form of theories in HOL theorem proving system [7].

We chose HOL as logic to formalize cycles and planar graphs by the following reasons.

- Cycles and planar graphs are defined inductively in this paper. However, unlike other inductively defined and well-known structures such as lists and trees, there is no standard way of constructing a given cycle or a planar graph. Therefore it turns out to be important to select a logic in which we can compare these structures in an extensional way, not by how they are constructed. As for this point, equality of functions or predicates in HOL is defined extensionally, and this feature is suitable for our purpose.
- Proofs of theorems in the field of discrete mathematics tend to contain lots of case analyses. Some of them can be proved simply by rewriting a goal with assumptions after splitting the previous goal into cases. Rewriting tactics in HOL are convenient in such situations.
- HOL system comes with two useful libraries, sets library and inductive definition package. They are indispensable tools for theories in this paper. We mention these libraries in the following subsections.

## 2.1   Sets Library

The sets library [4] provides a mechanism to deal with the polymorphic type of sets. A value $s$ of type $\alpha$ set corresponds to a predicate $P$ on type $\alpha$ where $x \in s$ if and only if $P\ x$ holds. In this library, several predicates on sets (e.g., membership, being a subset, disjointness, being a singleton, and finiteness) are predefined, and we can handle sets with predefined operations (union, intersection, difference, etc.) and theorems on them.

Almost all the structures in this paper are coded using sets, pairs, or their combinations. For example, a cycle, which is a main component of planar graphs, is represented as a set of pairs. Operations that construct cycles are coded by insertions and deletions on sets.

## 2.2 Inductive Definition Package

The inductive definition package [5] facilitates the introduction of relations that are defined inductively by a set of rules. We obtain two theorems on the relation defined by a set of specified rules; one asserts the relation satisfies these rules, and the other guarantees it is the least such relation. This package also provides functions to derive a theorem for the case analysis of the relation, and a tactic for proving theorems on it.

In this paper, we use this package to define predicates that characterize two notions, cycles and planar graphs. Thanks to this package, we can define them without taking trouble with proving several theorems such as the existence of the predicates.

## 2.3 Notations

We here introduce notations used in the later sections. By attaching the turnstile mark $\vdash$, we will specify that its following sentence is a theorem. The mark $\vdash_{def}$ is used for definitions and constant specifications similarly. We use Greek letters, $\alpha, \beta, \gamma, \ldots$, as type variables, **bold** font letters for type constants and sans serif font letters for term constants. Inside a definition or a constant specification, the defined term constant is written by an underlined symbol.

# 3 Preliminaries of Graph Theory

## 3.1 Graph

A *graph* consists of two finite sets, the set of vertices and the set of edges. Vertices are drawn from a fixed set, which is denoted by $\alpha$ in the following formalization. An edge is a two-element set of vertices. Therefore we do not allow multiple edges between vertices nor an edge comprising a cycle by itself.

Since edges are not directed, graphs we treat in this paper are so-called simple undirected graphs.

## 3.2 Planar Graph

A graph is called *planar* if it can be embedded into the two-dimensional plane.

In the following formalization, however, we explicitly add to a planar graph its structure of regions each of which is surrounded by a cycle of the graph. We therefore define a planar graph as a quadruple; the set of vertices, the set of edges, the set of regions and the outer cycle. Forgetting the last two components, we obtain a usual graph.

## 3.3 $n$-Connectivity

A graph is called *n-connected* if it keeps connected after any $n - 1$ vertices are removed from it. As will be explained in detail later, we restrict ourselves to planar graphs that are 2-connected.

# 4   Defining Planar Graphs

## 4.1   Embedding in $\mathbb{R}^2$ versus Inductive Definition

A *planar graph* is defined as a graph that can be embedded into two-dimensional Euclidean plane $\mathbb{R}^2$. However, this definition is not appropriate for the formal treatment of planar graphs by the following reasons.

- We have to formalize many advanced notions related to the two-dimensional plane, such as the real space $\mathbb{R}$, the topological space on $\mathbb{R}^2$, and embedding functions from graphs into $\mathbb{R}^2$ in terms of continuous functions.
- The definition by embeddings gives us naive intuition, and we can easily understand the explanations of properties of planar graphs by drawing diagrams. However, such intuition and explanations are of no use when dealing with planarity in a formal way.
- Turning our eyes to applications of planar graphs such as colorability and duality, we only use few of the properties of the two-dimensional plane. We only use properties such as separation of plane into regions by edges, and connectivity(or adjacency) of regions rather than those of $\mathbb{R}^2$ such as coordinates.

As a result of the above considerations, we do not adopt the embeddings into $\mathbb{R}^2$, but define planar graphs *inductively*. As a base case, a graph whose shape is just as a ring (we call such graphs *cycles* in this paper) is a planar graph. We then consider an operation that adds a sequence of edges to the outer region of a planar graph. A graph is defined to be planar if it is constructed repeatedly applying this operation to the base case. Since this inductive definition does not refer to any complex notion about the real space, and facilities that allows us to handle the inductive definition are already prepared(see Sect.2.2), it is suitable for formalization of planar graphs. The precise definition of planar graphs will appear in Sect. 6.

## 4.2   Why Introduce Cycle?

Instead of defining planar graphs directly, we prepare another theory called *cycle* and define planar graphs using cycles as main components. Intuitively speaking, cycles are similar to lists, but elements are distinct and the first element is treated as next to the last element. Regarding an element of a cycle as a vertex and adjacent elements as an edge, we can say a cycle is a special case of a planar graph.

As we mentioned slightly in the previous subsection, we define planar graphs by beginning with a cycle as a base case, and repeatedly adding a new sequence of edges. If cycles are used only in the base case of planar graphs, separating the theory of cycles from that of planar graphs is not so meaningful. However, we also use cycles when adding a new sequence of edges, regarding a cycle with a certain element of it as a list of distinct elements. Operations such as contracting two elements of a cycle and concatenating cycles are closed in cycle theory, and are used as fundamental operations when constructing planar graphs. For this reason, we prepare the theory of cycles independently.

## 5   Cycles

### 5.1 Definition and Basic Operations

A cycle is a circular list of distinct elements. In other words, the $k$th element is prior to the $(k+1)$th element and the last element to the first element. Although a cycle can be defined as an equivalence class of a list modulo circularity, it is too cumbersome to prove each property on cycles using this definition. Instead of making a definition by lists, we define a cycle as a special case of directed graphs. A cycle corresponds to its counterpart of a directed graph as follows:

- The set of elements of the cycle is equal to the set of vertices of the graph;
- The element $x$ is prior to $y$ in the cycle if and only if there is a directed edge from $x$ to $y$ in the graph.

Since an edge of a directed graph can be represented as an ordered pair of incident vertices in the order of direction, we represent a cycle as a value of $(\alpha \times \alpha)$ set which satisfies a predicate IS_CYCLE. The predicate IS_CYCLE : $(\alpha \times \alpha)$ set $\rightarrow$ bool is defined using the inductive definition package of HOL as:

$$\vdash_{def} \forall x.\, \text{IS\_CYCLE}\{(x,x)\}$$
$$\vdash_{def} \forall s\, \forall y\, \forall x\, \forall z.\, \text{IS\_CYCLE}\, s \wedge \neg(\exists w.(y,w) \in s) \wedge (x,z) \in s$$
$$\Longrightarrow \text{IS\_CYCLE}((s \setminus \{(x,z)\}) \cup \{(x,y),(y,z)\})$$

Figure 1 illustrates examples of cycles validated the above definitions. The left part of this figure shows a cycle of a self loop consisting of one element, which corresponds to a value $\{(x,x)\}$ of type $(\alpha \times \alpha)$ set. The right part corresponds to the above second formula. If IS_CYCLE $s$, there is an edge from $x$ to $z$ in $s$, and $y$ is not in $s$ (the left hand side of $\Rightarrow$), then we get another cycle by taking a side trip to $y$ on our way from $x$ to $z$ (the right hand side of $\Rightarrow$).

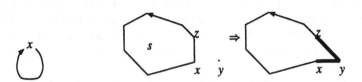

Fig. 1. Examples of cycles

The following figure shows a simple example of a cycle consisting of three elements and demonstration of its validity.

The set of its vertices is $\{u, v, w\}$ and the set of its edges $\{(u, v), (v, w), (w, u)\}$. The cycle drawn on the left is $\mathsf{ABS\_cycle}(\{(u, v), (v, w), (w, u)\})$ with validity of $\mathsf{IS\_CYCLE}(\{(u, v), (v, w), (w, u)\})$, which is inductively proved as follows:

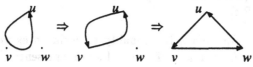

We also get an induction scheme on IS_CYCLE.

$$\vdash \forall P : (\alpha \times \alpha)\,\mathbf{set} \to \mathbf{bool}.(\forall x.P\,\{(x, x)\}) \wedge$$
$$(\forall s\,\forall y\,\forall x\,\forall z\,.P\,s \wedge \neg(\exists w.(y, w) \in s) \wedge (x, z) \in s \Longrightarrow$$
$$P\,((s \setminus \{(x, z)\}) \cup \{(x, y), (y, z)\})) \Longrightarrow$$
$$(\forall s.\,\mathsf{IS\_CYCLE}\,s \Longrightarrow P\,s)$$

Then we define a new type $\alpha\,\mathbf{cycle}$ using IS_CYCLE as a characteristic predicate. The bijection from $\alpha\,\mathbf{cycle}$ to $(\alpha \times \alpha)\,\mathbf{set}$ and its inverse automatically introduced in defining the type $\alpha\,\mathbf{cycle}$ are $\mathsf{REP\_cycle} : \alpha\,\mathbf{cycle} \to (\alpha \times \alpha)\,\mathbf{set}$ and $\mathsf{ABS\_cycle} : (\alpha \times \alpha)\,\mathbf{set} \to \alpha\,\mathbf{cycle}$, respectively.

$$\vdash_{def} (\forall a : \alpha\,\mathbf{cycle}.\,\mathsf{ABS\_cycle}(\mathsf{REP\_cycle}\,a) = a) \wedge$$
$$(\forall r : (\alpha \times \alpha)\,\mathbf{set}.\,\mathsf{IS\_CYCLE}\,r \Longleftrightarrow \underline{\mathsf{REP\_cycle}(\mathsf{ABS\_cycle}\,r) = r})$$

Basic operations on cycles are defined as follows. CYC_DOM returns a set of elements of a given cycle. FORW returns the next element of a given element in a given cycle.

$$\vdash_{def} \forall c : \alpha\,\mathbf{cycle}.\,\underline{\mathsf{CYC\_DOM}}\,c = \{x \mid \exists y : \alpha.(x, y) \in (\mathsf{REP\_cycle}\,c)\}$$
$$\vdash_{def} \forall c : \alpha\,\mathbf{cycle}\,\forall x : \alpha.x \in \mathsf{CYC\_DOM}\,c \Longrightarrow (x, \underline{\mathsf{FORW}}\,c\,x) \in \mathsf{REP\_cycle}\,c$$

CYC_BASE and CYC_INSERT are used for constructing a cycle. $\mathsf{CYC\_BASE}\,x$ creates a cycle of a self loop consisting of one element $x$. $\mathsf{CYC\_INSERT}\,c\,x\,y$ returns a cycle formed by deleting an edge from $x$ to $(\mathsf{FORW}\,c\,x)$ and adding two edges from $x$ to $y$ and from $y$ to $(\mathsf{FORW}\,c\,x)$.

$$\vdash_{def} \forall x : \alpha.\,\underline{\mathsf{CYC\_BASE}}\,x = \mathsf{ABS\_cycle}\{(x, x)\}$$
$$\vdash_{def} \forall c : \alpha\,\mathbf{cycle}\,\forall x : \alpha\,\forall y : \alpha.\,\underline{\mathsf{CYC\_INSERT}}\,c\,x\,y =$$
$$\mathsf{ABS\_cycle}((\mathsf{REP\_cycle}\,c \setminus \{(x, \mathsf{FORW}\,c\,x)\}) \cup \{(x, y), (y, \mathsf{FORW}\,c\,x)\})$$

Relations between CYC_DOM, FORW, CYC_BASE, and CYC_INSERT (we here call them four basic operations) are proved as follows:

$$\vdash \forall x.\,\mathsf{CYC\_DOM}(\mathsf{CYC\_BASE}\,x) = \{x\}$$
$$\vdash \forall c\,\forall x\,\forall y.y \notin \mathsf{CYC\_DOM}\,c \wedge x \in \mathsf{CYC\_DOM}\,c \Longrightarrow$$
$$(\mathsf{CYC\_DOM}(\mathsf{CYC\_INSERT}\,c\,x\,y) = \{y\} \cup \mathsf{CYC\_DOM}\,c)$$
$$\vdash \forall x.\,\mathsf{FORW}(\mathsf{CYC\_BASE}\,x)\,x = x$$
$$\vdash \forall c\,\forall x\,\forall y.y \notin \mathsf{CYC\_DOM}\,c \wedge x \in \mathsf{CYC\_DOM}\,c \Longrightarrow$$
$$(\forall z.z \in \mathsf{CYC\_DOM}(\mathsf{CYC\_INSERT}\,c\,x\,y) \Longrightarrow$$
$$(\mathsf{FORW}(\mathsf{CYC\_INSERT}\,c\,x\,y)\,z =$$
$$((z = x) \Rightarrow y \mid ((z = y) \Rightarrow (\mathsf{FORW}\,c\,x) \mid (\mathsf{FORW}\,c\,z)))))$$

The induction scheme on IS_CYCLE can be rewritten with CYC_BASE and CYC_INSERT as:

$$\vdash \forall P : \alpha\, \mathbf{cycle} \to \mathbf{bool}.(\forall x.P\,(\text{CYC\_BASE}\,x))\wedge$$
$$(\forall c\,\forall x\,\forall y\,.P\,c \wedge y \notin \text{CYC\_DOM}\,c \wedge x \in \text{CYC\_DOM}\,c \Longrightarrow$$
$$P\,(\text{CYC\_INSERT}\,c\,x\,y)) \Longrightarrow$$
$$(\forall c.P\,c)$$

In the following theorems, the equality of cycles is expressed in terms of CYC_DOM and FORW. Therefore, if two cycles are represented by CYC_BASE and CYC_INSERT, we can test the equality between them using these theorems and the relations between the four basic operations.

$$\vdash (c = c') \Longleftrightarrow$$
$$((\text{CYC\_DOM}\,c = \text{CYC\_DOM}\,c')\wedge$$
$$\forall x.(x \in \text{CYC\_DOM}\,c) \Longrightarrow (\text{FORW}\,c\,x = \text{FORW}\,c'\,x))$$
$$\vdash (c = c') \Longleftrightarrow$$
$$(\forall x.x \in \text{CYC\_DOM}\,c \Longrightarrow x \in \text{CYC\_DOM}\,c' \wedge (\text{FORW}\,c\,x = \text{FORW}\,c'\,x))$$

Using the above theorems, we get the following theorems about commutation of CYC_BASE and CYC_INSERT.

$$\vdash x \neq y \Longrightarrow$$
$$(\text{CYC\_INSERT}(\text{CYC\_BASE}\,x)\,x\,y = \text{CYC\_INSERT}(\text{CYC\_BASE}\,y)\,y\,x)$$
$$\vdash y \notin \text{CYC\_DOM}\,c \wedge x \in \text{CYC\_DOM}\,c \wedge y' \notin \text{CYC\_DOM}\,c \wedge y \neq y' \Longrightarrow$$
$$(\text{CYC\_INSERT}(\text{CYC\_INSERT}\,c\,x\,y)\,x\,y' =$$
$$\text{CYC\_INSERT}(\text{CYC\_INSERT}\,c\,x\,y')\,y'\,y)$$
$$\vdash y \notin \text{CYC\_DOM}\,c \wedge x \in \text{CYC\_DOM}\,c \wedge y' \notin \text{CYC\_DOM}\,c\wedge$$
$$x' \in \text{CYC\_DOM}\,c \wedge y \neq y' \wedge x \neq x' \Longrightarrow$$
$$(\text{CYC\_INSERT}(\text{CYC\_INSERT}\,c\,x\,y)\,x'\,y' =$$
$$\text{CYC\_INSERT}(\text{CYC\_INSERT}\,c\,x'\,y')\,x\,y)$$

We must prepare CYC_REV to define planar graphs. This returns the cycle whose order is the reverse of that of a given cycle.

$$\vdash_{def} \text{REP\_cycle}(\underline{\text{CYC\_REV}}\,c) = \{(y, x) \mid (x, y) \in \text{REP\_cycle}\,c\}$$

Although this definition is clear and simple, we must go back to $(\alpha \times \alpha)$ set when proving properties on CYC_REV, and this definition violates the data abstraction by the four basic operations. In the next subsection, we will present another definition of CYC_REV without resort to REP_cycle or ABS_cycle.

## 5.2 Primitive Recursive Functions on Cycles

Now fix a cycle $c$. The commutation theorem in Sect. 5.1 implies the way of constructing $c$ from CYC_BASE and by repeatedly applying CYC_INSERT is not unique. On the other hand, if you take a list $l$, the way of creating $l$ using NIL and CONS can be determined uniquely. This is one of the significant difference between **list** and **cycle**. This variety of construction of a cycle becomes an obstacle to prove the existence of primitive recursive functions on cycles. However, when we fix an element $x$ of $c$, the way of constructing $c$ from CYC_BASE $x$ and repeatedly applying CYC_INSERT to the position of $x$ is determined uniquely. For example, if $c$ is denoted by $[\ldots, y_n, x, y_1, y_2, \ldots]$ in the order of elements, the cycle $c$ can only be constructed as $[x] \rightarrow [x, y_n] \rightarrow \cdots \rightarrow [\ldots, y_n, x, y_1, y_2, \ldots]$. Then the cycle $c$ can be seen as a pair of an element $x$ and a list $[y_1, \ldots, y_n]$. Using this idea, we have another induction scheme on cycles and a theorem useful to prove the existence of primitive recursive functions on cycles.

$$\vdash \forall x : \alpha \,\forall P : \alpha \,\mathbf{cycle} \rightarrow \mathbf{bool}. P\,(\text{CYC\_BASE } x) \wedge$$
$$(\forall c. x \in \text{CYC\_DOM } c \Longrightarrow P\, c \Longrightarrow$$
$$(\forall y. y \notin \text{CYC\_DOM } c \Longrightarrow P\,(\text{CYC\_INSERT } c \, x \, y))) \Longrightarrow$$
$$(\forall c. x \in \text{CYC\_DOM } c \Longrightarrow P\, c)$$
$$\vdash \forall x : \alpha \,\forall b : \beta \,\forall i : \beta \rightarrow \alpha \,\mathbf{cycle} \rightarrow \alpha \rightarrow \beta.$$
$$\exists f : \alpha \,\mathbf{cycle} \rightarrow \beta. (f\,(\text{CYC\_BASE } x) = b) \wedge$$
$$(\forall c \,\forall y. y \notin \text{CYC\_DOM } c \wedge x \in \text{CYC\_DOM } c \Longrightarrow$$
$$(f\,(\text{CYC\_INSERT } c \, x \, y) = i\,(f\, c)\, c \, y))$$

The first theorem is an induction scheme on cycles in case that $x$ is contained in $c$. This induction has loose conditions than the one that appeared in the previous subsection. The second one is just like the theorem that is used for proving the existence of a primitive recursive function on lists. Using this theorem, we can treat a cycle $c$ as a list if a certain element $x$ of $c$ is fixed. In the rest of this subsection, we show some examples of function definitions where the existence of the each function are proved by the above theorem. All of the functions in these examples are indispensable to our definition of planar graphs.

The first example is a function CYC_INSERT. Let $c$ and $c'$ be cycles that have a common element $x$ and no other common elements. Then CYC_INSERTC $c \, x \, c'$ returns a cycle that is jointed at the position $x$. More precisely, if $c$ and $c'$ is denoted by $[\ldots, y_n, x, y_1, y_2, \ldots]$ and $[\ldots, z_m, x, z_1, z_2, \ldots]$ respectively, then CYC_INSERTC $c \, x \, c'$ is denoted by $[\ldots, y_n, x, z_1, z_2, \ldots, z_m, y_1, y_2, \ldots]$.

$$\vdash_{\text{def}} (\underline{\text{CYC\_INSERTC}}\, c \, x \,(\text{CYC\_BASE } x) = c) \wedge$$
$$(y \notin \text{CYC\_DOM } c' \wedge x \in \text{CYC\_DOM } c' \Longrightarrow$$
$$(\underline{\text{CYC\_INSERTC}}\, c \, x \,(\text{CYC\_INSERT } c' \, x \, y) =$$
$$\text{CYC\_INSERT}(\underline{\text{CYC\_INSERTC}}\, c \, x \, c')\, x \, y))$$

The next example is a function CYC_CONTRACT. Let $c$ be a cycle denoted by $[\ldots, y_n, x, y_1, y_2, \ldots, y_m, z, y_{m+1}, y_{m+2}, \ldots]$, then CYC_CONTRACT $c \, x \, z$ is denoted by $[\ldots, y_n, x, z, y_{m+1}, y_{m+2}, \ldots]$ as the following figure.

$$\vdash_{\text{def}} (\underline{\text{CYC\_CONTRACT}}(\text{CYC\_BASE } x)\, x \, z = \text{CYC\_BASE } x) \wedge$$
$$(y \notin \text{CYC\_DOM } c \wedge x \in \text{CYC\_DOM } c \Longrightarrow$$
$$(\underline{\text{CYC\_CONTRACT}}(\text{CYC\_INSERT } c \, x \, y)\, x \, z =$$
$$((z = y) \Rightarrow (\text{CYC\_INSERT } c \, x \, y) \mid (\underline{\text{CYC\_CONTRACT}}\, c \, x \, z))))$$

$$CYC\_CONTRACT\,c\,x\,z$$

The last example is another definition of CYC_REV. First, we define an auxiliary function CYC_REVX primitive recursively.

$$\vdash_{def} (\underline{CYC\_REVX}\, x\, (CYC\_BASE\, x) = CYC\_BASE\, x)\wedge$$
$$(y \notin CYC\_DOM\, c \wedge x \in CYC\_DOM\, c \Longrightarrow$$
$$(\underline{CYC\_REVX}\, x\, (CYC\_INSERT\, c\, x\, y) =$$
$$CYC\_INSERT(\underline{CYC\_REVX}\, x\, c)\, (FORW\, c\, x)\, y)))$$

Since we can prove $(x \in CYC\_DOM\, c \wedge x' \in CYC\_DOM\, c \Longrightarrow (CYC\_REVX\, x\, c = CYC\_REVX\, x'\, c))$, the first argument of CYC_REVX has no effect on the return value. Using this, we can prove the existence of CYC_REV satisfying the following condition:

$$\vdash_{def} (\underline{CYC\_REV}(CYC\_BASE\, x) = CYC\_BASE\, x)\wedge$$
$$(y \notin CYC\_DOM\, c \wedge x \in CYC\_DOM\, c \Longrightarrow$$
$$(\underline{CYC\_REV}(CYC\_INSERT\, c\, x\, y) = CYC\_INSERT(\underline{CYC\_REV}\, c)\, (FORW\, c\, x)\, y)))$$

## 6  Inductive Definition of Planar Graphs

In this section, we define planar graphs with the help of cycles. As we mentioned in Sect 3.3, we restrict ourselves to planar graphs that are 2-connected for simplicity. Although we place a limit on the definition of planar graphs as a result of this restriction, by the fact that any connected planar graph can be separated into the maximal 2-connected subgraphs (called *blocks*), cut vertices, and bridges, we can say the case of 2-connected planar graphs is an important issue enough to be studied independently. Moreover, 2-connected components in a planar graph are the essence of its planarity as stated in [8]:

> A simple observation shows that we can actually restrict the planarity test and, later, the design of a suitable drawing algorithm to the biconnected components of graphs: a graph is planar if and only if its biconnected components are. (This follows because the biconnected components of a graph can intersect in at most one node.)

Hence we here concentrate our attention on the case of 2-connected planar graphs and leave the remaining case as future work.

### 6.1  How to Define Planar Graphs

A planar graph consists of a set ($vs : \alpha\, set$) of vertices, a set ($es : \alpha\, set\, set$) of edges, a set ($rs : \alpha\, cycle\, set$) of regions, and an outer cycle ($oc : \alpha\, cycle$).

Vertices *vs* and edges *es* are as in the usual definition of a graph, though each edge is represented as a two-element set of incident vertices rather than an ordered pair of them, for here we formalize only undirected graphs. Regions *rs* are parts of the plane separated by edges. The outermost region is not included in *rs*, but treated specially as an outer cycle *oc*. Each region is represented by a cycle that encloses it, and is directed so that if two regions *r* and *r'* are adjacent with a common edge $\{x, y\}$, then *r* is directed(say, from *y* to *x*) in the opposite direction of *r'* (from *x* to *y*) as the following figure.

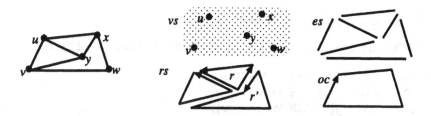

Planarity is characterized by a predicate IS_PGRAPH on quadruple $(vs, es, rs, oc)$. IS_PGRAPH is defined inductively[4] as follows:

- *Base case:* For each cycle *c* that contains more than two elements, we have IS_PGRAPH($vs, es, rs, oc$) where:
  - *vs* is a set of elements of *c*, i.e., $vs = \text{CYC\_DOM}\,c$;
  - $\{x, y\} \in es$ if and only if $x \in \text{CYC\_DOM}\,c$ and $y = \text{FORW}\,c\,x$;
  - *rs* is a singleton set of CYC_REV *c*;
  - *oc* is *c* itself.
- *Induction step:* Let *c* be a cycle. Suppose that the following conditions hold:
  - IS_PGRAPH($vs, es, rs, oc$);
  - Distinct elements *u* and *v* are contained in *oc*;
  - CYC_DOM *c* shares an element *u* with *vs* and no other elements;
  - If *c* is a cycle of one element, the edge $\{u, v\}$ is not contained in *es*.
  Then IS_PGRAPH($vs', es', rs', oc'$) holds satisfying that
  - $vs' = vs \cup \text{CYC\_DOM}\,c$;
  - $\{x, y\} \in es'$ if and only if one of the following holds:
    * $\{x, y\} \in es$;
    * $x \in \text{CYC\_DOM}\,c$, $y = \text{FORW}\,c\,x$, and $y \neq u$;
    * $x \in \text{CYC\_DOM}\,c$, $y = v$, and $\text{FORW}\,c\,x = u$.
  - $rs' = rs \cup \{\text{CYC\_INSERTC}(\text{CYC\_REV}\,c)\,u\,(\text{CYC\_CONTRACT}\,oc\,v\,u)\}$;
  - $oc' = \text{CYC\_INSERTC}(\text{CYC\_CONTRACT}\,oc\,u\,v)\,u\,c$.

The following figure illustrates the induction step of IS_PGRAPH.

---

[4] The inductive definition package is also used here.

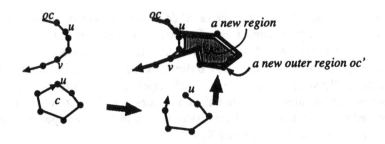

## 6.2 Basic Properties of Planar Graphs

The following basic properties are derived from the definition of Sect. 6.1.

- $vs$, $es$, and $rs$ are finite sets (easy);
- CYC_DOM $oc \subseteq vs$, $oc \notin rs$;
- $vs = \bigcup es = \bigcup \{$CYC_DOM $r \mid r \in rs\}$;
- $\{x, y\} \in es$ if and only if one of the following holds:
    - There exists $c$ and $c'$ in $rs$, FORW $c\, x = y$, and FORW $c'\, y = x$;
    - There exists $c$ in $rs$, FORW $c\, x = y$, and FORW $oc\, y = x$.

We must prepare theorems as above together with some theorems on the functions CYC_CONTRACT and CYC_INSERTC to prove Euler's formula, which we will discuss in Sect. 7.

## 6.3 Validity of Definition

By drawing diagrams, we can easily show that a graph constructed by our inductive definition is a planar graph in the sense that there exists an embedding into the two-dimensional plane. We here informally prove that any 2-connected planar graph can be constructed by our definition.

**Definition 1.** For any planar graph, we can classify edges that are incident to the outer region into sequences of adjacent edges so that the following conditions hold.

- The degrees of vertices at the end of a sequence are more than 2.
- For each sequence, vertices in it (except at the end) are of degree 2.

We call each sequence of adjacent edges as an *edge sequence*.

The following figure illustrates the classification into edge sequences. Each edge of the right hand side of the arrow corresponds to an edge sequence of the left hand side.

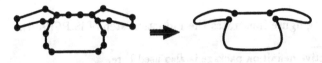

**Definition 2.** Let $r$ and $r'$ be two regions of a graph. We call $r$ is *adjacent to $r'$* if they share common edges. Two regions sharing vertices but not sharing edges are *not* called as adjacent regions in this paper.

**Lemma 3.** *For every 2-connected planar graph $g$, if the number of regions that are adjacent to the outer region is more than 1, then there exists an edge sequence $s$ of $g$ such that 2-connectivity is preserved when $s$ is eliminated from $g$. We write the graph made by eliminating $s$ from $g$ as $g \setminus s$.*

**Proposition 4.** *For every 2-connected planar graph $g$, we can construct $g$ by our inductive definition of planar graph.*

*Proof of proposition.* By induction on the number $n$ of regions of $g$. Suppose $n = 1$, then $g$ is a cycle, which is the base case of our definition. If $n > 1$, there is an edge sequence $s$ of $g$ and $g \setminus s$ is also 2-connected by Lemma 3. Since the number of regions of $g \setminus s$ is less than that of $g$ by 1, we can construct $g \setminus s$ by our definition using induction hypothesis. Adding $s$ to $g \setminus s$ is exactly the same operation as that of the induction step of our definition.  □

*Proof of lemma.* By induction on the number $n$ of regions that are adjacent to the outer region. First, there is no vertex whose degree is less than 2 by 2-connectivity. If all the vertices that is incident to the outer region are of degree 2, then since this graph is a cycle, we have $n = 1$ and this contradicts with the assumption. So we assume there is at least one vertex whose degree is more than 2 and which is incident to the outer region in the rest of the proof.

The situation that the elimination of an edge sequence that is incident to a region $r$ (not the outer region) results in a non-2-connected graph is characterized by the left part of the following figure,

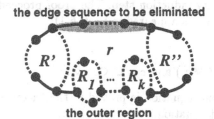 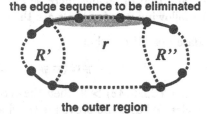

where $R'$, $R''$, and $R_1, \ldots, R_k$ are blocks of regions, which are not adjacent with one another. Since in case $k > 0$, we can eliminate an edge sequence in $R_1$ without producing a non-2-connected graph, we restrict ourselves to the case that $k = 0$ (the right part of the above figure). The induction on the number of regions adjacent to the outer one goes as follows.

- *Base case:* Suppose $n = 2$. We cannot construct a graph like the above figure. So this case is impossible.
- *Induction step:* We try to make $r$ not adjacent to the outer region by contracting the edges surrounding $r$ as follows.

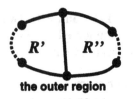

the outer region             the outer region

Consider the right part of the figure if the edges that $R'$ and $r$ have in common and those that $R''$ and $r$ do are made into just one edge. Otherwise consider the left part. By induction hypothesis, we have at least one edge sequence $s$ in the new graph such that the elimination of $s$ produces a 2-connected graph. Then elimination of the counterpart of $s$ in the original one also results in a 2-connected graph.

□

## 7  Euler's Formula

Euler's formula is one of the most fundamental theorems on planar graphs. This formula is a relationship among the numbers of vertices, that of edges, and that of faces as follows:

$$\#(vertices) - \#(edges) + \#(faces) = 2$$

where $\#s$ means cardinality of a set $s$.

In our formalization, Euler's formula is expressed as:

$$\#vs + \#(rs \cup \{oc\}) = \#es + 2$$

where IS_PGRAPH($vs, es, rs, oc$).

To prove this theorem, we must check how the number of vertices, that of edges, and that of regions increase at every induction step. The proof proceeds in the following way (symbols are the same as appeared in Sect. 6.1):

- *Base case:* Prove that

$$\#es = \#vs \; (= \#(\text{CYC\_DOM}\,c)) \text{ and } c \neq \text{CYC\_REV}\,c$$

  If $\#(\text{CYC\_DOM}\,c) > 2$, then the second equation is proved in the theory of cycles. What to be proved is the first equation.
  - Case [$\#(\text{CYC\_DOM}\,c) = 3$] Prove the first equation by giving concrete elements of $c$.
  - Case [$c = \text{CYC\_INSERT}\,c'\;x\;y$, and the first equation holds in case of $c = c'$] Prove that $\#es$ and $\#vs$ increase just by 1, i.e., the new vertex and the new edge are not included in the old one.
- *Induction step:* Prove $vs$ increases by $\#(\text{CYC\_DOM}\,c)-1$, $es$ by $\#(\text{CYC\_DOM}\,c)$, and $rs$ by 1. This is done by proving that newly added vertices, edges, and regions are not contained in the original graph (i.e., set union operations in the induction step of the definition of planar graphs are disjoint unions essentially) except one vertex.

# 8 Future Work

Although we have formalized graphs, cycles, and planar graphs in the previous sections and proved several properties and theorems, we leave many notions and properties on planar graphs unformalized and unproved: colorability, duality, and Kuratowski's theorem, which is the ultimate goal in our work, as we mentioned in the introduction section. We carefully examined the proof of Kuratowski's theorem and we informally checked that the theorem can be proved by using our formalization of planar graphs. We know that we must formalize several notions on graph theory before proving Kuratowski's theorem:

- Path, tree, cut point, degree, subgraph, graph minor, minimal forbidden minor, subdivision.
- Adding, splitting, deletion, contraction.

And the following seems to be key theorems for formalizing Kuratowski's theorem:

**Theorem 5.** *Any graph that has five or more vertices has a contractible edge.*

**Theorem 6.** *If a planar graph $G$ has four or more vertices, the following two conditions are equivalent.*

*1. $G$ is 3-connected.*
*2. All the regions that are adjacent to a vertex of $G$ are surrounded by a cycle.*

Besides these theorems, we must prove more fundamental properties of planar graphs. For example, any planar graph can be redrawn with any specified region as the outer cycle. In order to prove this kind of property, we must develop techniques for transforming a given way constructing a planar graph into an appropriate one.

# 9 Related Work

Wong formalized directed graphs and applied to the railway signalling system [9]. From among the notions in graph theory, he placed stress on the derivation of paths that plays an important role in railway signalling. Chou, a pioneer of formalization of undirected graphs, described a formal theory of undirected (labeled) graphs in higher-order logic [1]. A large number of notions in graph theory are formalized in his paper. In particular, he gave deep consideration to treatment of trees, for his work was motivated by the mechanical verification of distributed algorithms [2]. Unlike our approach to planar graphs, his strategy for defining graphs, which may not be planar and are permitted to include self loops and multiple edges, is by a non-inductive relation, not in a constructive way. Although there are a few gaps between his approach and ours, we conjecture that his method of formalizing trees, paths, bridges, and connectivity, will be greatly helpful when extending the definition of 2-connected planar graphs to that of connected planar graphs (i.e., connection of blocks by bridges and cut vertices).

# 10  Conclusion

In this paper, we formalized cycle theory, and described a formal definition of planar graphs using cycles. The cycle theory provides mechanism for constructing cycles, theorems that helps to prove equality of them, and facilities for defining primitive recursive functions on them. Planarity of graphs is defined inductively, not using embeddings into the two-dimensional plane. We have proved Euler's theorem, the relation among the numbers of vertices, edges, and faces of planar graphs. We also sketched a prospect of a formal proof of Kuratowski's theorem.

## Acknowledgements

The authors would like to thank Mary Inaba for her help and advice on graph theory, and are also grateful to anonymous referees for their constructive suggestions.

## References

1. Ching-Tsun Chou. A formal theory of undirected graphs in higher-order logic. In Thomas F. Melham and Juanito Camilleri, editors, *7th International Workshop on Higher-Order Logic Theorem Proving System and Its Applications*, volume 859 of *Lecture Notes in Computer Science*, pages 144–157. Springer-Verlag, 1994.
2. Ching-Tsun Chou. Mechanical verification of distributed algorithms in higher-order logic. In Thomas F. Melham and Juanito Camilleri, editors, *7th International Workshop on Higher-Order Logic Theorem Proving System and Its Applications*, volume 859 of *Lecture Notes in Computer Science*, pages 158–176. Springer-Verlag, 1994.
3. Frank Harary. *Graph theory*. Addison-Wesley series in mathematics. Addison-Wesley, London, 1969.
4. Thomas F. Melham. *The HOL sets Library*. University of Cambridge, Computer Laboratory, October 1991.
5. Thomas F. Melham. A package for inductive relation definition in HOL. In *Proceedings of the 1991 International Tutorial and Workshop on the HOL Theorem Proving System*, pages 27–30. IEEE Computer Society Press, August 1991.
6. Seiya Negami. *Discrete Structures*. Number 3 in Information Mathematics Lectures. Kyouritsu Shuppan, Tokyo, Japan, May 1993. (in Japanese).
7. University of Cambridge, Computer Laboratory. *The HOL System: DESCRIPTION*, March 1994.
8. Jan van Leeuwen. *Graph Algorithms*, volume A of *Handbook of Theoretical Computer Science*, chapter 10, pages 525–633. MIT Press, 1990.
9. Wai Wong. A simple graph theory and its application in railway signaling. In M. Archer *et al.*, editor, *Proc. of 1991 Workshop on the HOL Theorem Proving System and Its Applications*, pages 395–409. IEEE Computer Society Press, 1992.

# A Hierarchical Method for Reasoning about Distributed Programming Languages *

Cui Zhang, Brian R. Becker, Mark R. Heckman
Karl Levitt and Ron A. Olsson

*Department of Computer Science*
*University of California, Davis, CA 95616*
email: {zhang, beckerb, heckman, levitt, olsson}@cs.ucdavis.edu

**Abstract.** This paper presents a technique for specifying and reasoning about the operational semantics of distributed programming languages. We formalize the concept of "vertical stacking" of distributed systems, an extension of Joyce's, Windley's and Curzon's stacking methodologies for sequential systems and of the CLI "short stack" which stacks interpreters for object code, assembly code, and a high-level sequential language. We use a state transition model to account for the issues of atomicity, concurrency and nondeterminism at all levels in our stack. A correctness definition is given, which for each pair of adjacent language semantics and mappings between them, produces proof obligations corresponding to the correctness of the language implementation. We present an application of the method to a two-level stack: the microSR distributed programming language and a multi-processor instruction set, which is the target language for a compiler for microSR. We also present the development of a verified programming logic for microSR, using the same microSR semantic specification. The HOL system is used for the specification and the proofs.

## 1 Introduction

This paper presents a mechanized methodology for specifying and reasoning about the operational semantics of distributed programming languages, and for "stacking" them as layers in distributed systems.

There has been much research related to our work. CLI's work on the "short stack" has shown the feasibility of full sequential computer system verification using a vertical-layer proof technique [3]. Joyce's work has provided a HOL-based method to verify a sequential programming language implementation [10]. Windley has developed a method to specify instruction set architectures and a generic interpreter model for verifying microprocessors by layered proof [13, 14]. Curzon's work provides a method to combine a derived programming logic with a

---

* This work was sponsored by ARPA under contract USN N00014-93-1-1322 with the Office of Naval Research and by the National Security Agency's UR Program.

verified compiler for an assembly language [5]. von Wright has verified the correctness of refinement rules for developing programs from higher-level specifications into efficient implementations, based on the weakest precondition calculus [12]. Abadi and Lamport's work has shown a theoretical completeness result on the existence of a refinement mapping when one system specification implements another, under reasonable assumptions [1]. These works provide us with insights into the specification of semantics and the verification of implementations, using a compositional method.

Similar to [3, 5, 10, 13, 14], which provide mechanized methodologies and techniques that work very well for verifying the implementation of sequential languages (or instruction sets) but are not sufficient for distributed language implementation, we focus on the verification of programming language implementation ("compilation" or "translation"), taking a very general view of programming languages as interfaces. In particular, we present a mechanized methodology for "stacking" layers in a distributed system to verify, for example, a compiler for a distributed programming language that generates code targeted to a collection of machines that communicate by message passing. The compiler, including the "translation" of instructions and semantic domains, has to be specified as mappings between the two languages' semantic specifications. Its correctness has to be verified with respect to the pair of semantics.

Like most of the work in concurrent program verification [1, 4, 6, 11], our research is based on a state transition model. We claim that distributed programming languages, whether higher-level programming languages, lower-level instruction sets for multi-computer systems, or a network interface wherein processes communicate through the exchange of packets, can be formalized under the state transition model. We define a framework for specifying the operational semantics of distributed programming languages using this model. Our work is a generalization of Windley's result of treating each layer in a microprocessor system as a generic interpreter. Our technique not only makes the specification of distributed programming languages a step-by-step task, but also eases the verification of the language implementation and the verification of a prover for distributed programs, the latter an extension of Gordon's work [7].

As claimed, our technique works for any stacking of distributed programming languages. In this paper, we apply our methodology to two levels. MicroSR, a derivative of the SR language [2], is used as the higher-level distributed programming language. MP, a multiple processor instruction set architecture, is used as the lower-level implementation language. In addition to those constructs basic to common sequential programming languages, our current version of microSR includes: (1) the guarded *If* (conditional) statement and the guarded *Do* (loop) statement; (2) the asynchronous *Send* statement; (3) the synchronous *Receive* statement with an optional *Synchronization Expression* providing further control on the synchronization of communication; (4) the *Call* statement that provides either one-way or two-way communication with rendezvous between the caller and the process executing an *Input* statement; (5) the guarded communication *Input* statement, which is similar to the receiving part in Hoare's CSP;

and, (6) the *Co* (co-begin) statement for specifying the concurrent execution of processes that communicate via message passing but do not share memory. The MP machine is an abstraction of an interface to multiple simple processors. Each individual processor (called a Virtual Machine) is a model of a very simple sequential microprocessor containing a basic set of instructions, but no communication primitives. Two communication mechanisms are provided by the MP machine: (1) the *SEND* instruction, which asynchronously sends a message to a specific message queue over an abstract network; and, (2) the *RCV* instruction, which synchronously receives a message from a specific message queue. The MP machine is also responsible for scheduling the execution of instructions by the various VMachines. At present, this is a very simple scheduler but will be made more realistic in future work.

Our work has been performed in HOL [8]. In Section 2, we describe our state transition model with support for atomicity, concurrency, and nondeterminism. Section 3 presents our framework, clarifying the definitions that need to be made to specify distributed programming languages in general and how to define them in HOL. The specification of microSR provides an example for describing our technique. Section 4 describes the application of our technique to the verification of a microSR implementation. Section 5 presents a verified HOL prover for microSR. The final section discusses future work.

## 2 The State Transition Model

### 2.1 For Any Given Level

In our model, all *primitive* statements are atomic but with different granularities at different levels. At all levels, once a process starts executing a primitive statement, whether an internal (i.e., intra-process) statement or a communication primitive, no other process can influence that statement's execution or observe intermediate points of its execution. Thus, if two primitive statements, say C1 and C2, are executed concurrently in processes P1 and P2, the net effect, as modeled in our state transition system, is either that of C1 followed by C2, or of C2 followed by C1. Although we model the concurrent execution of two statements by two processes as a linearly ordered sequence of state transitions, the actual order in which selectable (i.e., eligible to execute) statements are executed is nondeterministic. With this view of atomicity, the behavior of a distributed program is modeled as a sequence of state transitions, each of which is accomplished by an atomic step. As we indicate below, this view of atomicity models the actions of distributed processes whose actions change their own local state or change a shared message pool. It also models all execution interleavings of processes which are allowed by the language semantics.

The abstract syntax of a statement set can be defined recursively, using the *type definition package* in HOL. Below is the abstract syntax of the current version of microSR. By structural induction, the execution of a composite statement is actually an interleaving of the execution of its atomic components and the execution of atomic primitives of other processes.

```
⊢ def Stmt = Assign Var IExp          % v := iexp %
         | Send Op IExp               % Send op(iexp) %
         | Receive1 Op Var            % Receive op(v) %
         | Receive2 Op Var BExp       % Receive op(v) suchthat bexp %
         | Call1 Op IExp              % Call op(iexp) %
         | Seq Stmt Stmt              % stmt1; stmt2 %
         | If BExp BExp Stmt Stmt     % If b1 → stmt1 [] b2 → stmt2 fi %
         | Do BExp BExp Stmt Stmt     % Do b1 → stmt1 [] b2 → stmt2 od %
         | In1 Op Op Var Stmt Stmt    % In op1(v)→stmt1 [] op2(v)→stmt2 ni %
         ...
```

## 2.2  For Any Two Adjacent Levels

We say two languages are at two adjacent levels when the relatively higher-level
language is implemented by the lower-level language. Careful attention has to
be given to what will happen between the two language systems when both support concurrency and when both are modeled using state transitions. Because
the atomicities of the two different levels have different granularities, a single
atomic state transition at the higher language level corresponds to multiple
state transitions at the lower language level. Therefore, each higher-level execution interleaving is likely to have multiple corresponding lower-level execution
interleavings. Among them, only those interleavings which exhibit equivalent effects on the externally visible state will be allowed by a correct implementation
of the higher-level language by the lower-level language.

For the simple microSR program below, there are multiple possible interleavings at the microSR level, reflecting concurrency in the microSR semantics
and the nondeterministic execution order of simultaneously selectable (i.e., eligible to execute) statements in our state transition model. In these possible
valid interleavings, the execution of *Send op1(msg31)* in process P3 can occur
either earlier or later than the execution of statements in other processes. This is
because the asynchronous messages from the same sender to the same message
channel (which is called an operation in SR's terminology) have to be well ordered, but the order of messages from different senders is nondeterministic. Since
a single state transition step at any given level represents a completed primitive
statement, the execution of synchronous *Receive op2(v)* cannot be selectable until at least one message has been sent to the operation op2. When a microSR
program is compiled to MP code, the MP code for *Receive op2(v)* may begin its
execution before the code for a microSR *Send* finishes execution. For illustration purpose, figure 1 shows the implementation of microSR *Send* and *Receive*
each requiring three MP instructions. The first one stands for the message and
operation preparation, the second is a *SEND* labeled *S* or a *RCV* labeled *R* for
accessing an operation, and the third one is for the clean-up. Figure 1 shows the
execution of the MP instruction "21", which corresponds to the "preparation"
for the instruction *R*, beginning before the execution of instruction "13", which
corresponds to the "clean-up" after the instruction *S*. We allow this overlapping
because of the finer atomicity and the finer interleavings at the MP level. However, the instruction *R* can be selectable only when at least one message has
been sent to the operation by the instruction *S*.

MicroSR Program: (Process P1) ... Send op2(msg11); Send op2(msg12); ...
(Process P2) ... Receive op2(v) ...
(Process P3) ... Send op1(msg31) ...

Some possible interleavings:

... Send op2(msg11) Receive op2(v) Send op2(msg12) Send op1(msg31) ...

... Send op2(msg11) Send op1(msg31) Receive op2(v) Send op2(msg12) ...

... Send op1(msg31) Send op2(msg11) Send op2(msg12) Receive op2(v) ...

Note that a given interleaving at the microSR level can have multiple corresponding MP level interleavings because compiled code has finer granularity of atomicity at the MP level, as indicated by Figure 1. However, only interleavings which have an equivalent effect on the externally visible state (i.e., the shared message pool) are allowed by a correct implementation. Therefore, the specification of the system-wide and intra-process sequencing, defining the allowable or valid interleavings of state transitions, has to be a part of the semantic specification at any given level under our model. Also, an equivalence with respect to the impact of interleavings at two adjacent levels on their externally visible states has to be specified, in order to identify allowable interleavings and to guarantee that no others are generated by the implementation.

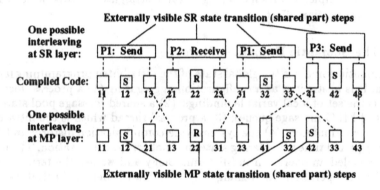

**Fig. 1.** Interleavings at microSR Level and MP Level

# 3   The Framework for Semantic Specification

A semantic specification of a given distributed programming language can be modeled by a state transition system:

*Specification (Syntax, State, Continuation, Selection, Meaning, System)*

The *Syntax* is the abstract syntax of the language. The *State* represents the semantic domain of the language. The other four components in the framework are definitions of relations which are necessary to specify concurrency and nondeterminism. Similar to [6], the *Continuation* is the relation that formalizes intra-process syntactic continuation, representing the "rest" of the computation that still has to be executed in the process at a given state. It can also be viewed as the definition of intra-process sequencing, because the execution of a composite statement (i.e., *If*, *Do*, or *In*) is the intra-process sequential execution of its atomic primitive components interleaved with the execution of other processes. We believe the continuation model is appropriate to formalize the recursive decomposition of composite statements' execution into primitive atomic execution steps. The concept of *program counters* or *statement lines* is not sufficient to represent this decomposition. The *Selection* is the relation that formalizes the execution eligibility of any statement in a state. By defining the eligibility of statements to be executed, this relation actually defines the system-wide sequencing of valid interleavings in the system. The *Meaning* is the relation that specifies "what can happen" in the semantic domain (i.e., changes to the state) by executing a selectable statement. The *System* is the relation that formalizes a valid state transition sequence for the defined language.

The following subsections describe how to define these relations and the state. We emphasize that the framework is generic because it aggregates all top-level definitions that have to be made for the specification of distributed languages. When two adjacent languages are defined according to this framework, the relation concerning the equivalent execution effects and the proof obligations for verifying the implementation can be generated straightforwardly, which can simplify the verification effort.

## 3.1 The Semantic Domain – State

The semantic domain is the global state of the distributed programming language system. It is abstracted as a configuration aggregating (1) a process' local state, which is the set of local variable bindings; (2) a shared message pool state which is a set of FIFO message queues; (3) a process' thread which is the intra-process syntactic continuation; (4) the system-wide communication history; and (5) the auxiliary global time for ordering states but not for synchronization. The message queue is called an operation in SR terminology and we use the terms "message queue" and "operation" interchangeably. The *State* is defined in HOL as follows.

State: (Proc_local_state)list#Pool_state#(Thread)list#History#Global_time
Proc_local_state: Var → Value         Pool_state: Op → Opvalue
History: (Snd_rcv_flag#Message)list     Message: (Data#Proc_id#Flag)
Opvalue: (Message)list#Sent_count#Received_count    Thread: (Stmt)list

Auxiliary operations are needed in the specification to access the components of the global state, such as *get_thread*, *get_local_state*, *get_pool_state*, *get_history*, etc. In our example, the state at the microSR level and the state at the MP level are defined with similar basic structures but with different definitions for basic domain elements such as the Value, Message, etc. However, the definition of auxiliary operations at both levels are very similar.

## 3.2 The System Relation

The relation *System ((State)list → Bool)* characterizes the state sequences which are allowed in the state transition system modeling a given language. Since the *Meaning* relation encapsulates all valid single-step transitions, the definition of the relation *System* actually defines all possible valid interleavings of state transitions. We use this definition generically for both the microSR level and the MP level.

$\vdash$ *def* System (sseq:(State)list) =
  $\forall$ (i:num). ($\exists$ (p:proc_id)(j:num).
  Meaning (current_stmt(get_thread(EL i sseq)p)) (EL i sseq) (EL j sseq) p)

## 3.3   The Meaning Relation

The relation *Meaning (Stmt $\rightarrow$ State $\rightarrow$ State $\rightarrow$ Proc_id $\rightarrow$ Bool)* specifies "what can happen" in the semantic domain, i.e., the complete effect on any given state of executing a selectable statement in a process. The most important feature of this specification is that all possible interleavings are taken into account, through the following definition steps. Clearly, nondeterminism is permitted.

– The *Continuation* and *Selection* relations are defined in the next two sections.
– Relations on the execution effect of selectable atomic statements are specified. *m_Skip* indicates no effect on the state. *m_Assign* indicates an effect on the process' local state only. *m_Send* and *m_Call1* have only an effect on the shared pool state by adding a single message into a given message queue. *m_Receive1* and *m_Receive2* define the effect on both the process' local state and the pool state by receiving a message from a given operation and assigning the data of the received message to the appropriate variable.
– The effect on a state by the intra-process sequence of state transitions, possibly interleaved with system-wide valid state transitions by other processes, is defined. The key relation defined for this purpose is *m_proc_Seq*. We use this definition generically for both the microSR and the MP specifications.
– The effect on the state of executing a selectable composite statement in a process where interleavings with other processes are possible is defined by the relations *m_If, m_Do, m_In1*, etc. By structural induction, the state transitions caused by executing these statements are reduced to their component statements when the appropriate conditions are satisfied. The definition of these relations depends on a nondeterministic guard checking, the relations of the atomic state transitions, and the relation of the intra-process sequencing.

$\vdash$ *def* m_proc_Seq(m_stmt1,m_stmt2: State$\rightarrow$State$\rightarrow$Proc_id$\rightarrow$Bool)
          (s1,s2:State)(p:proc_id)=
  $\exists$(s3:State)(s4:State) .
  ((s3 = s4) $\wedge$ m_stmt1 s1 s3 p $\wedge$ m_stmt2 s3 s2 p) $\vee$
  ($\neg$(s3 = s4) $\wedge$ ((get_local_state s3 p) = (get_local_state s4 p)) $\wedge$
  m_sys_Seqn s3 s4 $\wedge$ m_stmt1 s1 s3 p $\wedge$ m_stmt2 s4 s2 p)
$\vdash$ *def* m_sys_Seqn (s1:State) (s2:State) =
  $\exists$ (m_stmts:(State$\rightarrow$State$\rightarrow$Proc_id$\rightarrow$Bool)list) (sl:(State)list)(pl:(Proc_id)list).
  ((EL 1 sl) = s1) $\wedge$ ((EL (LENGTH sl) sl) = s2) $\wedge$
  ($\forall$ (i:num). (EL i m_stmts) (EL i sl) (EL (i+1) sl) (EL i pl))

The relation *Meaning* for microSR has to be specified recursively on the syntactic structure of microSR statements. A state transition by a process is allowed only if its current statement satisfies relations on valid interleavings. The state transition accomplished by a composite statement is reduced recursively to state transitions accomplished by its component statements which must satisfy relations on valid interleavings as well. In this way, the concurrency and nondeterminism in the state transition model are completely specified.

⊢ *def* (Meaning (Seq stmt1 stmt2) s1 s2 p =

           ((CurrentStmt (get_thread s1 p)) = (Seq stmt1 stmt2)) ∧

           Selection s1 (Seq stmt1 stmt2) p ∧

           Continuation (get_thread s1 p)(get_thread s2 p)

                    (Seq stmt1 stmt2)(get_local_state s1 p) ∧

           m_proc_Seq (Meaning stmt1)(Meaning stmt2) s1 s2 p) ∧

   (Meaning (Call1 op e) s1 s2 p =

           ((CurrentStmt (get_thread s1 p)) = (Call op e)) ∧

           Selection s1 (Call1 op e) p ∧

           Continuation(get_thread s1 p)(get_thread s2 p)

                    (Call op e)(get_local_state s1 p) ∧

           m_Call1 op e s1 s2 p) ∧

   (Meaning (In1 op1 op2 v stmt1 stmt2) s1 s2 p =

           ((CurrentStmt (get_thread s1 p)) = (In op1 op2 v stmt1 stmt2)) ∧

           Selection s1 (In op1 op2 v stmt1 stmt2) p ∧

           Continuation (get_thread s1 p)(get_thread s2 p)

                    (In op1 op2 v stmt1 stmt2) (get_local_state s1 p) ∧

           m_In1 op1 op2 v (Meaning stmt1) (Meaning stmt2) s1 s2 p) ∧

   ...

## 3.4 The Selection Relation

In any given state, some statements are always selectable (such as *Skip, Assign,* and *Send*). Some are only conditionally selectable (e.g. *Receive*), reflecting the synchronous message receiving. Some are only conditionally selectable due to rendezvous communication (such as *Call* and *In*). As shown below, the relation *Selection (State → Stmt → Proc_id → Bool)* for microSR, is also defined recursively on the abstract syntactic structure of the statement. *Receive op(v)* is selectable only when there is at least one unreceived message in the given operation *op*. *Receive op(v) suchthat bexp* is selectable when the synchronization expression *bexp* is true in the given state by substituting the data of the first unreceived message of *op* for any occurrences of the variable *v* in *bexp*. *Call* is selectable, when its message will be the first unreceived message in the given *op*, and when there exists a process in the system whose current statement (the first statement in its thread) is an *In* statement where the *op* in the *Call* equals either *op1* or *op2* in the *In*. *In* is selectable, when there is only one unreceived message sent by a *Call* in either message queue *op1* or *op2*. Thus, rendezvous communication is modeled by two successive state transitions.

⊢ *def*

(Selection (s:State) (Send op iexp) (p:Proc_id) = T) ∧

(Selection s (If bexp1 bexp2 stmt1 stmt2) p =

$((\neg(\text{Mbexp bexp1}(\text{get\_local\_state s p})) \wedge \neg(\text{Mbexp bexp2}(\text{get\_local\_state s p})))\vee$
$(\text{Mbexp bexp1 (get\_local\_state s p)} \wedge \text{Selection s stmt1 p}) \vee$
$(\text{Mbexp bexp2 (get\_local\_state s p)} \wedge \text{Selection s stmt2 p})) \wedge$
$(\text{Selection s (Seq stmt1 stmt2) p} =$
$(\text{Selection s stmt1 p}) \wedge (\exists \text{ s'. (Selection s' stmt2 p)})) \wedge$
$(\text{Selection s (Receive2 op var bexp) p} =$
$((\text{get\_sent\_count op(get\_pool\_state s)})>(\text{get\_received\_count op(get\_pool\_state s)}))\wedge$
$(\text{get\_first\_msg\_tag op (get\_pool\_state s)} = \text{send\_by\_Send}) \wedge$
$(\text{Mbexp(SubstValue bexp var (get\_first\_msg op}$
$(\text{get\_pool\_state s)}))(\text{get\_local\_state s p}))) \wedge$
$(\text{Selection s (Call1 op iexp) p} =$
$((\text{get\_sent\_count op(get\_pool\_state s)})=(\text{get\_received\_count op(get\_pool\_state s)}))\wedge$
$(\exists \text{p2 op1 op2 var1 var2 stmt1 stmt2}.$
$(\text{CurrentStmt(get\_thread s p2)} = (\text{In1 op1 op2 var1 var2 stmt1 stmt2})) \wedge$
$((\text{op1=op}) \vee (\text{op2=op}))) \wedge$
$(\text{Selection s (In1 op1 op2 var stmt1 stmt2) p} =$
$(((\text{get\_sent\_count op1 (get\_pool\_state s)}) =$
$(\text{get\_received\_count op1 (get\_pool\_state s)}) + 1) \wedge$
$(\text{get\_first\_msg\_tag op1 (get\_poo\_state s)} = \text{sent\_by\_Call})) \vee$
$(((\text{get\_sent\_count op2 (get\_pool\_state s)}) =$
$(\text{get\_receive\_count op2 (get\_pool\_state s)}) + 1) \wedge$
$(\text{get\_first\_msg\_tag op2 (get\_pool\_state s)} = \text{sent\_by\_Call}))) \wedge$
...

## 3.5 The Continuation Relation

The relation *Continuation (Thread → Thread → Stmt → Proc_local_state → Bool)* for microSR, is recursively defined as well. Primitive microSR statements (such as *Assign, Send, Receive, Call*) correspond to atomic state transition steps, while composite statements (such as nondeterministic *If, Do, In*) have to be decomposed recursively as an intra-process sequence of state transition steps interleaved with other processes. The definition indicates how an atomic primitive is "popped off" from the thread and how to decompose the composite statement recursively to generate a new thread that represents the syntactic continuation. When the *oldthread* is an empty statement list, *TL oldthread* returns an arbitrary unknown list in HOL. It will not affect the correct evaluation of the *Meaning* relation, however, because the *Meaning*, as well as relations *CurrentStmt* and *Selection* in the *Meaning* definition are only satisfied when the *oldthread* is a non-empty list. We are also not bothered by the finiteness limitations of HOL lists, because the semantic specification is not used for the symbolic execution of programs, but for the verification.

$\vdash def$
$(\text{Continuation oldthread newthread (Send op iexp) (ls:Proc\_local\_state)} =$
$(\text{newthread} = (\text{TL oldthread}))) \wedge$
$(\text{Continuation oldthread newthread (Seq stmt1 stmt2) ls} =$
$(\exists \text{ls'. (Continuation (APPEND[stmt1;stmt2](TL oldthread))}$
$(\text{APPEND[stmt2](TL oldthread)) stmt1 ls}) \wedge$
$(\text{Continuation(APPEND[stmt2](TL oldthread)) newthread stmt2 ls'})) \wedge$
$(\text{Continuation oldthread newthread (If bexp1 bexp2 stmt1 stmt2) ls} =$

$((\neg(\text{Mbexp bexp1 ls})\wedge\ \neg(\text{Mbexp bexp2 ls})) \Rightarrow (\text{newthread} = (\text{TL oldthread}))|$
$((\text{Mbexp bexp1 ls}\ \wedge$
$(\text{Continuation}(\text{APPEND}[\text{stmt1}](\text{TL oldthread}))\ \text{newthread stmt1 ls}))\ \vee$
$(\text{Mbexp bexp2 ls}\ \wedge$
$(\text{Continuation}(\text{APPEND}[\text{stmt2}](\text{TL oldthread}))\ \text{newthread stmt2 ls})))))\ \wedge$

## 3.6 The Language Implementation

As mentioned previously, we focus on the distributed programming language implementation with the specification of mappings serving as the language "compilation" or "translation". In order to prove the correct implementation of a higher-level distributed language by a lower-level language, we have formalized the concept of the correct implementation relationship of two adjacent language levels, both of which are specified in our semantic framework. As shown below, it is necessary to specify three mappings between two adjacent language levels to represent the language implementation. Its correctness has to be verified with respect to the pair of operational semantics, i.e., the "execution" of the lower-level instructions "generated", with respect to its start and final states at the lower level, have to be proved to correctly implement the meaning of the corresponding higher-level instruction, with respect to its start and final states at the higher level. The proof obligation *Stmt_implemented_correct* for the implemen-

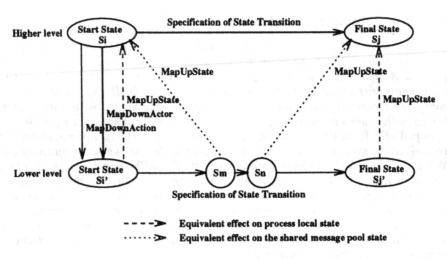

**Fig. 2.** The Correct Implementation Relationship Between Two Adjacent Levels

tation correctness and the relation *Equivalent_interleaving* for the equivalence of interleavings at two adjacent levels are specified below. Figure 2 shows a schematic of the proof obligation. The equivalence of interleavings at two levels is specified with respect to their equivalent effects on states, because the lower level has a finer granularity of atomicity and, therefore, more possible interleavings. The equivalence of effects on the local states of processes is defined with respect to start and final states at both levels, while the equivalence of effects on the shared message pool states is defined with respect to start and final states at the higher level and two intermediate states at the lower level. The intermediate states at the lower level allow for the finer atomicity and interleavings. Since interleavings at both levels are taken into account in their semantics, the generic definition of the proof obligations is relatively straightforward.

MapDownAction: high_Stmt → low_Stmt
MapDownActor: high_Proc_id → low_Proc_id
MapUpState: Low_State → high_State

⊢ *def* Stmt_implemented_correct (statei,statej:high_State)
        (statei',statej':low_State)(stmt:high_Stmt)(p:high_Proc_id) =
        (low_m_proc_Seq(MapDownAction stmt) statei' statej' (MapDownActor p)
          ⇒ Equivalent_interleaving statei statej statei' statej' p)
        ⇒ high_Meaning stmt statei statej p
⊢ *def* Equivalent_interleaving si sj si' sj' p =
        (high_get_local_state si p = high_get_local_state(MapUpState si')p) ∧
        (high_get_local_state sj p = high_get_local_state(MapUpState sj')p) ∧
        ((high_get_history si = high_get_history sj) ∨
          (¬(high_get_history si = high_get_history sj) ∧
          (∃ (sm':low_State)(sn':low:state). ordered si' sm' sn' sj' ∧
          (high_effect_on_pool si sj p =
          high_effect_on_pool (MapUpState sm')(MapUpState sn') p) ∧
          (high_get_history si = high_get_history (MapUpState sm')) ∧
          (high_get_history sj = high_get_history (MapUpState sn')))))

# 4 The Verification of the MicroSR Implementation

We have applied our methodology to the verification of a microSR implementation. Using the generic framework, we specify both the microSR semantics and the MP semantics. Like the specification for microSR outlined in Section 3, several specific items must be defined at the MP layer: the State, and the Continuation, Selection, and Meaning relations. Furthermore, we must define mappings between the two layers and verify the implementation of microSR on an MP machine.

## 4.1 The MP Machine State

The definition of the state in the MP machine is

mp_STATE:
(vm_State)list#mp_POOL#(mp_Thread)list#mp_MSG_HISTORY#mp_CLOCK
mp_Code: (mp_Inst)list    mp_PC:num    mp_Thread: mp_Code#mp_PC

which is similar to the definition of the microSR state (see section 3.1). The significant differences between layers are hidden by the definitions of basic domain elements within each layer. An example of such a difference between the microSR layer and the MP machine layer can be found in the definition of the thread. Because of the relatively high level of abstraction at which microSR operates, it is convenient to define the thread as a list of statements, where the first statement on the list represents the next one to execute. This is not possible at the MP layer because, for example, there is no well-defined concept of a loop. Only individual instructions, including branch instructions, are modeled. Thus, the MP machine requires that a program counter be associated with each thread, reflecting a more concrete view of the system than the microSR level.

## 4.2 Continuation, Selection, and Meaning Relations

In the MP layer, the Continuation and Selection relations are very much simpler than their counterparts at the microSR layer, reflecting MP's relative simplicity. Whereas, with microSR code, there are various conditions which determine what part of the code will be executed next, the MP instruction set only has to concern itself with branches. If the instruction is not a branch, then all we are required to do to find the continuation state is to increment the program counter for the current thread. Otherwise, the program counter should (conditionally) be set to the value specified in the branch instruction. Likewise, the Selection relation is greatly simplified because, of all the instructions in the MP instruction set, only the RCV instruction has conditional eligibility (a message must exist on a queue in order to receive it). The MP meaning relation is defined much like microSR's.

## 4.3 microSR–MP Mappings

There must be mapping functions between any two adjacent layers, as shown in Figure 2. These mappings must exist between microSR and the MP, for the State, the Actor (process), and the Action (code) and have the following signatures:

mp_to_sr_MapUpState:     mp_STATE → State
sr_to_mp_MapDownActor:   Proc_id → mp_PID
sr_to_mp_MapDownAction:  Stmt → (mp_Thread)

Of these three mappings, the first two are relatively straightforward, involving primarily projections. However, in order to map the code, we must give a "compiler" specification from the microSR language to the MP instruction set.

The compiler (called sr_to_mp_MapDownAction) is a collection of simple recursive definitions which specifies an implementation of a compiler of microSR code into MP machine code. This is really a transformation of the microSR abstract syntax tree into a list of MP machine instructions. The transformation of a sample of representative statements is shown below:

```
⊢ def (sr_to_mp_MapDownAction (state:compile_State) Skip =
        (state,[VM (vm_INST NOP (ARGS0))]])) ∧
    (sr_to_mp_MapDownAction state (Send op expr) =
        let expr_res = compile_IExpr state expr 0 in
        let op_res = compile_Op (FST expr_res) op in
        (FST op_res, (APPEND SND expr_res)
```

$$[\text{MP SEND (ARGS2 (SND op\_res) 0)]})))\wedge$$
$$(\text{sr\_to\_mp\_MapDownAction state (Seq s1 s2)} =$$
$$\text{let first} = \text{sr\_to\_mp\_MapDownAction state s1 in}$$
$$\text{let second} = \text{sr\_to\_mp\_MapDownAction (FST first) s2 in}$$
$$(\text{FST second, APPEND (SND first) (SND second)})) \wedge$$

...

Because, at present, microSR has no variable declarations, and, thus, variables are implicitly declared, the compilation process must carry around a "compile state" which shows the bindings between variables and assigned memory locations. This is the first element returned by each of the compilation functions; the compiled code is the second element. Also, in the compiled code shown above, there are two MP-level instructions shown. The first, VM (vm_INST NOP (ARGS0)) actually represents a VM (hardware) instruction—in this case, the NOP instruction—which takes zero arguments. The second, MP SEND (ARGS2 (SND op_res) 0) represents a true MP instruction—here, the SEND instruction— which takes two arguments: the compiled representation of the destination operation, and a register number which contains the value of the data to be sent.

## 4.4 Verification of the Implementation

Finally, we use the complete definitions of microSR and the MP machine to verify that the microSR compiler is correct. In order to do this within the general framework, we show that executing the MP code which results from compiling the microSR code yields the effect on the global state that the semantics of microSR requires.

One of the theorems which we prove concerning the implementation relationship between microSR programs and the corresponding MP programs is the correct implementation of the *Send* statement. Other statements yield similar results.

$$\vdash \forall \text{ hstatei hstatej lstatei' lstatej' p op e .}$$
$$(\text{mp\_m\_proc\_Seq (sr\_to\_mp\_MapDownAction (Send op e))}$$
$$\text{lstatei' lstatej' (sr\_to\_mp\_MapDownActor p)}$$
$$\Rightarrow \text{mp\_sr\_Equivalent\_interleaving hstatei hstatej lstatei' lstatej' p)}$$
$$\Rightarrow \text{Meaning (Send op e) hstatei hstatej p}$$

# 5 The Verified Programming Logic for MicroSR

In this section we present our proof of the soundness of a programming logic for the microSR language with respect to the same semantic specification we captured above. All axioms and inference rules in the logic are formally proved as theorems in HOL. Compared with our early exercise on mechanizing a smaller programming logic for an early version of microSR in HOL [15], this work differs mainly in the following aspects, besides the enhancements to both the language and the mechanized programming logic.

- Reflecting our new framework, concurrency and nondeterminism have been completely handled in the semantic specification for microSR, of which the mechanized programming logic is the logical implication.
- Not only the abstract syntax and the semantics of the microSR are specified in HOL, but also the abstract syntax of an assertional language is defined by *Type Definition* in HOL, and substitution, which is the major operation on assertions in the programming logic, is defined recursively on the syntactic structure. [9] has shown how to formalize, in HOL, an assertional language and the substitution for a programming logic of a *while_loop* language. Because of the message passing constructs and the related concurrency and nondeterminism in the microSR, our assertional language includes assertions on a process' local state as well as assertions (called global invariants) on the shared message pool especially on counters of messages that have been sent into or received from operations. This made the abstract structure of assertions more complicated and the proofs about the substitution nontrivial. However, it is still syntactically manipulable. Abstracting the pool state as the function as shown in the state abstraction in our framework made the definition more expressive and the proof more tractable.
- Because the complete theory of microSR is deeply embedded in HOL, the proved theorems about the substitution on assertions including global invariants, and the proved theorems of axioms and inference rules for language constructs now comprise a verified HOL prover for microSR programs.

The following are representative proved theorems on substitutions which are necessary for the soundness proof of the programming logic. $P$ represents an assertion on a process' state, $GI$ an assertion on the shared message pool, and $SB$ the synchronous expression used in the *Receive* statement. The notation $A[v \leftarrow exp]$ represents the substitution of the expression $exp$ for the variable $v$ in the assertion $A$, $ls[v \Leftarrow va]$ represents the state change in the process local state $ls$ by giving the variable $v$ a new value $va$. Similarly, $GI[op \leftarrow op+msg]$ is the substitution of $op+msg$ for the operation name $op$ in the assertion of the global invariant $GI$, and $pool[op \Leftarrow op+msg]$ is the state change of the message pool by binding a new message queue with $op$ after adding one message $msg$ into the message queue.

M_as:     Assertion $\rightarrow$ Proc_local_state $\rightarrow$ Bool
M_exp:    Exp $\rightarrow$ Proc_local_state $\rightarrow$ Value
M_gi:     GIAssertion $\rightarrow$ Proc_local_state $\rightarrow$ Pool_state $\rightarrow$ Bool
M_bexp:   BExp $\rightarrow$ Proc_local_state $\rightarrow$ Bool

$\vdash$ M_as (P[v$\leftarrow$exp]) ls = M_as P (ls[v$\Leftarrow$(M_exp exp ls)])
$\vdash$ M_gi (GI[op$\leftarrow$(op+msg)]) ls pool = M_gi GI ls (pool[op$\Leftarrow$(op+msg)])
$\vdash$ M_gi (GI [v$\leftarrow$first_msg(pool op)][op$\leftarrow$(op-msg)]) ls pool =
         M_gi GI (ls[v$\leftarrow$first_msg(pool op)]) (pool[op$\Leftarrow$(op-msg)])
$\vdash$ M_bexp (SB[v$\leftarrow$first_msg(pool op)]) ls = M_bexp SB (ls [v$\Leftarrow$first_msg(pool op)])

Representative proved axioms and inference rules are listed below. The axioms listed have weakest pre-conditions. Theorems with stronger pre-conditions for these constructs can be deduced by the mechanized rule for strengthening pre-conditions. We have the *Intra-Process Sequencing Rule* because the intra-process sequencing has interleavings with other processes as specified by $m\_proc\_Seq$. The *statement1 ;; statement2* in the *Rendezvous Rule* is defined as two successive state transitions without any interleavings after executing *statement1* in one process and before executing *statement2* in another process.

Send Axiom:

$$\vdash \{P \wedge (GI[op \leftarrow (op + msg(e))])\} \ Send \ op(e) \ \{P \wedge GI\}$$

Call Axiom:

$$\vdash \{P \wedge (GI[op \leftarrow (op + msg(e))])\} \ Call \ op(e) \ \{P \wedge GI\}$$

Receive Axiom1:

$$\vdash \{(Q[v \leftarrow e]) \wedge (GI[v \leftarrow e][op \leftarrow (op - msg)])\} \ Receive \ op(v) \ \{Q \wedge GI\}$$

Receive Axiom2:

$$\vdash \{(Q[v \leftarrow e]) \wedge (GI[v \leftarrow e][op \leftarrow (op - msg)])\} \ Receive \ op(v) \ suchthat \ B \ \{Q \wedge B \wedge GI\}$$

Intra–Process Sequencing Rule:

$$\vdash \frac{\{GI \wedge P\} \ SL1 \ \{GI \wedge R\}, \ \{GI \wedge R\} \ SL2 \ \{GI \wedge Q\}, \ GI \ is \ true \ in \ other \ process}{\{GI \wedge P\} \ SL1; \ SL2 \ \{GI \wedge Q\}}$$

In Rule:

$$\vdash \frac{\{P \wedge GI\} \ Receive \ op1(v)\{R1 \wedge GI\}, \ \{R1 \wedge GI\} \ stmt1 \ \{Q \wedge GI\},}{\{P \wedge GI\} \ Receive \ op2(v) \ \{R2 \wedge GI\}, \ \{R2 \wedge GI\} \ stmt2 \ \{Q \wedge GI\}}{\{P \wedge GI\} \ In \ op1(v) \rightarrow stmt1 \ [] \ op2(v) \rightarrow stmt2 \ nI \ \{Q \wedge GI\}}$$

Rendezvous Rule :

$$\vdash \frac{\{Pproci \wedge GI\} \ Call \ op(e) \ \{Pproci \wedge GI\}, \ ((op = op1) \vee (op = op2)),}{\{Pprocj \wedge GI\} \ In \ op1(v) \rightarrow stmt1 \ [] \ op2(v) \rightarrow stmt2 \ nI \ \{Qprocj \wedge GI\}}{\{GI \wedge Pproci \wedge Pprocj\} \ Call \ op(e); ;}$$
$$In \ op1(v) \rightarrow stmt1 \ [] \ op2(v) \rightarrow stmt2 \ nI \ \{GI \wedge Pproci \wedge Qprocj\}$$

# 6  Discussion

We have developed a method for specifying and reasoning about the operational semantics of distributed programming languages. Our research has shown how to formalize, in HOL, distributed programming languages under the state transition model; how to structure the semantic specification by specifying necessary relations in our framework; and how to express proof obligations for the verification of language implementation when both the higher-level language and the lower-level language are specified by our framework. Processes in our distributed languages have their own local states and interact only through message passing, and our framework models all such interleavings of processes' actions. We apply our methodology to the verification of a compiler for a simple distributed language (microSR) which generates code for a multiprocessor which provides a simple assembly language and an operating system interface. This interface provides message passing through system calls. This simple example illustrates the methodology for a two-level system. Work in progress is on a still lower level, whose "distributed language" model is a network interface where processes communicate by exchanging packets; for this system it is necessary to verify that the operating system code that makes calls to the network interface correctly implements the system calls that effect message passing.

# References

1. M. Abadi and L. Lamport, The Existence of Refinement Mappings, Theoretical Computer Science, Vol. 82, pp.253-284, 1992.
2. G.R. Andrews and R.A. Olsson, The SR Programming Language: Concurrency in Practice, Benjamin/Cummings Publishing Company, Inc. Redwood City, CA, 1993.
3. W.R. Bevier, W.A. Hunt, J.S. Moore, and W.D. Young, An approach to systems verification, Journal of Automated Reasoning, 5 (1989) 411–428.
4. M. Chandy and J. Misra, Parallel Program Design: A Foundation of Programming Logic. Addison-Wesley Publishing Company, Inc. 1988.
5. P. Curzon, Deriving Correctness Properties of Compiled Code, in Higher Order Logic Theorem Proving and Its Applications, pp327–346, IFIP Transactions, A-20, North-Holland, 1993.
6. N. Francez, Program Verification, Addison-Wesley publishing Company Inc, England, 1992.
7. M. J. C. Gordon.: Mechanizing Programming Logics in Higher Order Logic. In: Current Trends in Hardware Verification and Automated Theorem Proving. Springer-Verlag, New York, 1989.
8. M. J. C. Gordon and T. F. Melham, Introduction to HOL: A theorem proving environment for higher order logic, Cambridge University Press, Cambridge, 1993.
9. P. V. Homeier and D. F. Martin, Trusworthy Tool for Trustworthy Programs: A Verified Conditional Generator, in Higher Order Logic Theorem Proving and Its Applications, No.859 in LNCS, pp269-284, Springer-Verlag, 1994.
10. J.J. Joyce, Totally Verified Systems: Linking verified software to verified hardware. In M. Leeser and G. Brown, Eds., Specification, Verification and synthesis: Mathematical Aspects, Springer-Verlag, 1989
11. A. U. Shankar, An Introduction to Assertional Reasoning for Concurrent Systems, ACM Computing Surveys, Vol.25, No.3, pp225-262, September 1993.
12. J. von Wright, J.Hekanaho, P. Luostarinen and T. Langbacka, Mechanising some Advanced Refinement Concepts, in Higher Order Logic Theorem Proving and Its Applications, pp307-326, IFIP Transactions, A-20, North-Holland, 1993.
13. P. J. Windley, A theory of generic interpreters, in Correct Hardware Design and Verification Method, No. 683 in LNCS, pp122-134, Springer-Verlag, 1993.
14. P. J. Windley, Specifying Instruction-Set Architectures in HOL: A Primer, in Higher Order Logic Theorem Proving and Its Applications, No.859 in LNCS, pp440-455, Springer-Verlag, 1994.
15. C. Zhang, R. Shaw, R. Olsson, K. Levitt, M. Archer, M.Heckman, and G. Benson, Mechanizing a Programming Logic for the Concurrent Programming Language microSR in HOL, in Higher Order Logic Theorem Proving and Its Applications, No.780 in LNCS, pp31-44, Springer-Verlag, 1994.

# Springer-Verlag
# and the Environment

We at Springer-Verlag firmly believe that an international science publisher has a special obligation to the environment, and our corporate policies consistently reflect this conviction.

We also expect our business partners – paper mills, printers, packaging manufacturers, etc. – to commit themselves to using environmentally friendly materials and production processes.

The paper in this book is made from low- or no-chlorine pulp and is acid free, in conformance with international standards for paper permanency.

# Lecture Notes in Computer Science

For information about Vols. 1–903

please contact your bookseller or Springer-Verlag